THE DESIGN INFERENCE

THE DESIGN INFERENCE

ELIMINATING CHANCE THROUGH SMALL PROBABILITIES

2ND EDITION, REVISED AND EXPANDED

WILLIAM A. DEMBSKI AND
WINSTON EWERT

SEATTLE   DISCOVERY INSTITUTE PRESS   2023

Description

A landmark of the intelligent design movement, *The Design Inference* revolutionized our understanding of how we detect intelligent causation. Originally published twenty-five years ago, it has now been revised and expanded into a second edition that greatly sharpens its exploration of design inferences. This new edition tackles questions about design left unanswered by David Hume and Charles Darwin, navigating the intricate nexus of chance, probability, and design, and thereby offering a novel lens for understanding the world. Using modern concepts of probability and information, it exposes the inadequacy of undirected causes in scientific inquiry. It lays out how we infer design via events that are both improbable and specified. Amid controversial applications to biology, it makes a compelling case for intelligent design, challenging the prevalent neo-Darwinian evolutionary narrative. Dembski and Ewert have written a groundbreaking work that doesn't merely comment on contemporary scientific discourse but fundamentally transforms it.

Cover

Design by Nathan Jacobson. Photo by Galyna Andrushko, "Rushmore," Adobe Stock, standard license.

Library Cataloging Data

The Design Inference: Eliminating Chance Through Small Probabilities, 2nd Edition by William A. Dembski and Winston Ewert
583 pages, 6 x 9 inches
Library of Congress Control Number: 2023945861
ISBN: 978-1-63712-033-0 (hardcover), 978-1-63712-034-7 (paperback), 978-1-63712-035-4 (EPUB), 978-1-63712-036-1 (Kindle)
BISAC: MAT029000 MATHEMATICS / Probability & Statistics / General
BISAC: COM031000 COMPUTERS / Information Theory
BISAC: SCI027000 SCIENCE / Life Sciences / Evolution

DISCOVERY
INSTITUTE
PRESS
Academic

Publisher Information

Discovery Institute Press, 208 Columbia Street, Seattle, WA 98104
Internet: https://www.discoveryinstitutepress.com
Published in the United States of America on acid-free paper.
Second Edition, October 2023

Advance Praise

Ever since Darwin, most scientists have adopted a principled view by which they reject out of hand any non-naturalistic explanations. This works perfectly well in the physical sciences, but less so in biology where, due to the incredible complexity of biological systems, appearance of design is overwhelming. Yet, by appealing to this metaphysical principle, intelligent design (ID) ideas are automatically rejected; natural selection and random mutations are viewed as the only acceptable explanation of the mechanism by which biological evolution takes place.

Though I take an agnostic position on ID, I have no doubt that its main proponents, Behe and Dembski, have brought to light important challenges to the reigning neo-Darwinian version of evolutionary biology. This second edition of Dembski's classic *The Design Inference* is well argued and eminently readable. The appendix provides the reader with a short, effective, introduction to the probabilistic and statistical methods used throughout the book. The authors give plenty of well-motivated, non-biological examples on how specified events of small probability lead to a convincing inference of intelligent design. The same arguments become controversial only when applied to biology! I don't see how any open-minded scientist can ignore this important book.

— **Sergiu Klainerman**, Higgins Professor of Mathematics, Princeton University, member of the National Academy of Sciences

When technology guru George Gilder described William Dembski as the Isaac Newton of the information age, he was in no way guilty of hyperbole. In *The Design Inference*, Dembski cracked a profound philosophical and scientific problem that had persisted unsolved for 2,500 years.

Western philosophers and scientists from Plato, Aristotle, and Cicero to Maimonides, Aquinas, and Kant to Boyle, Paley,

Maxwell, and even Newton himself have long perceived evidence of design in nature. Yet none of these great thinkers were able to explicate objective criteria by which the intuition of design could be justified.

In the first edition of *The Design Inference*, published with Cambridge University Press in 1998, Dembski explicated joint criteria of improbability and specification by which rational agents can reliably detect the activity of other rational agents. He also showed that human beings as rational agents routinely make such design inferences even if they are unaware of, or unable to, articulate the criteria by which they make them.

Now in the second edition of *The Design Inference*, writing with computer scientist Winston Ewert, Dembski trades on the same powerful concepts developed in the first edition, while providing additional analytical rigor and an updated account of the key notion of specification. In so doing, he explains how the concept of specification can be legitimately applied to a wider variety of cases.

Consequently, his work has now even more obvious and profound implications for the debate about design in biology, physics, and cosmology. Specifically, Dembski's updated account of specification makes it easier to explain why the digital code stored in the DNA molecule, and the fine tuning of the fundamental parameters of physics, exemplify both small probability events *and* specifications—and, thus, why they rightly trigger an awareness of intelligent activity or design.

This updated edition of *The Design Inference* shows the enduring power of Dembski's ideas and itself reveals the work of a profound intelligence. Clearly written, rigorous, and intellectually compelling. A work of genuine genius.

— **Stephen C. Meyer**, Director of Discovery Institute's Center for Science and Culture, author of *Signature in the Cell, Darwin's Doubt*, and *Return of the God Hypothesis*

Appearing a quarter century after the first edition, this second edition of *The Design Inference* is bolder, richer, and without the burden and dictates of a doctoral dissertation. More than double in size, it is a testimony to Dembski's abiding commitment to elucidation of design in all its myriad forms. The second edition could not have come into existence without the work of the designed minds (Dembski and Ewert) who have now added a new layer to the discourse by inviting readers to reflect on the processes of evolution in addition to the products of evolution. This extension to the refreshingly restated arguments of the first edition makes this second edition a compelling refutation of the neo-Darwinian narrative; it will be a gamechanger in the discourse on whether or not life has been designed.

> — **Muzaffar Iqbal**, Founder-President of the Center for Islamic Sciences, past Director of the Pakistan Academy of Sciences

Prepare to be dazzled. This new edition of *The Design Inference* is a *tour de force* of thinking and explaining—a veritable feast. If you are serious about understanding fundamental reality, evidence, and reasoning, read this book.

> — **Gale Pooley**, Associate Professor of Business Management, Brigham Young University-Hawaii, co-author of *Superabundance*

Richard Dawkins famously commented that, with Charles Darwin and his theory of evolution, it became possible to be an intellectually fulfilled atheist. The "design" in nature is merely apparent, and natural selection acting on random variation explains why, according to Dawkins. Despite the likes of Michael Faraday, Gregor Mendel, James Clerk Maxwell, and Arthur Eddington, Dawkins considers modern science and traditional religious conviction to be incompatible.

Not so, and with the help of William Dembski and Winston Ewert, it has now become even more possible to be an intellectually fulfilled theist. The second edition of *The Design Inference* makes a compelling case that the "design" in nature is real and can be scientifically inferred. As they show, the specified complexity of the information contained in DNA and RNA—its small probability and conformity to a complex pattern—cannot plausibly be attributed to unguided natural processes; logically and causally, it requires an intelligent designer.

As Dembski and Ewert realize, their work does not get us to the benevolent Deity of the Bible, and it will probably not satisfy die-hard fans of Darwinian selection and the quantum multiverse, including Dawkins. I highly commend *The Design Inference*, however, to anyone who wants to be intellectually informed about probability, reason, and faith.

> — **Timothy P. Jackson**, the Bishop Mack B. and Rose Stokes
> Professor of Theological Ethics, Candler School of Theology,
> Emory University

Darwinists have long asserted any appearance of design in life is the result of natural unintelligent processes that had no end in mind. Any suggestion that there is a designer is merely a primitive "god of the gaps" argument from ignorance. But Drs. Dembski and Ewert show in an accessible and testable way that intelligent design is not an argument from ignorance—life itself contains empirically, verifiable evidence *for* design. And the evidence is mathematically overwhelming. Some will use their designed minds to continue to resist the conclusions of this brilliant *tour de force*, but given the evidence I suspect any resisters will be either stubborn ideologues or really bad at math.

> — **Frank Turek**, President of CrossExamined.org, author and
> speaker

This second, and expanded, edition benefits greatly from the significant advances in understanding the design inference over the last twenty-five years. Over half a century ago, in *Chance and Necessity*, Jacques Monod admitted that there is a "fundamental epistemological contradiction" when we attempt to understand teleological objects "objectively." This contradiction arises because he dogmatically insisted that objective science has no place for design. *The Design Inference* shows how meaningful events can be, and indeed are, recognized using a simple criterion: *specified complexity*. In this second edition, Dembski and Ewert carefully explain this criterion, showing how it not only helps us understand and analyze a wide range of decisions that we make as part of everyday life, but also extends the reach of science.

— **Fred Skiff**, the Harriet B. and Harold S. Brady Chair in Laser Physics, University of Iowa

This new and expanded edition of *The Design Inference* follows in a long line of books over the last quarter century by distinguished mathematician William Dembski, renowned as the leading intelligent design (ID) specialist in the world. This book is another important step along the way to validating intelligent design as a mainstream and scientifically robust alternative to Charles Darwin's nineteenth-century philosophy of natural selection.

A key question this book addresses is: Does natural selection have sufficient creative power to account for the immense complexity and information-richness of life? To date, mainstream science has failed to answer this question. An appeal to faith in natural selection's information-creating powers—in the continued absence of clear confirming evidence of such—remains the current leading answer for information creation. But in some countries, such as Brazil, ID is making dramatic inroads as a sub-discipline within biology. In other countries where a more traditional ruling scientific orthodoxy holds sway, ID is a target of scientific censorship.

It is a well-known adage, "Where there's smoke, there's fire." There is certainly a lot of smoke surrounding this ID-versus-natural-selection controversy. This book gives us the clearest picture yet of the fire, and thereby takes the science of ID one step closer to validation as a rigorous and scientifically robust explanation for the immense information-richness of life.

— **Andrew Ruys**, Professor of Biomedical Engineering (Retired), University of Sydney

The Design Inference, 2nd Edition, is a much needed (and much appreciated) update to Bill Dembski's classic work, this time co-authored with Winston Ewert. In the twenty-five years since the first edition, the work has been strengthened and substantially extended, and in the process the ideas have been made more accessible to less technical audiences. Among its many other benefits, this work provides a solid foundation for future research into the exquisite and astounding design of living systems.

— **Steve Laufmann**, enterprise systems architect, Program Chair of the Conference on Engineering in Living Systems (CELS), leader of Discovery Institute's Engineering Research Group (ERG), and co-author with Howard Glicksman of *Your Designed Body*

The first edition of *The Design Inference*, published by Cambridge University Press, was welcomed with academic accolades because in many disciplines we need a robust capacity to rule out chance, and Dembski's design-inferential method confers that capacity. But when it was realized that this method, based on sound statistical and information-theoretic concepts, may trigger a design inference once applied to biological complexity, the storm clouds gathered. It is not that Dembski's design inference was discredited, but rather that, for many, the rule of naturalism must remain absolute.

In this expanded second edition we are shown how the design inference makes design part of the very fabric of science. Dembski and Ewert make a strong case for regarding specified complexity as the key to design detection and as a normal tool for everyday inquiry. They carefully define specification and complexity mathematically and illustrate these concepts with helpful examples. Given the solid theoretic foundation for design inferences provided here, researchers will be better equipped to answer whether biology is solely the domain of unguided processes or instead reveals the activity of a designing mind.

— **Mark Fitzmaurice**, MD, Sydney, Australia

Mathematical probabilists like myself happily work deductively with axioms proposed by A. N. Kolmogorov some ninety years ago. By and large we leave to statisticians the daunting task of combining probability theory with inductive evidence. And I don't know that even statisticians tend to be eager to ply their trade on the controversial, ultimate questions of existence. It is natural that the lot of us would feel convicted by Pascal's rebuke, "Those who do not love the truth take as a pretext that it is disputed, and that a multitude deny it."

Dembski and Ewert, on the other hand, don't shy away from discussing the evidence for a designing intelligence, framing their arguments in a systematic and general way. In *The Design Inference* they write clearly and irenically, which makes the book a pleasure to read. I believe their work is worthy of attention and respect.

— **Christopher P. Grant**, Associate Professor of Mathematics, Brigham Young University

This second edition is many things—humble conceptual remodeling, quarter-century labor of love, and testament to true intellectual partnership. Dembski and Ewert remind us of something we already know: debate drowns in shallow water and thrives in logical depth. Layperson and expert alike are encouraged to explore the endnotes and remarkable appendices. The authors leave no stone unturned, and neither should the reader.

— **Tristan Abbey**, President, Comarus Analytics LLC

In his book, *Six Great Ideas*, the philosopher Mortimer Adler stated: "There would be stars and atoms in the physical cosmos with no human beings or other living organisms to perceive them. But there would be no ideas as objects of thought without minds to think about them."

Stars and atoms are the venue of physics. During the last century, the vast majority of physicists have made peace with the notion that the universe had a beginning, and that any attempt to assign a cause or mechanism or prior state to that beginning lies beyond the reach of the natural sciences.

Darwinists, however, persist in the hubris of believing that they have fully resolved how the chance assemblage of an exponentially lengthy sequence of statistically impossible events could produce life in all its variations. Any objections to this conclusion are met with censorship, derision, and a profound obliviousness to the mathematical hurdles confronting the Darwinian view. We have sadly reached an anti-scientific point in many circles where even openly thinking about an alternative explanation is viewed as heresy.

In this second edition of *The Design Inference*, Dembski and Ewert present a formidable probabilistic and information-theoretic method for determining whether design, rather than chance, was the cause of an observed event. They then apply this method to the intricate forms we find in biology. With devastating mathematical precision, the book demonstrates that any complex event having both a briefly described specification and a small probability of occurrence—that is, small in light of all available probabilistic resources—must logically be attributed to design rather than chance.

This edition also incorporates further mathematical refinements, particularly in the account of *specified complexity*. It updates many of the references. And it convincingly refutes the various objections raised since the publication of the original version.

It is remarkable that the question of design, ubiquitous in everyday experience, is met with such ferocious resistance when it comes to thinking about the origin of living organisms, which represent the ultimate in specified complexity. Minds open to the issues raised in this book will be able to fruitfully engage the debate over biological origins. In this greatly revised and expanded edition, opponents of design have a new and unenviable challenge to surmount.

> — **Terry Rickard**, PhD, Engineering Physics, University of California, San Diego

The first edition of *The Design Inference* in 1998 cemented the validity of ID as the first bona fide challenge to natural evolution, whose foundational tenets had stood largely unchanged, and whose position in biology was not seriously threatened for over a hundred years. In this second edition, authors Dembski and Ewert provide even more incontrovertible evidence that luck has its limits, and the devil is truly in the details. Meanwhile, the Darwinists still have no effective counterarguments save that their blind watchmaker did indeed manage to make the watch you wear.

— **Kenneth Poppe**, EdD, Southern Nazarene University, professor and author of books on educational reform

Rare are the truly original, landmark books that have deeply impacted both science and the philosophy of science. For biology, one such work was Charles Darwin's *Origin of Species* (1859), whose thesis was that the design features of the biological world are mere illusion. Another was Jacques Monod's *Chance and Necessity* (1970), which reduced life to Darwin's natural selection and chance variations interpreted as the blind ruthlessness of the laws of nature working on the blind randomness of genetic mutations.

But in 1998, just when we thought the last nail had been driven into the coffin of intelligent goal-directedness in the natural world, Dembski's epochal, paradigm-shifting *The Design Inference* shook the foundations of materialistic reductionism, giving new vigor to a seemingly moribund teleology. Consequently, a reanimated *teleological realism* is now a great poker hand to hold in an intellectual wager where a designing mind appears to have stacked the deck in its favour.

The Design Inference launched the intelligent design movement as a truly novel *scientific* theory within natural philosophy and placed it on a solid intellectual footing with its notion of specified complexity. Dembski and Ewert's 25th-anniversary edition adds many new silvery pearls to the intellectual necklace of the first. This book applies to all areas of science. But for biology, and especially for "organs of extreme perfection" like the eye, it implies that after 160 years, "Darwin's dangerous idea" has met its match in "Dembski's pivotal postulate."

— **Marc Mullie**, MD, Ophthalmologist, Montreal, Canada

It is increasingly understood that the random-mutation selection mechanism is a woefully inadequate explanation for the origin and development of all of life. Probability arguments strongly support the conclusion that a random search algorithm (which the Neo-

Darwinian mutation/selection mechanism is) on the space of possible arrangements of nucleotides, amino acids, saccharides and glycerols is far too limited to invent or discover genetic codes, molecular machines and metabolic networks, to name a few.

But the unresolved question is if a design hypothesis can be given credence.

Sometimes, but not always, artists put their signature on a work of art. But signature or not, there are usually telltale signs that identify the object as having been designed. In the language of Bill Dembski and Winston Ewert, these are that the object or event has very low probability of occurring "naturally" and has evidence of being specified. It is these two characteristics that Dembski identifies as a design filter, and in this book, a revision of Dembski's groundbreaking, but controversial, *The Design Inference*, he gives a more precise definition of what these terms mean. With this amplification and clarification, significant steps have been taken to better quantify the likelihood that something is designed. Since it uses probabilistic arguments, there can be no rigorous mathematical proof that life is designed. But in my opinion, in no small part informed by Dembski's arguments, it would take a truly foolish person to bet against it.

I highly recommend you give this book a serious read.

— **James P. Keener**, Distinguished Professor of Mathematics, University of Utah

To our parents,
Bill and Ursula Dembski,
Ken and Micki Ewert,
Proverbs 1:8–9

The same Arguments which explode the Notion of Luck, may, on the other side, be useful in some Cases to establish a due comparison between Chance and Design: We may imagine Chance and Design to be, as it were, in Competition with each other, for the production of some sorts of Events, and may calculate what Probability there is, that those Events should be rather owing to one than to the other.

—Abraham de Moivre, *Doctrine of Chances*, 1718

TABLE OF CONTENTS

Table of Contents

FOREWORD

IN 1998 PHILOSOPHER AND MATHEMATICIAN WILLIAM DEMBSKI published a book with Cambridge University Press that would forever change the debate about design and purpose in biology. *The Design Inference: Eliminating Chance Through Small Probabilities* provided a powerful conceptual framework for understanding the origin of complexity and purpose in living things. The second edition of this seminal work, co-authored with software engineer Winston Ewert, is a profound and long-awaited reflection on the design inference and its relevance to biological complexity, specification, and information.

For millennia, philosophers and scientists ascribed biological complexity and purpose to design. In the age of theistic faith, the awe-inspiring purposefulness and complexity of living things seemed as convincing an argument for divine providence as could be imagined. With the publication in 1859 of Darwin's *Origin of Species*, and the accompanying tsunami of atheist ideology, a new age of atheistic faith inundated the scientific world. It seemed that biological design could be explained away by invoking Darwinian random heritable variation and natural selection.

In his book *The Blind Watchmaker*, Richard Dawkins famously summed up the Darwinian perspective. "Biology," he admitted, "is the study of complicated things that give the appearance of having been designed for a purpose."[1] But, he quickly added, any such appearance of design is illusory: "Natural selection is the blind watchmaker, blind because it does not see ahead, does not plan consequences, has no purpose in view. Yet the living results of natural selection overwhelmingly impress us with the appearance of design as if by a master watchmaker, impress us with the illusion of design and planning."[2] Because all biological design is thereby explained away, Dawkins concluded, "Darwin made it possible to be an intellectually fulfilled atheist."[3]

Of course, even after Darwin, perceptive scientists continued to point out that the complexity and purpose evident in living things still left atheists a bit short of intellectual fulfillment. Darwinian faith was de rigueur in the twentieth century, and this despite the discovery of a computer code in DNA and astonishingly elegant molecular nanotechnology in living things, including cellular organelles like the bacterial flagellum that work according to obvious engineering principles.[4] Nonetheless, it took a brave soul to dare question the Darwinian paradigm in biology. Those scientists who did question atheist dogma tended to become unfulfilled in the sense of "unemployed."

For a few scientists, truth mattered more than sinecure. Yet the scientific truth was hard to come by. The problem was that the mountains of evidence both for and against Darwinian theory were largely mountains of anecdotes. Darwinists pointed to anecdotal hunches that an imagined sequence of variation and selection could produce the genetic code, the beating heart, and even the brain by which an improbably evolved species of apes could ask such questions. Design scientists pointed to functional biological complexity that seemed beyond what even the most fanciful Darwinian stories could account for—think of the camera eye, which gave even Darwin sleepless nights.[5]

What was needed to settle the issue was not merely a rehash of the conflicting anecdotes for and against design, but a conceptual framework on which to organize and probe this trove of disparate evidence. What was needed was a *scientific theory* of design detection—a theory that could detect its presence, confirm its absence, and was falsifiable and thus testable by observation and experiment. This was a big ask. The reason it was so difficult to make the design question a genuinely scientific question is that a cogent theory of design—a method by which to detect its presence and confirm its absence—requires deep understanding of biology and mathematics, and a subtle and clear understanding of the treacherous philosophical terrain on which any scientist probing design in biology must walk. Particularly necessary was skill with the philosophy and mathematics of probability theory, which is an indispensable part of the process of testing the design inference. Few thinkers, no matter how astute and how motivated, had the diligence and resources to apply such insight to the study of living things.

Bill Dembski, fortunately, did. The brilliance of Dembski's work is that it provides a rubric—the Explanatory Filter—by which design can be confidently inferred. The Filter can be applied to a host of scientific inquiries—forensic science, origin-of-life research, SETI, cryptography, and archeology, among many others. It provides mathematical and logical rigor to an enduring human intuition that design can be inferred in nature. The Explanatory Filter, and the conceptual framework that undergirds it, describes a quantitative method, which by eliminating chance and necessity infers design. In particular, design becomes detectable by identifying a specifiable pattern in a highly improbable event. Although *The Design Inference* does not invoke Aristotle, its method of design detection, when applied to biology, dovetails nicely with Aristotle's definition of life as substance that seeks intrinsic goals.[6]

This second edition of *The Design Inference* is a fascinating in-depth exploration of the scientific and mathematical issues Dembski introduced twenty-five years ago. The authors carefully explain the conceptual development of the Explanatory Filter and its theoretical underpinning, specified complexity. Though they view biology as the ultimate proving ground for the design inference, they also show its applicability to a wide variety of human endeavors. They even discuss "bad" design, a topic frequently raised by Darwinists to undermine the design inference.[7]

Dembski's and Ewert's discussion in the epilogue on conservation of information, a topic not developed in the first edition, is enthralling. They make the profound observation that biological *processes*, and not simply biological *products*, can exhibit design. Just as the design inference is an analytic tool for products such as biomolecules, conservation of information is an analytic tool for the processes that originate biological products.

Biology is not merely the study of the structure of living things; it is the study of the biological processes that build, orchestrate, and transmit life, including the evolutionary process itself. And as mind-boggling as the structure of the DNA double helix and the intricate anatomy of the brain and camera eye are, the processes by which these components of living things arise and function and integrate with other living components are, in my view, even more fascinating and deeply beautiful. In this second edition of *The Design Inference*, the authors devote a few pages to conservation of information as a framework for the study of biological processes. The topic is so important, however, that it warrants a separate book, which they report is forthcoming.

The design inference provides two fundamental insights for scientific research that are quite different: a theoretical design inference and a methodological design inference. The theoretical design inference uses the tools of probability, information, and complexity theory. It infers design in nature via objective quantitative methods. Its ability to detect design by means of these tools has great scientific, philosophical, and theological salience. The methodological design inference, by contrast, is a heuristic method. It assumes design for the fruitful insights that this assumption is likely to yield. It provides a powerful tool to guide scientific research, independent of its theoretical implications. In my own experience, both the theoretical and the methodological design inference have been very important.

I converted from agnosticism (atheism, really, on my bad days) to Christianity about twenty years ago. I have always loved science, but I have long suspected that the Darwinian explanation for the marvelous workings of living things fell short of reality and very short of good science. Even in high school, I noticed that while physics, chemistry, and general biology were solid sciences, evolutionary biology seemed to be a collection of dogmatic just-so stories. My suspicions were affirmed in college when, as a biochemistry major, I was very uncomfortable with the inference that intricate molecular pathways arose without planning or design.

How did the Krebs cycle or DNA transcription and translation happen without intelligent agency? The Darwinian explanation—that such elegant molecular arrangements were merely the consequence of hundreds of millions of years of (biotic or prebiotic) natural selection—struck me as implausible in the extreme. And I noticed that my professors of biochemistry and evolutionary biology were always short on specifics—they couldn't account quantitatively for a single molecular structure or pathway using Darwinian mechanisms. I was told, implicitly, to take it by faith. If I believed, I would come to understand. So, through college and medical school I stuck with Darwinian explanations. I assumed that my doubts were the result of my ignorance of Darwinian theory, and when I understood evolution well enough, it would all make sense.

My conversion to Christianity is a complex story, but the design inference and the intelligent design movement played a big role. I came to see that my doubts about Darwinian explanations were well founded—my doubts were in fact based on good and rigorous science. Reading and understanding *The Design Inference* helped me to clearly see the evidence for God's unmistakable creativity and wisdom in living things. I was also deeply impressed by Dembski's application of rigorous mathematics to the question of design. It mirrors, I believe, the landmark work of computer scientist Judea Pearl in the development of causal analysis.[8] The design inference is the application of probability and information theory to the detection of purpose in living things. Like Pearl's groundbreaking work, Dembski's research provides us with the opportunity to move beyond mere statistical correlation and to detect causation and design in nature. What an exciting and powerful approach to biological science!

But the importance of the design inference goes well beyond theoretical science. The design inference is also heuristic—a powerful tool for biological research as well as a landmark in theoretical biology. Of course, the inference to design has been the cornerstone of biological science since Aristotle. Scientists take a living thing apart, just as engineers would take apart a manufactured device of unknown provenance, to understand the design principles by which it was made and by which it works. Biological research is quite literally the reverse-engineering of living things. Nothing in biology really makes sense without the inference to design.

The fundamental questions a biologist must ask include the following: What is the purpose of this structure? What does it accomplish in the organism? Why is it designed in this way and not another? Traditionally, these questions were framed within a broader theological question, namely, Why did God make it that way? Nowadays, in the shadow of the atheism that constrains modern science, scientists are careful to hide their design inferences (if they wish to remain employed). They generally infer design implicitly, not explicitly. But the design inference is *always* there, for the atheist as well as the theist. Imagine doing research on the heart without acknowledging that it is a pump, or on DNA without admitting that it contains a code, or on ribosomes or mitochondria without knowing they are protein factories or power plants inside cells.

I am a pediatric neurosurgeon and a research professor at Stony Brook University, and the design inference has been essential to my own research. One of the most common disorders I treat is hydrocephalus, which is the accumulation of cerebrospinal fluid in the brain. It afflicts millions of children and adults, and often threatens life. I have known for many years that our traditional understanding of its cause is woefully inadequate, and I have worked for twenty years to better understand it. For a century, neurosurgeons have believed that hydrocephalus is caused by a blockage of the cerebrospinal fluid absorption sites in the brain (e.g., due to infections or hemorrhage or congenital deformities), but many of us in the hydrocephalus research community knew that this understanding of the disease was inadequate.[9] The cornerstone of our new approach to understanding hydrocephalus was the work of Dan Greitz, a neuroscientist in Stockholm who used MRI studies to show that hydrocephalus is intimately connected to abnormalities of the way the brain pulses in response to the arterial pulse from the heart.[10]

Following on Greitz's research, we noted two remarkable things about the coupling of the heart pulse and the brain pulse.[11]

First, the brain selectively suppresses the energy of the arterial pulse around the frequency of the heartbeat, thereby protecting the delicate brain capillaries from damage due to the high energy of the heartbeat. This phenomenon has been found in other organs of the body and is called the *windkessel effect*. Second, we found that the brain pulse *precedes* the heartbeat by about 150 milliseconds—that is, the pulse in the brain happens *before* the pulse from the heart reaches the brain, which supposedly causes it! Clearly, a new understanding of intracranial dynamics was needed.[12] I searched the physiological literature for an explanation of these findings. There was none—in fact, previous medical researchers weren't even aware of these remarkable characteristics of the pulse in the brain.

To understand the dynamics of the arterial pulse and the wind-kessel effect in the brain, I applied the Explanatory Filter from the design inference. These characteristics of the brain pulse were clearly not due to a regularity or necessity, because in nature the selective suppression of forced vibrations does not regularly occur at the frequency of the forcing pulse (it can occur at any frequency naturally) and the timing of the vibrations varies considerably. The characteristics of the brain pulse were certainly not due to chance, because the finding of frequency-specific suppression and phase lead of the brain pulse is quite consistent across all experiments and is found in all animals and humans.[13] It is also unimaginably improbable that chance mutations, whether or not coupled to natural selection, could have engineered such an exquisite system in the step-by-functional-step process required of natural selection, creating a functional brain pulse in a common ancestor of the various living forms that possess it. So that left us with one explanation—the pulsatile dynamics in the brain are designed.

For me, as for others, that inference was not the end of the investigative process, but the beginning. I turned to the study of design, which is engineering. I bought every engineering book I could find on harmonic motion and vibration suppression. I studied how human designers build systems to suppress pulses in machines, in electrical circuits and in pipes carrying water and gas. I learned that there was a design for vibration suppression called a band stop filter that specifically suppresses frequencies near the fundamental frequency of the harmonic force causing the vibration (i.e., the heart rate), just as we had found happens in the cerebral windkessel mechanism.

Together with my engineering colleagues at Stony Brook University, we simulated the windkessel effect that we measured in a series of experimental dogs on a simple band stop filter electrical circuit.[14] The voltage in the circuit corresponded to the pressure in the brain, the current in the circuit corresponded to the motion of fluids and tissues, and the charge in the circuit corresponded to the displacement of intracranial fluids and tissues. The output of the circuit was almost identical to the pressure tracings from the brain. A simple electrical circuit designed according to data from pulsatility in the brain simulated the dynamics of the pulse in the brain with remarkable accuracy. Both systems— the designed circuit and the pulsing brain—worked on the same dynamic principles.

We carefully studied the dynamics of the simple circuit and came to understand how the windkessel effect in the brain worked—the rhythmic expansion and relaxation of the brain suppressed the arterial pulse in a manner analogous to the rhythmic loading and unloading of the capacitor in the electrical circuit. This shed light on hydrocephalus, which we now understand as a disorder of the cerebral windkessel mechanism caused by excessively high resistance in the pulsatile motion (the pulsating "current") of cerebrospinal fluid in and around the brain.

Furthermore, our design-based research provided an answer to the perplexing observation that the pulse in the brain *precedes* the arterial pulse entering the brain by about 150 milliseconds. In a steady-state oscillating system hampered by appreciable resistance to harmonic motion (which is certainly the case with the brain), the optimal pulsation suppression occurs when the mass of the pulse is increased, which results in a leading phase between the pulse that is being suppressed (the brain pulse) and the forcing pulse (the arterial pulse). Our design-based research explained the perplexing lead of the brain pulse with respect to the arterial pulse— it optimizes the windkessel mechanism in order to protect brain capillaries. The cerebral windkessel mechanism is not merely designed. When functioning normally it is *optimally* designed and carefully tuned!

Our design research is ongoing, but the design inference is transforming our understanding of blood and cerebrospinal fluid flow in the brain and providing novel insights into treatment of disorders such as hydrocephalus, stroke, and brain trauma. Our work has been predicated on the reverse engineering of intracranial dynamics—all of which is based on the design inference.

This book, *The Design Inference*, is ideologically neutral. Its design inferential logic is a tool for research, not an assertion of faith. It confers scientific rigor on the inescapable everyday intuition that we can discern design in many of the designed objects around us. This logic can provide compelling evidence for design and, as an impartial arbiter, it can be used to raise legitimate doubts in cases where someone may have erroneously ascribed design. It is as relevant to the Darwinist as to the design scientist. Both intelligent design theory and Darwinian theory key off of the design inference, as opposite sides of the same coin.

Oddly, the Darwinian reaction to the design inference has not been analytic. It has been allergic. The quest for the truth about design in biology seems to be the farthest thing from the Darwinian mind. This book's generous gift of a scientific methodology that could convincingly confirm or disconfirm Darwinian theory has been met by Darwinists not by relief and gratitude but by evasion and contemptuous silence punctuated occasionally by outright malice.[15] Some of the explanation for this unscientific response is undoubtedly the challenging mathematics inherent to the detection of specified complexity and the discomfort evolutionary biologists understandably feel over the quantitative analysis of their own claims. Much of the Darwinian opposition to the design inference is ideological.

What is needed in biology is a willingness by biologists to subject Darwinian evolutionary claims to objective quantitative testing. The time for "mountains" of anecdotes has passed. In this superb new edition of *The Design Inference*, Dembski and Ewert set design science on a solid scientific foundation and provide scientists with the opportunity to test their theories for and against design using objective quantitative methods. It's hard to escape the conclusion that the Darwinists' striking refusal to engage this work thoughtfully and honestly is motivated by the strength, not the weakness, of the challenge the design inference poses to modern evolutionary theory.

Only by applying the design inference will biologists be able to decisively confirm their theories about the origin of biological complexity, specification, and information. The risk, of course, is that doing so will instead disconfirm them. But such are the hazards of any scientific enterprise, whose primary concern must always be the rigorous pursuit of truth. Full scientific rigor in the biological sciences requires the logic of the design inference as laid out in this book.

Michael Egnor
Director of Pediatric Neurosurgery
Renaissance School of Medicine
Stony Brook University

INTRODUCTION TO THE SECOND EDITION

THE FIRST EDITION OF *THE DESIGN INFERENCE* APPEARED IN 1998 with Cambridge University Press. It evoked strong reactions, both pro and con. Subtitled *Eliminating Chance Through Small Probabilities*, it offered a statistically rigorous method for eliminating chance and detecting design. Published by a major university press in a statistical monograph series (Cambridge Studies in Probability, Induction, and Decision Theory), it promised to make the investigation of design in nature scientifically rigorous. Its immediate effect was to energize the then newly emerging intelligent design movement, giving credibility to its scientific aspirations. In the monograph's wake, the intelligent design (or ID) movement has grown and in places even flourished.

Many have sought to conscript, analyze, and refute the first edition's method of design detection, or what came to be called simply *the design inference*.[1] That method, as will be argued in this second edition, remains valid. Yet the need for simplifications, clarifications and, above all, extensions to the first edition became evident quickly after it was published. For instance, the concept of specification, so central to this entire project, admitted a significant simplification: it proved possible to dispense entirely with one of its defining conditions (the conditional independence condition) by more effectively deploying the other defining condition (the tractability condition). Or take the Explanatory Filter, a flowchart offering a quick intuitive reconstruction of the design inference. It required further clarification to ensure that its terminal nodes covered all explanatory options adequately and consistently.

Several emendations along these lines made it into the sequel to *The Design Inference*, a book titled *No Free Lunch: Why Specified Complexity Cannot Be Purchased without Intelligence* (2002), as well as into *The Design Revolution: Answering the Toughest Questions About Intelligent Design* (2004). Yet none of this additional work supplanted the original monograph, which remained the primary focus of attention among supporters, sympathizers, and critics. As critics raised objections, I [WmD] would typically write rejoinders, sometimes responding with full-fledged articles. Some of the objections proved baseless. Others rightly identified room for improvement. Where they were baseless, I attempted to show why. Where they called for improvement, I attempted to offer the improvement. Throughout all this, I held that the design inference as developed in the first edition was, in broad strokes, sound and that any faults in it were peripheral and could be fixed. I still maintain that.

The Connection with Information

After the publication of the first edition, the question of how design inferences connected with information theory became increasingly urgent. In the epilogue to the first edition, I explicitly made that connection. There I sought to identify what was "the main significance of the design inference for science." My answer was that "the design inference detects and measures *information*."[2] I then sketched in the few final paragraphs of the first edition how the relation between specifications and small probability events mirrored the transmission of messages along a communication channel from sender to receiver.

By 2004, I had made some progress in detailing the connection between design inferences and information. But a pressing concern about the concept of specified complexity remained. Specified complexity is a term that does not appear in the first edition, but is there conceptually in the form of specified improbability, where it is presented as the key to design detection. Complexity, as used in the term "specified complexity," thus becomes equivalent to improbability. That equivalence makes good sense. Take the tossing of a fair coin. The more it is flipped, the greater the improbability of the sequence of heads and tails that is observed, but also the greater the complexity of that sequence as measured in bits. The probability of the sequence thus coincides precisely with the number of bits needed to describe it.

Yet even with such information-theoretic analogues, specified complexity did not seem to constitute a clearly defined informational concept. Specified complexity was, in the first edition, ascertained by going down a checklist and confirming that certain logical, statistical, and informational conditions were met. Information was there in the mix, but it also seemed that other non-informational things were happening. Moreover, specified complexity's precise connection with Shannon information, algorithmic information, or any other well-established mathematical theory of information remained less than obvious.

My friend and colleague Jay Richards, guest-editing a special issue of the philosophy journal *Philosophia Christi*, asked me to address this concern. I did so in an article titled "Specification: The Pattern That Signifies Intelligence," which appeared in that journal in 2005. There I showed how specified complexity could be defined as a unified information measure constructed by suitably combining Shannon information, as applied to an event, with algorithmic information, as applied to a pattern. This definition resolved the problem of how specified complexity could be a form of information. It was conceptually satisfying in that it encapsulated into a single information measure the diverse elements of the design inference.

An Argument from Ignorance?

None of these efforts to articulate the informational underpinnings of design inferences, however, assuaged the critics. Tacitly in the first edition of *The Design Inference* and explicitly in its sequel, *No Free Lunch*, I argued that natural selection and random variation could not create the sort of complexity we see in living things. My approach in applying the design inference to biology was to piggyback on the work of design biologists such as Douglas Axe and Michael Behe. They had identified certain subcellular systems (e.g., bacterial flagella and beta-lactamase enzymes) that proved highly resistant to Darwinian explanations.

Our joint task was to put plausible numbers to these systems so that even factoring in Darwinian natural selection, the probability of these systems arising was exceedingly small. Note that the specification of these systems, as in their exhibiting the right sort of pattern for a design inference, was never in question. The issue was always whether the probabilities were small enough. In using specified improbability to draw a design inference for biology, I therefore needed to argue that the probabilities for Darwinian processes producing certain biological systems, such as those identified by Axe and Behe, were indeed small.

As far as Darwinists were concerned, however, all attempts to show such biological systems to be vastly improbable were misguided and irrelevant. Any design inferences meant to defeat Darwinian evolution were, according to them, arguments from ignorance. For them, unidentified Darwinian pathways could never be decisively ruled out, so their mere possibility invalidated any design inference applied to biological evolution. In short, no calculated improbability could ever convince the Darwinian critics that the probabilities were actually small.

It didn't matter that Darwinists were ignorant of any detailed evidence for such Darwinian pathways, and thus had no counter-probabilities to offer. It was enough for them merely to gesture at the possibility of such pathways, as though raising a possibility could itself constitute evidence for an argument from improbability. To ID proponents critical of Darwin's theory, the argument-from-ignorance objection seemed to apply more aptly to the Darwinists themselves for positing unsubstantiated Darwinian pathways that offered no nuts and bolts, no nitty-gritty, just hand-waving.

No matter. For Darwinists to refute ID, they merely needed to postulate unidentified, and perhaps forever unidentifiable, indirect Darwinian pathways in which structure and function coevolved and led to the complex biological features in question. Brown University biologist Kenneth Miller led the way. Michael Behe had defined a system (biological or otherwise) to be *irreducibly complex* if its function was lost by removing key parts. He argued that such systems resisted Darwinian explanations. Miller countered that Behe's concept of irreducible complexity was ill-conceived be-cause removing parts from, or otherwise simplifying, a biological system could always yield a system with a different function. To convinced Darwinists like Miller, design in biology was therefore a nonstarter. Darwinian pathways to all complex biological systems had to exist, and any inability to find them simply reflected the imperfection of our biological knowledge, not any imperfection in Darwin's theory.

Richard Dawkins, better than anyone, has publicly championed the dogma that Darwinian pathways can and must always exist for any biological system. In a 1990s television interview he memo-rably took Behe to task for claiming that irreducibly complex biochemical machines, of the sort Behe popularized in *Darwin's Black Box*, were beyond the reach of Darwinian processes. Dawkins charged Behe with being "lazy" (yes, he used that very word) for seeing in the irreducible complexity of these machines a reason to conclude design, and thus to rule out any further effort to discover how Darwinian processes could have formed, say, a bacterial flagellum. That is, instead of concluding that these systems were designed by a real intelligence, Behe should get back into the lab and redouble his efforts to discover how Darwinian evolution could have produced them apart from design.[3]

The reaction of the ID community to Dawkins' "laziness challenge" was that he might just as well have recommended to physicists that they keep trying to construct a perpetual motion machine. Yet why did one task seem futile (constructing a perpetual motion machine) but not the other (discovering Darwinian pathways to irreducibly complex biochemical machines)? Physicists had the second law of thermodynamics to rule out the charge of laziness. That's why Dawkins would never have said to a physicist, "You're just being lazy for giving up on inventing a machine that can run itself forever."

Even so, Dawkins' "laziness challenge" was and remains misguided because Behe's skepticism is based not on ignorance but on careful study of the obstacles that Darwinian evolution must overcome and its consistent failure to do so. To seal the deal, however, the ID research community still needed something like the second law for biology. We found it in the *law of conservation of information*. This law logically completes the design inference. We'll address this law in the epilogue. The law of conservation of information is a vast topic and requires a book of its own, a book we intend to finish after this second edition appears in print.

Natural Selection as the Great Designer Substitute

Darwinian critics, however much they were willing to permit design inferences in other contexts, reflexively ruled them out as soon as they impacted biology or cosmology or anyplace where a non-natural designer might be implicated. They thereby gutted the design inference of any larger worldview significance, ensuring that it could never be applied to humanity's really big and important questions.

Early in *The Blind Watchmaker*, Dawkins stated that life is special because it exhibits a "quality" that is "specifiable in advance" and "highly unlikely to have been acquired by random chance alone."[4] All the elements of specified complexity are there in Dawkins' characterization of life. Yet Dawkins, along with fellow Darwinians, did not see in specified complexity a marker of actual design but rather the outworking of natural selection, naturalism's great designer substitute. For Dawkins, natural selection removes the small probabilities needed to make the design inference work. As he remarks, the "belief, that Darwinian evolution is 'random,' is not merely false. It is the exact opposite of the truth. Chance is a minor ingredient in the Darwinian recipe, but the most important ingredient is cumulative [i.e., natural] selection, which is quintessentially nonrandom."[5] *Nonrandom* here means, in particular, *not all that improbable.*

For a design inference to properly infer design, two conditions must be met:

1. an observed outcome matches an independently identifiable pattern, or what we call a specification (what Dawkins means by "specifiable in advance"); and

2. the event corresponding to that pattern has small probability (think of the pattern as a target and an arrow landing anywhere in it as the corresponding event).

With these conditions satisfied, the design inference ascribes such an observed outcome to design. Dawkins finds no fault with this form of reasoning provided the probabilities are indeed small. He even admits that scientific theories are only "allowed to get away with" so much "sheer unadulterated miraculous luck" but no more.[6] Dawkins is here expressing the widespread intuition that certain events are within the reach of chance but that others are not. He's right that people widely embrace this intuition, and he's right that this intuition applies to science.

Given his view that scientific theorizing can only permit a limited amount of luck (a view ID proponents share), Dawkins would be forced to concede that if randomness were operating in the evolution of life, the resulting probabilities would be small, and a design inference would be warranted. As evidence that Dawkins does indeed make this concession, consider the way he commends William Paley's design argument. In a remarkable moment of candor, Dawkins writes, "I could not imagine being an atheist at any time before 1859, when Darwin's *Origin of Species* was published... [A]lthough atheism might have been logically tenable before Darwin, Darwin made it possible to be an intellectually fulfilled atheist."[7] According to Dawkins, but for Darwin, we would be stuck with Paley and compelled to be theists.

Darwin, in positing natural selection as the driving force behind evolution, was thus seen as breaking the power of classical design arguments. Natural selection, with its ability to heap up small incremental improvements, would allow evolution to proceed gradually, baby step by baby step, overcoming all evolutionary obstacles. In proceeding by baby steps, Darwinian evolution is supposed to mitigate the vast improbabilities that might otherwise constitute insuperable obstacles to life's evolution, substituting at each step probabilities that are eminently manageable (not too small).

Darwinian processes, by overcoming probabilistic hurdles in this way, are thus said to banish design inferences from biology. The actual small probabilities needed for a valid design inference, according to Dawkins and fellow Darwinian biologists, thus never arise. Indeed, that was Dawkins' whole point in following up *The Blind Watchmaker* with *Climbing Mount Improbable*. Mount Improbable only seems improbable if you have to scale it in one giant leap, but if you can find a gradual winding path to the top (baby step by baby step), getting there is quite probable.

Dawkins never gets beyond such a broad-brush description of how vast improbabilities that might otherwise dog evolution can be mitigated. As it is, there are plenty of probabilistic situations in which each step is reasonably probable but the coordination of all these reasonably probable events contributes to an outcome that is highly improbable. Flip a coin a hundred times, and at each flip the coin is reasonably likely to land heads. But getting a hundred heads in a row is highly improbable, and we should not expect it to happen by chance. Dawkins doesn't just need reasonably sized probabilities at each step, but a kind of coordination or ratcheting that locks in prior benefits and keeps striving for and accumulating future benefits. Showing that natural selection possesses this power universally goes well beyond what he, or any other Darwinian biologist, ever established probabilistically.

In short, Darwinian critics of the design inference conflate *apparent* specified complexity with *actual* specified complexity.

Darwinists like Dawkins grant that actual specified complexity warrants a design inference. But they view the Darwinian mechanism of natural selection as a probability amplifier, making otherwise improbable events probable and thus rendering them no longer complex. As a consequence, it does not matter that specified complexity, as a matter of statistical logic, warrants a design inference because, according to Darwinists, life does not actually exhibit specified complexity. Darwinists will, to be sure, claim that the Darwinian mechanism creates specified complexity. But what they really mean is that the Darwinian mechanism causes life to exhibit the illusion of specified complexity. Living systems only seem to be highly improbable, but they're not once you understand how Darwinian evolution brings them about. In this way, the majority of evolutionary biologists, insofar as they understand the design inference at all, rationalize it away.

Since the publication of the first edition of this book, the debate over the design inference and its applicability to evolution has centered on whether such gradual winding paths exist and how their existence or non-existence would affect the probabilities by which Darwinian processes could originate living forms. Design theorists have identified a variety of biological systems that resist Darwinian explanations and argued that the probability of such systems evolving by Darwinian means is vanishingly small. They thus conclude that these systems are effectively unevolvable by Darwinian means and that their existence warrants a design inference. In this book, we recap that debate and contend that intelligent design has the stronger argument.

Reading the Second Edition

This second edition firmly supersedes the first edition. The first edition may remain of historical interest, but this new edition is now the canonical place for anyone who wants to assess whether, and to what degree, small-probability design-inferential reasoning has merit. The first edition was notation heavy and terminologically hard going. The second edition, while more thorough and rigorous, is also conceptually clearer and cleaner. Readers with no exposure to the first edition are invited simply to dive into this second edition. Readers of the first edition, on the other hand, may first want to read Appendix C.5, "Transitioning from the First to the Second Edition."

The first two chapters of this new edition ("The Challenge of Small Probabilities" and "A Sampler of Design Inferences") are readily accessible to the general reader. They show just how widespread design inferences are and how deeply they pervade human thought and experience. Like the American Express Card, we can't leave home without design inferences in our mental toolchest. We use them everywhere. And even if we haven't explicitly articulated to ourselves the core concepts of specification and small probability, we understand and employ these concepts intuitively.

The next two chapters ("Specification" and "Probabilistic Resources") lay out the twin pillars on which design inferences rest, namely, the type of patterns conducive to identifying the effects of intelligence (i.e., the specifications) and the relevant opportunities for an event to happen by chance (i.e., its probabilistic resources). These chapters, as well as the subsequent chapters, are more technical than the first two. To help in understanding these later chapters, we've included Appendix A ("A Primer on Probability and Information"). Readers are advised to look at this appendix after reading the first two chapters. Readers who want to delve still deeper into specification should look at Appendix B ("Select Related Topics"), and specifically at Appendices B.1–B.4.

Chapter 5 ("The Logic of the Design Inference") ties together the work on specification and probabilistic resources from the two previous chapters. It lays out the formal logical apparatus for small-probability chance-elimination arguments in general and for design inferences in particular. In motivating this formal apparatus, Chapter 5 also includes a detailed account of the suspicious ballot line selections by New Jersey clerk Nicholas Caputo (who was "immortalized," as philosopher Charles Chastain put it, in the first edition). This chapter also includes the Explanatory Filter, a flow chart that supplies a user-friendly reconstruction of design inferential reasoning. If a reader remembered just one thing from the first edition, it tended to be the Filter.

Chapter 6 ("Specified Complexity") provides a mathematically rigorous account of specified complexity. It thus fills in some key technical details in our account of specification, namely, what it means to say, with full rigor, that a description of an event has short length. It then shows how specification combines with small probability to form a unified information measure capable of underwriting design inferences. This chapter draws on theoretical computer science (notably, algorithmic information theory) to convincingly answer certain technical concerns that were inadequately addressed in the first edition. This chapter also provides some detailed examples illustrating specified complexity.

Chapter 7, the last chapter ("Evolutionary Biology"), provides the ultimate proving ground for this book's key claims. If the design inference were confined to the effects of human agency, it would be uncontroversial. The problem is that when applied to certain features in nature (biological and cosmological), it suggests design where no human or alien intelligence could have been involved. Naturalistic researchers therefore reflexively attempt to invalidate such design inferences. We argue, in contrast, that the design inference legitimately extends to these cases. Specifically, certain aspects of biological systems convincingly warrant full-fledged design inferences. Appendices B.5–B.8 expand on this chapter, containing much interesting material, some of which is quite accessible. We close with an epilogue that briefly discusses conservation of information, which complements the design inference and is the topic for a sequel to this book.

Interspersed at a handful of places in the text are excursus that provide helpful elaborations. Most of these started as endnotes that became unwieldy because of their length; yet they also seemed important enough to include in the main body of the text lest they get overlooked in the endnotes. There are nine such excursus: "The Need for BOTH Specification AND Small Probability" (Section 1.3), "The Metaphysics of Chance" (Section 1.4), "David Hume versus Thomas Reid" (Section 2.8), "Probabilistic Complexity and

Descriptive Simplicity" (Section 3.1), "The Metaphysics of Design and Information" (Section 5.5), "The Reason for Specified Complexity's Demotion" (Section 6.1), "Variations on the WEASEL" (Section 7.1), "Who Is Arguing from Ignorance?" (Section 7.2), and "The Breakdown of Evolvability" (Section 7.6). Each excursus is signaled before and after with a double-helix icon.

Like the first edition, this second edition remains a research monograph. But also like the first edition, this second edition is intended to address a wide range of interests, engaging technical and non-technical readers alike. We therefore encourage all readers to begin at Chapter 1 and read sequentially through Chapter 7 and then the Epilogue, referring to the appendices as needed. Readers with a technical background should plough directly through the material from start to finish. General readers, however, should skip the parts they find impenetrable, yet realizing that they will still glean plenty of value by doing so. In particular, general readers should avoid giving up the first time they hit one of the more daunting technical/mathematical sections. In this book, something readily comprehensible always follows something less comprehensible.

A Long and Winding Road

This second edition might have been completed a decade ago (though the additional gestation time has made it a better book). Several obstacles intervened. For one, I [WmD] had received a Templeton Foundation book award in 1999. In applying for the award, I had to submit a detailed book proposal. In winning it, I had agreed to write a book on the science and metaphysics of information. I decided to turn this into two books, one on the science of information, which appeared in 2002, titled *No Free Lunch*, and the other on the metaphysics of information, titled *Being as Communion*.[8]

I kept delaying the latter book because it was meant to bridge information theory with evidence of purpose in nature, and that very connection was the subject of intense research by me, at first solo and then later in collaboration with colleagues at the Evolutionary Informatics Lab (especially on the topic of conservation of information). I finally felt in a position to write *Being as Communion* in 2013, wrote it that year, and then published it with Ashgate's science and religion series in 2014. *Being as Communion*, many years overdue, finally made good on my Templeton award.

With the publication of *Being as Communion*, the way was clear to begin work on the second edition. So why didn't I start right away? There were other reasons for the delay. Cancel culture is in the news these days, but I experienced it two decades earlier as an academic working on intelligent design trying to keep his head above water. By the early 2010s I had had enough and decided to leave academia for good. A friend of mine, who had done well with online businesses, took me under his wing and showed me the ropes. Through his guidance, I became an entrepreneur developing educational websites and technologies. In 2013, I left college and university life, never to look back. It was the right move for me to become a businessman, and my life has been much happier since. But it also meant that I now had a full-time non-academic day job and that my attention to intelligent design would henceforth be much reduced.

It may seem strange, but the main obstacle for me personally to doing this second edition came from feeling at odds with my publisher, Cambridge University Press. My editor at Cambridge, Terence Moore, had initially been quite happy with the first edition of *The Design Inference*. It had been the bestselling Cambridge philosophical monograph over the previous five years.[9] I therefore approached him about whether Cambridge would be open to publishing a sequel that made explicit the connection between design inferences and biological evolution.

But by that point, my own views in favor of intelligent design and critical of Darwinism were public knowledge. Pulling back the curtain on the inner workings of the press, Moore told me that members of the Cambridge Syndicate (especially one or two biologists) opposed my work on intelligent design, and that even if such a sequel were approved on this side of the Atlantic, the Syndicate would likely deep-six it once it went for approval on the other side of the Atlantic. Normally, the Syndicate rubberstamps books recommended on this side of the Atlantic. In my case, the Syndicate was going to make an exception.

Terence Moore was a stand-up guy, and I'm grateful for his plain speaking. Sadly, he died of cancer shortly after this exchange. As it is, I ended up publishing *The Design Inference*'s sequel, *No Free Lunch*, with another press, Rowman and Littlefield. For many years, I didn't look back. Instead, I simply kept extending the ideas and technical machinery inspired by the first edition of *The Design Inference*, without any thought of revising and upgrading it in a second edition. And on leaving the academy to become a tech entrepreneur, I was even less incentivized to do a second edition.

But the thought that my publisher, Cambridge University Press, was not in my corner kept nagging me. Finally, in 2020, on impulse, I contacted Cambridge University Press and asked for the rights to the book back. Within a week, I had the rights back and now could do with the book whatever I wished. The sense of inertia I had felt about doing a second edition now vanished. To pursue the work in earnest, I therefore solicited a grant from Seattle's Discovery Institute. They had supported the work on the first edition, and now generously provided a grant for the second. And so, the path was cleared for this second edition of *The Design Inference*.

Acknowledgments for the Second Edition

When Winston Ewert agreed to come on board as a full co-author of this second edition of *The Design Inference*, I [WmD] sensed that this would be a fruitful collaboration, as has now been confirmed by this finished manuscript. The two of us have collaborated since 2009 on the information-theoretic underpinnings of intelligent design, many aspects of which trace back to the first edition of this book. Winston's background in computer science complements mine in pure mathematics, making this second edition a far richer and more stimulating volume than the first.

The first edition of *The Design Inference* was praised as well as criticized from its inception. Many scholars over the course of more than two decades offered insights, clarifications, and objections to ideas expressed in the first edition. Their engagement with those ideas has helped make this second edition a much better book than the first, and so we want to acknowledge at least some of them here. Twenty-five years is a long time, and memories grow fuzzy. Of course, all the people acknowledged in the first edition deserve still to be acknowledged, which is why we have included the acknowledgment to the first edition in an appendix to this volume (Appendix C.4).

Our good friend and colleague Robert Jackson "Bob" Marks II stands out for thanks and commendation. As a key collaborator of mine [WmD] since 2004 and as Winston's doctoral supervisor in electrical and computer engineering at Baylor University, he has played a crucial role in taking the ideas articulated in the first edition and bringing them to their present maturity. Here it's necessary also to mention George Montañez, who did graduate studies under Bob's supervision at Baylor at the same time as Winston, and then went on for a doctorate in computer science at Carnegie Mellon. The four of us have contributed to a growing list of peer-reviewed articles on the conservation of information, which takes the design inferential apparatus developed in this book to its logical conclusion (a topic touched on in the epilogue and deferred to a follow-up book—stay tuned!).

Seattle's Discovery Institute was pivotal in bringing the first edition of this book to life back in the late 1990s, and it is likewise pivotal in bringing this second edition to life now. Both authors are grateful for Discovery's financial support of this second edition, but also for its continuing support of the entire ID research program. Its fellows and staff are the heartbeat of the ID movement. Warm thanks to all of them! Looking over the people acknowledged twenty-five years ago in the first edition, we find most still alive, actively contributing to ID, and cheering on the work of this second edition. Stephen Meyer, Paul Nelson, Jonathan Wells,

Michael Behe, David Berlinski, Bruce Gordon, Jay Richards, Douglas Axe, Walter Bradley, and the late Phillip Johnson (1940–2019) stand out in this regard.

In their leadership of Discovery Institute, Stephen Meyer, John West, Casey Luskin, Steven Buri, and Bruce Chapman have been visionaries in advancing intelligent design. We are especially grateful to them for seeing the importance of this project and positioning it to succeed. Also to be commended is the team at Discovery Institute Press—including Jonathan and Amanda Witt—whose editing and production skills make good books better—a lot better. In this regard, we also want to acknowledge publisher Jon Buell (1939–2020), whose Foundation for Thought and Ethics was assimilated into Discovery Institute Press. In the 1970s and '80s Buell was crucial to launching the ID movement, notably by arranging the publication of *The Mystery of Life's Origin* by the mainstream academic press Philosophical Library.

Others we would like to acknowledge for their contribution to this book include the following: Jake Akins, the American Scientific Affiliation, Eric Anderson, Barry Arrington, Joe Audi, Pam Bailey, James Barham, Jonathan Bartlett, Chris Bishop, John Bracht, Linda Montgomery Buell, Ashby Camp, Gregory Chaitin, David Chiu, the Christian Scientific Society, Sal Cordova, Paul Davies, Michael Denton, Daniel Andrés Díaz-Pachón, Kirk Durston, Michael Egnor (special thanks for his wonderful foreword!), Mark Fitzmaurice, Michael Flannery, Ann Gauger, Mike Gene, Howard Glicksman, Guillermo Gonzalez, Niels Henrik Gregersen, Charles Harper, Joseph Holden, Cornelius Hunter, Muzaffar Iqbal, Stuart Kauffman, David Klinghoffer, Rob Koons, Erik Larson, Steve Laufmann, John Leslie, Rick Martinez, Sean McDowell, Tim and Lydia McGrew, Jonathan McLatchie, Angus Menuge, Brian Miller, Todd Moody, the late Fr. Richard John Neuhaus, Denyse O'Leary, the late Willis Page, Nancy Pearcey, J. Brian Pitts, Martin Poenie, Mary Poplin, Alex Pruss, Terry Rickard, the late Noel Rude, Douglas Rudy, Michael Ruse, Andrew Ruys, John Sanford, James Shapiro, Rob Sheldon, Rupert Sheldrake, Michael Shermer, Edward Sisson, Fred Skiff, Wolfgang Smith, David Snoke, Micah Sparacio, Richard Sternberg, John Stonestreet, Iain Strachan, the

Templeton Foundation, James Tour, Ide Trotter, Frank Turek, Fr. Pasquale Vuoso, Rob Ward, and Bill Wilberforce.

Finally, we want to remember and thank our families. Winston's wife Elyse and Bill's wife Jana deserve special thanks for their patience, support, and love throughout the editing and writing. Winston is now at the age that Bill was when he published the first edition of this book. Happily, Winston's parents are still alive to enjoy seeing this work come to press. Bill's parents passed away in the intervening years. The first edition was dedicated to Bill's parents, Bill and Ursula. In this second edition, Winston's parents, Ken and Micki, are added to the dedication.

William A. Dembski
Winston Ewert

CHAPTER 1

THE CHALLENGE OF SMALL PROBABILITIES

1.1 Historical Backdrop

ELIMINATING CHANCE THROUGH SMALL PROBABILITIES HAS A LONG history. In his dialogue on the nature of the gods, written over two millennia ago, Cicero remarked, "If a countless number of copies of the one-and-twenty letters of the alphabet, made of gold or what you will, were thrown together into some receptacle and then shaken out on to the ground, [would it] be possible that they should produce the *Annals* of Ennius?... I doubt whether chance could possibly succeed in producing even a single verse!"[1] In this passage, Cicero contemplates the chance assembly of words by randomly tossing what are essentially Scrabble pieces. Cicero concludes that even a single verse of poetry is beyond the reach of chance. Yet it is within the reach of poets, who obviously act by design.

Eighteen centuries later, the Marquis Pierre Simon de Laplace would question whether Cicero's method of randomly tossing letters could by chance produce even a single word: "On a table we see letters arranged in this order, *Constantinople*, and we judge that this arrangement is not the result of chance, not because it is less possible than the others, for if this word were not employed in any language we should not suspect it came from any particular cause, but this word being in use among us, it is incomparably more probable that some person has thus arranged the aforesaid letters than that this arrangement is due to chance."[2] A whole book, a single verse, even a long word are so unlikely that we attribute the arrangement of their letters to some cause other than chance. For Laplace, that cause was a person acting by design.

Thomas Reid, in his 1780 *Lectures on Natural Theology*, argued against the absurdity of asserting chance in the face of small probabilities: "Can any thing done by chance have all *the marks of design*? If a man throws dice and both turn up aces, if he should throw 400 times, would chance throw up 400 aces? Colors thrown carelessly upon a canvas may have some rude appearance of a human face, but would they form a picture beautiful as the Coan Venus? A hog grubbing the earth with his snout may turn up something like the letter A, but would he turn up the words of a complete sentence?"[3] These are rhetorical questions, and the answer to each is, obviously, No. Moreover, in referring to "the marks of design," Reid clearly intends the alternative to chance to be design.

In the preface to his classic treatise on gambling, Abraham de Moivre contrasts chance with a mode of explanation he calls design:

> The same Arguments which explode the Notion of Luck, may, on the other side, be useful in some Cases to establish a due comparison between Chance and Design: We may imagine Chance and Design to be, as it were, in Competition with each other, for the production of some sorts of Events, and may calculate what Probability there is, that those Events should be rather owing to one than to the other. To give a familiar Instance of this, Let us suppose that two Packs of Piquet-Cards being sent for, it should be perceived that there is, from Top to Bottom, the same Disposition of the Cards in both packs; let us likewise suppose that, some doubt arising about this Disposition of the Cards, it should be questioned whether it ought to be attributed to Chance, or to the Maker's Design: In this Case the Doctrine of Combinations decides the Question; since

it may be proved by its Rules, that there are the odds of above 263130830000 Millions of Millions of Millions of Millions to One, that the Cards were designedly set in the Order in which they were found.[4]

Nor has inferring design through small-probability chance-elimination arguments lost its force in our own day. When the twentieth-century statistician Ronald Fisher charged Gregor Mendel's gardening assistant with data falsification, it was because Mendel's data matched Mendel's theory too closely. In fact, the probability of so close a match was so small that Fisher could not in good conscience countenance ascribing the match to chance.[5] Obviously, in charging Mendel's gardening assistant with data falsification, Fisher was drawing a design inference. Data falsification is an area where design inferences prove themselves especially useful (see Section 2.3).

When biologist Richard Dawkins takes creationists to task for misunderstanding Darwin's theory of natural selection, it is for failing to appreciate that natural selection renders the emergence and development of living forms sufficiently probable to remove the need for a designer. As a matter of general principle, however, Dawkins concedes that eliminating chance through small probabilities constitutes a valid mode of reasoning. It's just that in the case of biological origins he does not deem the probabilities to be small enough. As he writes in *The Blind Watchmaker*:

> We can accept a certain amount of luck in our explanations, but not too much. The question is, *how* much? The immensity of geological time entitles us to postulate more improbable coincidences than a court of law would allow but, even so, there are limits. Cumulative selection [i.e., a selection process capable of cumulating advantages, as with natural selection in biology] is the key to all our modern explanations of life. It strings a series of acceptably lucky events (random mutations) together in a nonrandom sequence so that, at the end of the sequence, the finished product carries the illusion of being very very lucky indeed, far too improbable to have come about by chance alone,

even given a timespan millions of times longer than the age of the universe so far... [W]e are asking *how* improbable, how *miraculous*, a single event we are allowed to postulate. What is the largest single event of sheer naked coincidence, sheer unadulterated miraculous luck, that we are allowed to get away with in our theories and still say that we have a satisfactory explanation of life?[6]

Dawkins' reference to "coincidence" here is on point. It's not just sheer improbability that makes for unsatisfactory chance explanations. Indeed, plenty of highly improbable events happen by chance all the time. The precise sequence of heads and tails in a long sequence of coin tosses, the precise configuration of darts thrown randomly at a dart board, and the precise seating arrangement of people at a cinema are all highly improbable events that, apart from any further information, can reasonably be explained by appealing to chance.

But what happens when the precise sequence of heads and tails coincides with one recorded in advance, when the precise configuration of the darts coincides with the target's bullseye, and when the precise seating arrangement at the cinema coincides with seats people were assigned on their ticket stubs? Clearly, we would deny that these events occurred by chance. But why? It's not just the sheer improbability of each of these events, but also its coincidence with or conformity to a *pattern* that compels us to look beyond chance.

"Eliminating chance through small probabilities" must therefore be interpreted as an incomplete expression that includes an implicit assumption. Sheer improbability by itself is not enough to eliminate chance. Rather, to eliminate chance, we also need to know whether an event conforms to a pattern. But what sort of pattern? The easiest way to eliminate chance through small probabilities is to designate a pattern prior to an event, and then eliminate chance in case the event conforms to that pattern. This method of eliminating chance is common in statistics, where it is known as setting a *rejection region* prior to performing an experiment. In certain types of statistical inference, if the outcome of an experiment (= event) falls within the rejection region (= pattern), the chance hypothesis supposedly responsible for the outcome is rejected (i.e., chance is eliminated). A significant aspect of this book is generalizing the rejection regions by which chance may be eliminated (see especially Sections 3.5 to 3.8 and also Appendix B.3).

Patterns given before an event are great when they can be had. Yet patterns need not be given before an event to justify eliminating chance. Consider, for instance, Alice and Bob celebrating their fiftieth wedding anniversary. Their six children show up bearing gifts. Each gift is part of a matching set of china. There is no duplication of gifts, and together the gifts form a complete set of china. Suppose Alice and Bob were satisfied with their old set of china and had no inkling before opening their gifts that they might expect a new set. Alice and Bob are therefore unaware of any relevant pattern that might characterize the upcoming distribution of gifts apart from actually receiving them from their children. Nevertheless, Alice and Bob will not attribute the gifts to random acts of kindness (i.e., to chance). Rather, Alice and Bob will attribute the new set of china to the cooperation of their children (i.e., to design). In receiving the gifts, Alice and Bob discern a pattern that—though discerned after the fact—cannot be reasonably explained apart from the planning and cooperation of their children.

When a small-probability event matches a pattern identified before it, chance may be rightly eliminated. When a small-probability event matches a pattern identified after it, chance may or may not be rightly eliminated. It depends on the type of pattern. Alice and Bob were faced with the right sort of pattern and thus were justified in eliminating chance after the fact. But suppose someone flips a coin a thousand times and simultaneously records the sequence of coin tosses on paper. The sequence flipped (= event) conforms to the sequence recorded on paper (= pattern). Moreover, the sequence flipped is vastly improbable (the probability is roughly 10^{-300}). Nevertheless, it's clear that the pattern to which these coin flips conform was merely read off the event and thus, as it stands, cannot legitimately justify eliminating chance.

Patterns may therefore be divided into two types, those that in the presence of small probabilities justify eliminating chance, and those that despite the presence of small probabilities do not justify eliminating chance. The first type of pattern will be called a *specification*, the second a *fabrication*.[7] Specifications, as we will see, are patterns that arise out of a unified conception and thus can be succinctly described (e.g., "an epic poem" or "a matching set of china"). Fabrications, on the other hand, are patterns that tend to be so variegated that many different aspects all need separate attention, and thus the pattern as a whole requires a long description (imagine describing the precise locations of a thousand distinct marbles randomly tossed on a floor).

Conventional statistical theory doesn't draw this distinction between specifications and fabrications with full generality. Typically, it focuses on how to eliminate chance in the simplest case where a pattern is given prior to an event, or what may be called a *prespecification*. These are the most straightforward specifications because they depend on a clear temporal ordering in which the pattern is identified first and the corresponding event happens next. A mature statistical theory needs to understand specifications in their full sense, especially after-the-fact specifications. Specifications are a prime focus of this book.

1.2 The Reach of Chance

We all have intuitions about what is and is not within the reach of chance. If you get out a fair coin and flip it 3 times, you may well witness 3 heads in a row. No surprise here. You might even flip 10 heads in a row if given an hour or so to toss the coin. Again, no surprise. But getting 100, to say nothing of 1,000, heads in a row by chance seems completely absurd, and not just for you but for all humans across history doing nothing throughout their lives but flipping coins. The mathematical theory of probability begins with a 1654 correspondence between Pierre Fermat and Blaise Pascal, in which they determined the fair distribution of gambling stakes if a game had to be ended prematurely. This work was soon followed up and extended by that of the Bernoullis, an illustrious family of mathematicians. But it didn't take formal mathematics to convince people across history about what is and is not within the reach of chance.

People are inveterate gamblers. Archeologists have found dice in what today is southeastern Iran dating back to almost 3000 BC. Archeologists have found dice from ancient Rome displaying the same numerical markings and same cubic shape that we use today. The casting or drawing of lots was common in the ancient world, and appears throughout the Old and New Testaments of the Bible. Besides their use in gambling, lots were also used in divination. The *I Ching*, for instance, describes a Chinese approach to divination in which lots are cast with both sticks and coins. Leaving aside the legitimacy of gambling or divination, the behavior of a coin, a die, or a lot of any kind, when repeatedly cast or drawn, quickly approaches a chance boundary from which we instinctively shrink back and beyond which we disbelieve that certain patterns can without interruption continue to appear by chance.

Consider the case of roulette. It is widely reported that for roulette, 32 reds in a row is the most that has ever been witnessed at any casino (supposedly at an American casino back in 1943).[8] American roulette has 18 red slots, 18 black slots, and 2 green slots marked 0 and 00. The chance of getting 32 reds in a row is thus on the order of 1 in 25 billion. From the branch of probability theory focused on recurrent events, this probability in turn requires about 46 billion spins of the roulette wheel, on average, for 32 reds to occur in a row.[9]

There are over 2,000 casinos in the United States.[10] At a typical casino, roulette wheels are spun 55 times an hour.[11] If we imagine a roulette wheel in each casino operating 24 hours a day, 365 days a year, over the course of 100 years, then we would have 96 billion spins. Some casinos may not have roulette wheels, or may not operate 24 hours, or may not have been in operation for 100 years. However, others will have been operating for longer and have multiple roulette wheels. Additionally, there are casinos in other countries besides the United States—close to 10,000 casinos worldwide.[12] As such, we are in our rights to think that the number of spins far exceeds the 46 billion spins that we calculated as needed on average to reach 32 reds in a row. Moreover, a long sequence of reds (or blacks) approaching 30 in a row would clearly be salient enough to make the news. The actual occurrence of 32 reds in a row at roulette is therefore plausible.

But what about 100 reds in a row? As it is, we don't find reports about roulette claiming anything close to 100 reds (or blacks) in a row. Why? Because, unless there is some sort of tampering with the roulette wheel or other deception (in other words, design), that many repetitions of the same color are completely beyond the reach of chance. Consider that if all 100 billion humans estimated ever to have existed each lived 100 years and did nothing but spin roulette wheels throughout their entire lives, they would need to achieve a rate of roughly 1.7 trillion spins per second to have an even chance of seeing even one run of 100 reds in a row.

Skeptics of intelligent design tend to trivialize startling correlations and coincidences, relegating them as much as possible to chance. They rationalize this dismissive attitude toward unusual connections, correlations, and coincidences by arguing that these really aren't all that improbable once we factor in all the events that happen but that don't draw our attention. The late skeptic Martin Gardner put it this way: "The number of events in which you participate for a month, or even a week, is so huge that the probability of noticing a startling correlation is quite high, especially if you keep a sharp outlook."[13] But not all startling coincidences are created equal. Some may be ascribed to chance, others not.[14]

How many events do you participate in during the course of your life? Let's be supremely generous and say you live 100 years and that each second of your life you participate in 100 events that could be salient enough to be part of a startling coincidence. With 100 years, 365.25 days per year, 24 hours per day, 60 minutes per hour, 60 seconds per minute, and 100 events per second, that comes to just under 316 billion events. If you're keeping a sharp lookout at all possible connections among the events you are witnessing, then there are 316 billion times 316 billion ways all these events could pair up and thus coincide, which comes to around 10^{23} (or 100 sextillion). If we now factor in the 100 billion humans estimated to have ever existed, that raises the number of coincidences that might ever register as salient on the consciousness of people as 10^{34}. It follows that humans should never expect to observe coincidences with an improbability more extreme than 1 in 10^{34}.[15] Such coincidences could thus legitimately be regarded as beyond the reach of chance.

People throughout history have intuited what's within and beyond the reach of chance. The tools of modern probability theory, however, have allowed us to put numbers on that intuition, making it more precise. Conventional statistical theory, with its use of rejection regions to eliminate chance hypotheses, goes some distance toward rigorously capturing that intuition. But conventional statistical theory typically puts alternative chance hypotheses in competition with each other and then lets "smallish" probabilities decide which hypothesis is to be rejected. For instance, probabilities in the range of .05 or .01 are typical in social science research for rejecting chance hypotheses. Yet probabilities in that range are not small enough to eliminate chance across the board.

In rejecting one chance hypothesis, statistical theory typically keeps other chance hypotheses alive. But when, as in the previous gambling examples, we think the reach of chance has been convincingly exceeded, it's because the calculated probabilities are not just jaw-droppingly small but also exhaustive of the relevant chance hypotheses on which those probabilities might be based. Accordingly, in such examples, small probabilities don't just eliminate a single chance hypothesis but sweep the field clear of all chance hypotheses that might apply. With a fair coin, for instance, we know the causal and probabilistic factors involved in its tossing. Thus, if we witness, say, 1,000 heads in a row, we would conclude not that some other chance hypothesis was involved, but that chance was not involved at all.[16]

The French mathematician Emile Borel grappled with how the reach of chance can be exceeded and thus the field swept clear of chance hypotheses when probabilities become too small. To answer this question, he proposed a principle governing small probabilities that he called the *Single Law of Chance*. According to this principle, *prespecified events of small probability do not occur by chance*.[17] In place of Borel's principle, we will propose and justify a more general principle governing inferences with small probabilities. We call it the *Law of Small Probability* (LSP). According to this principle, *specified events of small probability do not occur by chance*. The Law of Small Probability thus extends Borel's Single Law of Chance from prespecifications to specifications in general. The LSP is rigorously stated and justified in Section 5.4.

Borel's law and ours raise the need for a threshold beyond which a probability may legitimately be regarded as small. A probability small enough to fall below this threshold, in combination with a (pre)specification, can then be used to justify the elimination of chance. Such a threshold will need to be a concrete number, since otherwise there will be no clear objective way to determine whether the threshold has actually been crossed. Such a threshold gives Borel's Single Law of Chance and our own Law of Small Probability teeth. Accordingly, to use small probabilities to eliminate chance, we need a probability bound or threshold α (Greek lower-case alpha) so that any probability p less than or equal to α may rightly be regarded as small.

The need for such a probability bound is evident in the Single Law of Chance as Borel originally formulated it, and in the Law of Small Probability as subsequently formulated and generalized in this book. The key question facing these two laws is this: How small is small enough? In other words, how small does a probability need to be to effectively eliminate chance? This question demands a concrete numerical answer, because, without it, eliminating chance through small probabilities becomes subjective (what seems small to you might not seem small to me) and thus cannot claim the rigor of science. And to be clear, our aim is to make the design inference, along with its theoretical underpinnings, such as the Law of Small Probability, applicable across the sciences.

Borel was sensitive to the need to put concrete numbers on the probabilities that might legitimately be regarded as small enough to eliminate chance (as always, for specified, rather than unspecified, events). In 1930, he proposed 10^{-1000} as a bound below which probabilities could be neglected universally (i.e., neglected across the entire universe). Let's call such probabilities that eliminate chance across the entire scale of the universe *universal probability bounds*. In Borel's view, prespecified events at this level of improbability could for him always be rejected as not being the product of chance. Later, in 1939, by focusing on Earth-based human observers looking at the earth and across the universe, he proposed a less stringent universal probability bound of 10^{-50}.[18] Borel, however, never convincingly justified these bounds. Take 10^{-50}, the probability bound on which he ultimately settled. Borel argued for this bound as follows:

> If we turn our attention, not to the terrestrial globe, but to the portion of the universe accessible to our astronomical and physical instruments, we are led to define the negligible probabilities on the cosmic scale... We may be led to set at 10^{-50} the value of negligible probabilities on the cosmic scale. When the probability of an event is below this limit, the opposite event may be expected to occur with certainty, whatever the number of occasions presenting themselves in the entire universe. The number of observable stars is of the order of magnitude of a billion, or 10^9, and the number of observations which the inhabitants of the earth could make of these stars, even if all were observing, is certainly less than 10^{20}. [An event] with a probability of 10^{-50} will therefore never occur, or at least never be observed.[19]

This passage has a dated feel given the astrophysics on which Borel relied. Thanks to advances in telescope technology, we now know that the number of stars in the observable universe is many orders of magnitude greater than a billion. But Borel's outdated astrophysics is not the real problem here. Deeper probabilistic concerns exist with his case for making 10^{-50} a universal probability bound. With each of these concerns, the fault is not that Borel goes wrong but that he doesn't go far enough. First, Borel does not adequately distinguish the occurrence of an event from the observation of an event. There is a difference between an event never occurring and never being observed. For instance, in a large enough universe, the origin of life might, given the laws of chemistry, have a near-certain probability of occurring spontaneously even if its probability on Earth were less than Borel's 10^{-50}. In that case, we should never expect to observe life originate spontaneously on Earth even though we should expect its occurrence on the scale of the universe (in which case Earth might just be one of those lucky places where life did originate spontaneously by chance). Is it, therefore, that (pre)specified events of small probability are occurring all the time, but that we're just not observing them? Or are they not occurring at all? Borel doesn't elaborate.

Second, in setting his small probability bound, Borel neglects an inquirer's interests and context. In most contexts of inquiry, 10^{-50} is far too stringent; in others, it may not be stringent enough. When, for instance, Ronald Fisher charged Mendel's gardening assistant with data falsification, what elicited this charge was a specified event whose probability was around 4 in 100,000 (.00004).[20] What counts as a small probability depends on an inquirer's interests and context for eliminating chance. Social scientists are less stringent than criminal courts that establish guilt to a moral certainty beyond reasonable doubt, and these in turn are less stringent than cosmologists who work with small probabilities on a cosmic scale. Borel admits that what counts as a small probability on "the human scale" differs from what counts as a small probability on the "cosmic scale."[21] But he never clarifies how context of inquiry determines what may count as a small probability.

Universal probability bounds are meant to hold at the level of the entire universe, and thus be in some sense absolute (albeit dependent on our conception of the size of the known physical universe). Most probability bounds in practice, however, are *relative* or *local* probability bounds, relative to the interests and needs of the persons or groups who depend on the bounds not being breached by chance, and localized to the inquirers and contexts needing to assess whether chance was indeed breached.

Consider, for instance, why Visa issues 16-digit credit card numbers (though the first number is always a 4 and thus redundant), adding an expiration date as well as a 3-digit CVV number (card verification value). The reason for all these digits is to ensure that random guessing of Visa card information is highly unlikely to be successful in obtaining someone's credit card credentials. With 343 million Visa cards in circulation worldwide and 8 billion people on the planet, a lot of random guesses are possible. But with over a billion billion different possible numbers to uniquely identify a Visa card, random guessing is highly unlikely to identify any of those in circulation.

As a consequence, when someone not owning a Visa card is found with its credentials, the natural inference is not that it was randomly guessed but that it was hacked or stolen (the natural inference is therefore a design inference). In this way Visa ensures that however many random guesses might exist practically speaking, it should still be highly unlikely for any of them to succeed. Note that Visa could increase the security of its cards still further by adding more digits to its cards (e.g., by making the CVV ten digits rather than just three). But there's a downside to adding digits, which is the inconvenience to the user, who has to keep track of more and more digits. So, with local probability bounds, as opposed to universal probability bounds, the aim is to get good enough (rather than perfect or ideal) protection against chance.

Bounds less stringent than Visa's are common in business. Internet users these days, when trying to log into a website, often face a two-step verification. Thus, after you put in your login credentials, you may get a text message on your cell phone with a 6- or 8-digit code. You then have only a few minutes to enter the code, and you may only have a fixed number of attempts to enter the right code before the code becomes invalid. The implicit probability bounds here with the second step of this verification are on the order of 1 in a million or 1 in 100 million. These are not small probabilities at the scale of the universe, and thus chance can, at such a universal scale, be readily expected to overturn these probabilities. The probabilities are also bigger than the 1 chance in 292,201,338 of winning the Powerball jackpot. Yet the probabilities behind these 6- or 8-digit codes, when combined with the security of a username-password login combination, seem adequate for a reasonably secure two-step verification.

The third and last difficulty with Borel's characterization of small probability is that he never makes clear how his universal probability bound depends on the number of opportunities for an event to occur. Obviously, there is a connection. If, for instance, there are 10^{50} independent opportunities for an event of probability 10^{-50} to occur, then there is a slightly better than even chance that the event will occur. The actual probability is about .632, and the calculation to show this is straightforward. That same connection between opportunities and probabilities helped us to decide what's within and outside the reach of chance for the gambling examples that started this section.

With 10^{50} independent opportunities, an event of probability 10^{-50} will therefore occur with probability roughly .632. So, suppose we give Borel the benefit of the doubt, granting that the universe is such that no event has anywhere near 10^{50} opportunities to occur. Suppose instead that some event of probability 10^{-50} has at most 10^{30} independent opportunities to occur. Such an event therefore has probability around 10^{-20} of occurring. This last probability will strike most people as absurdly small. It may therefore seem that all we've done is substitute one ludicrously small probability (i.e., 10^{-20}) for another (i.e., 10^{-50}), and that in consequence we haven't actually explained what constitutes a small probability. There is a regress here, and even though it's not infinite, Borel does not point the way out of it. In fact, an easy way out of this regress exists and will be addressed in Chapter 4 on probabilistic resources.

1.3 Life in the Short Run

The opening scene of Tom Stoppard's 1990 film *Rosencrantz and Guildenstern Are Dead* illustrates the Theatre of the Absurd. Inspired by Shakespeare's *Hamlet*, the film begins with Hamlet's sidekicks Rosencrantz and Guildenstern on horseback. Rosencrantz finds a coin by the wayside and begins to toss it. Heads, heads, heads, heads, over and over again. Rosencrantz, played by Gary Oldman, takes the uninterrupted sequence of heads as a matter of course, anomalous to be sure, but not worldview-shattering. Guildenstern, played by Tim Roth, on the other hand, sees in this uninterrupted sequence of heads reason to question the nature of reality. Guildenstern muses:

> A weaker man might be moved to re-examine his faith, for nothing else at least in the law of probability... Consider: One, probability is a factor which operates within natural forces. Two, probability is not operating as a factor. Three, we are now held within sub- or supernatural forces... I think I have it. Time has stopped dead. The single experience of one coin being spun once has been repeated. A hundred and fifty-six times. On the whole, doubtful. Or, a spectacular vindication of the principle that each individual coin spun individually is as likely to come down heads as tails and therefore should cause no surprise each individual time it does.[22]

The scene ends with Rosencrantz and Guildenstern witnessing 157 heads in a row.

What should we make of the claim that someone flipped 157 heads in a row? With a fair coin, the probability of this result is 1 in 2^{157}, or less than 1 in 5 trillion trillion trillion trillion (that's a denominator with 48 zeros after a nonzero leading digit). It's estimated that there are about 7.5 million trillion grains of sand on the earth, so flipping this sequence of heads is vastly more improbable than finding one particular grain of sand among the whole lot of them by blind search. Consequently, if some persons reported that they, like Rosencrantz, had flipped a fair coin (i.e., a coin that's evenly balanced with distinguishable sides) and had witnessed 157 heads in a row, we would be inclined to disbelieve them. But what exactly justifies such a conclusion?

The laws of probability, insofar as they characterize the tossing of a fair coin, function differently from laws as we ordinarily understand them. We typically think of laws as conditionals: *if this, then that*. In other words, if a particular antecedent condition is satisfied, then a particular result will be the consequence. But with laws of probability, if a particular probability distribution operates, multiple distinct and even mutually exclusive events could happen so long as each has positive probability. Moreover, if an event has probability greater than a half, then we should expect it to happen; but unless the event has probability equal to one, and is thus certain, we cannot count on it happening. Ordinary laws have single consequences (jump in a pool and get wet). But probabilistic laws have multiple consequences (throw a die and get any one of six outcomes).

Nonetheless, the laws of probability also say that things should, in the aggregate, average out and become increasingly predictable to the point of certainty. Flip a coin, and any particular outcome will be completely uncertain. Flip a coin indefinitely, and the proportion of heads should, in the long run, converge upon a half with total certainty. Experience seems to confirm this. The laws of probability theory guarantee it. Uncertainty in the short run, certainty in the long run.

The economist John Maynard Keynes elaborated on this point memorably when he remarked that the "*long run* is a misleading guide to current affairs. *In the long run* we are all dead. Economists set themselves too easy, too useless a task if in tempestuous seasons they can only tell us that when the storm is long past the ocean is flat again."[23] In other words, we need to pay attention to the short run and not just to the long run. No one of us lives in the long run. We all live in the short run.

So, what exactly leads us to disbelieve that a coin tossed 157 times could land heads each time? Rosencrantz is certainly right when he remarks that "each individual coin spun individually is as likely to come down heads as tails and therefore should cause no surprise each individual time it does." But it's the aggregate behavior of the coin tosses that unsettles us. It's not any one particular toss that creates difficulty. It's their sum total, and the pattern they display. And so, we may be inclined to invoke Borel's Single Law of Chance or our generalization of it, the Law of Small Probability (to be described later in this chapter and then taken up earnestly in Section 5.4), to rule out that a (pre)specified event of small probability like this could happen by chance. But then the question becomes, What justifies such a principle of probabilistic rationality?

Rosencrantz's 157 heads in a row has probability close to and yet still greater than Borel's universal probability bound of 1 in 10^{50}. But even if Rosencrantz had managed to toss a few more heads in a row, putting the probability below 1 in 10^{50}, the resulting sequence would still have positive probability. As an event of positive probability, it would not be logically or even physically impossible. It would just be vastly improbable. Moreover, the laws of probability, notably the Strong Law of Large Numbers, guarantee that any event, however small its probability, is certain to happen in the long run. In fact, any event of positive probability, if given an unlimited number of opportunities to occur, will occur not just once but infinitely often.

Let's therefore imagine a universe with infinitely many people tossing infinitely many coins. Inflationary cosmology, quantum many worlds, and extreme modal realism all allow such over-stuffed universes (multiple worlds or multiverses of infinite variety). In that case, a Rosencrantz tossing 157 heads in a row is bound to happen over and over again. In fact, getting 157 heads in a row is peanuts for the long run. If you flip a coin long enough, you will see a sequence of coin tosses that, if interpreted as Unicode text (0 for tails, 1 for heads), will spell out the entire works of Shakespeare. But no need to stop with Shakespeare. With enough coin tosses, eventually you'll spell out the entire Library of Congress, the entire content of the World Wide Web, indeed, any bit string of any length whatsoever. Thus there will come an occasion when you witness a trillion trillion trillion heads in a row, all with a fair coin flipped fairly.

So how do we know that with the chance events we are witnessing in this life, we are not coming in, as it were, on coin tosses that are completely uncharacteristic of their "normal" chance behavior? When we look at nature, how do we know, speaking metaphorically, that we are not seeing a trillion trillion trillion heads in a row by chance even though chance would "ordinarily" present a roughly equal proportion of heads and tails? To say that the laws of probability favor equal proportion because it is "expected" or will happen "normally" or is "likely on average" begs the question, for why should chance behave in accord with our expectations just when we happen to be watching and taking notes (as with Rosencrantz and Guildenstern)? Why shouldn't chance behave anomalously, with extreme improbability in the short run, but making up the difference in the long run (which is all that the laws of probability can guarantee)?

Allowing chance to behave anomalously by deviating sharply from short-run expectation discredits all probabilistic inference and thereby *ruins practical reason*.[24] The fact is that we all draw design inferences based on small probability arguments whose input data are drawn entirely from the short run. Small probability acts as a safeguard for practical reason. When, in the presence of specifications, probabilities get too small, they give us reason to conclude that chance was not operating, and often that design was operating. Conversely, if we were to cast off small probabilities as a safeguard on practical reason, we would find ourselves attributing all manner of things to chance regardless of circumstances. Indeed, parity of reasoning would break down catastrophically if we could no longer use small probabilities to infer design. In such an alternative universe, chance would seem capable of virtually anything. The following examples make this point:

1. Someone seems to be a consummate pianist (Martha Argerich caliber), but in fact knows nothing about music and simply by chance, whenever she puts her fingers on the keyboard, sends forth fabulous music.

2. Someone seems to be a master investor (Warren Buffet caliber), but in fact knows nothing about business or

investing, simply by chance assigning dollar amounts to investment decisions, and thereby in trade after trade gaining a fabulous return.

3. Someone has the most reliable basketball shot ever witnessed (step aside Stephen Curry), never missing a basket, but is in fact a terrible athlete and simply by chance keeps making basket after basket.

4. Someone is a psychopath intent only on harm and mayhem (worse than the worst monsters of history), and yet every time he tries to do something bad, by chance it turns out good, bringing hope and joy to the world.

5. Someone is a saint, at least by intention (on the order of a Mother Teresa), and yet every time she tries to do something kind and loving, she happens by chance to kill, maim, and destroy human lives.

6. Someone seems to be a first-rate poet (a modern-day Keats), and yet is illiterate and simply by chance composes sublime poetry by randomly typing away at a keyboard.

7. Someone tosses a "magical" penny that acts as an oracle (far more impressive than Pythia at Delphi), assigning 0 to tails and 1 to heads, and uses ASCII or Unicode as the symbol convention for interpreting the strings of 0s and 1s. The penny operates purely by chance, but it answers all of humanity's pressing questions fully and succinctly (the cure for cancer, the solution to climate change, the key to world peace, completion of Schubert's unfinished symphony, etc.).

Such thought experiments are easily multiplied. Those listed here have been devised with reference to human behavior. Granted, aspects of human behaviors often follow well-defined probability distributions. Actuaries, for instance, keep track of accident statistics that neatly conform to probability distributions and yet ultimately rest on human intention and decision (such as the probabilities connected with automobile accidents). We will address how chance can be an outgrowth of intelligent activity in the next section. Notwithstanding, there are human behaviors that in any guise cannot rightly be ascribed to chance, such as sitting at a piano and skillfully playing a difficult composition. Small probability underlies our refusal to ascribe to chance such performances.

Similar thought experiments could as well have been devised with reference to the behavior of animals, rocks, oceans, stars, galaxies, and even space aliens. In each such thought experiment, a small probability would be seen to push against a proposed (and outlandish) chance explanation. Imagine, for instance, that the large-scale structure of some galaxy happened, by chance, clearly to spell out the phrase "Made by Yahweh." Imagine your dog talking philosophy (in English no less) with you over the course of an hour, and then reverting back to his usual barking. Such examples show that in the absence of some constraint on how we ascribe chance, chance can explain anything. Our ascriptions of chance must therefore be disciplined. Otherwise, chance is no better than the god of the gaps. Invoking chance, like invoking a deity, can easily degenerate into an argument from ignorance.

The key to disciplining chance, and thus refusing to ascribe 157 heads in a row to chance, is to recognize that we live in the short run and adapt our probabilistic reasoning to this fact. Thus, if probabilities are going to help us make sense of our short-run experience, we must assume, as a *regulative principle* guiding our use of probabilities, that our short-run experience of chance ought to be representative of our probabilistic expectation of chance. And this is just another way of saying that wild deviations from short-run probabilistic expectations ought not to be ascribed to chance. Wild deviations of this sort will always be highly improbable. What we're really saying, therefore, is that events of small probability, other than those that are unspecified, should not be ascribed to chance. Alternatively, it is to restate what we are calling the Law of Small Probability.

A sober science committed to understanding the world as it actually is must make its peace with life in the short run. Science, properly conceived, disciplines its use of probabilities and ascriptions of chance. Take Monte Carlo simulations, in which scientists model the probabilistic behavior of a system with a computer program and then run the program to determine the likely behavior of the system. Some strategies for playing poker with multiple players, for instance, admit no exact mathematical solution (at least none that is known), but these situations can be modeled with computer programs, and the programs can then be run to yield the likely outcomes of poker play if the strategies in question are adopted.

It's common these days for applied math and computer science undergraduates to solve problems like this by writing a Python program and running, in usually under a minute on current laptops, 10 million or so samples. Given Moore's Law (while it lasts), the number of samples taken in a minute will only increase over time, facilitating better accuracy and smaller standard errors (i.e., likely deviations from the true values). But the point to realize is that for most such problems, 10 million samples will be plenty to determine the behavior of the system to within a very small standard error.

What this means is that if a closed-form expression for the behavior of the system exists and can be accurately calculated (in other words, if there's an explicit formula that calculates the probability exactly), it will conform very closely with the behavior as indicated in the simulation (unless there is some problem with the random number generator on which the simulation depends). It also means that if a proposed closed-form expression doesn't match up even approximately to the behavior indicated by the simulation, then either the simulation doesn't adequately model the phenomenon in question, or the closed-form expression is in error and needs to be corrected.

To sum up, we discipline chance by refusing to ascribe our short-run experiences to chance if they deviate too sharply from probabilistic expectations. Such a refusal constitutes a regulative principle for life in the short run. This principle requires making sense of specifications, namely, those patterns that, in the presence of small probability, justify the elimination of chance. Moreover, it requires making sense of probabilistic resources, namely, the range of opportunities that exist for events like the one in question to happen by chance. We shall address these matters in subsequent chapters.

The key point to take away from this section is that without a regulative principle for eliminating chance through small probabilities, we have no way to rule out the assaults on practical reason that would lead us to ascribe anything and everything to chance. If ascribing events to chance is going to be useful and meaningful, then so must ascribing events to non-chance. We cannot reasonably ascribe everything to chance. That means we need a robust capacity to rule out chance. The design inference confers that capacity. Yet all such distinctions between chance and non-chance require recognizing that we live life in the short run and ensuring that our use of probabilities respects this fact.

The Need for BOTH Small Probability AND Specification

Ever since the publication of the first edition of this book, various critics have failed to understand or even recognize the significance of specification in eliminating chance. As a result, they suggest that arguments for eliminating chance, especially those forming design inferences, rely exclusively on the concept of improbability. Their mistake cannot be attributed to a lack of effort or clarity by design theorists in underscoring the crucial role specification plays in design inferential reasoning. In fact, we have bent over backwards to stress its importance.

This book's central claim, clearly articulated here and in the first edition, states that chance elimination requires *both* small probability *and* specification. This means—and we've said it till the cows come home—that for unspecified events, their low probability is irrelevant. Arguments for eliminating chance do not apply to such events because unspecified events with small probabilities can readily occur by chance.

It's therefore puzzling to see, a decade after the publication of the first edition, Bayesian probabilist Colin Howson write the following in his atheism-promoting book *Objecting to God*: "The intelligent-design advocate William Dembski also informs us that a *p*-value below a certain magnitude (in Dembski's reckoning 10^{-150}) corresponds in effect to a physical impossibility. To see why this is wrong consider a screen which randomly displays an integer between 0 and 9 inclusive, and let it do so 151 times. It will produce a physical impossibility according to Dembski's reckoning."[25]

Howson utterly mischaracterizes the design inference, rendering it unrecognizable. To see this, suppose the 151 digits on his screen are the ISBNs of Howson's books and those of his atheist colleagues. Would he attribute such a sequence to chance, or to design? Obviously, to design, and precisely because it joins to small probability that second essential feature for triggering a design inference, specification, a feature absent from Howson's 151 random digits. How could a trained philosopher who specializes in statistical reasoning miss something so basic? It calls to mind Upton Sinclair's dictum that it's hard to get people to understand something when their livelihood depends on not understanding it. The livelihood of professional atheists like Howson depends on blocking design inferences.

Atheist YouTube personality David Farina ("Prof. Dave") is another case in point. Farina, ignoring specification, offers the following analogy to argue against the design inference:

> Let's say 10 people are having a get-together, and they are curious as to what everyone's birthday is. They go down the line. One person says June 13th, another says November 21st, and so forth. Each of them have a 1 in 365 chance of having that particular birthday. So, what is the probability that those 10 people in that room would have those 10 birthdays? Well, it's 1 in 365 to the 10th power, or 1 in 4.2 times 10 to the 25, which is 42 trillion trillion. The odds are unthinkable, and yet there they are sitting in that room. So how can this be? Well, everyone has to have a birthday.[26]

But Farina's pattern of birthdays is completely unspecified. Imagine, instead, that each of these ten people had reported that their birthday is January 1. Such a coincidence would have a short description, such as "everyone has the same birthday" or "everyone was born New Year's Day." It would therefore constitute a specification. By combining small probability and specification, this coincidence would therefore have called for an explanation other than chance. It would not, in that case, be enough to say, as Farina did, "Well, everyone has to have a birthday."

Some critics, like Kenneth Miller, know better, but will nonetheless leave the impression that design theorists reject chance and infer design simply on the basis of brute improbability apart from specification. He and I debated—both in print and on stage—in the years immediately following the 1998 publication of the first edition of this book, so he knows the importance I attach to specification. Consider, in this light, the following exchange from a 2006 episode of BBC *Horizon*:

> BBC Commenter: In two days of testimony [at the Dover trial] Miller attempted to knock down the arguments for intelligent design one by one. Also, on his [i.e., Miller's] hit list, Dembski's criticism of evolution, that it was simply too improbable.
>
> Miller: One of the mathematical tricks employed by intelligent design involves taking the present-day situation and calculating probabilities that the present would have appeared randomly from events in the past. And the best example I can give is to sit down with four friends, shuffle a deck of 52 cards, and deal them out and keep an exact record of the order in which the cards were dealt. We can then look back and say, "My goodness, how improbable this is. We can play cards for the rest of our lives and we would never ever deal the cards out in this exact same fashion." You know what: that's absolutely correct. Nonetheless, you dealt them out, and nonetheless you got the hand that you did.

BBC Commentator: For Miller, Dembski's math did not add up. The chances of life evolving just like the chance of getting a particular hand of cards could not be calculated backwards. By doing so, the odds were unfairly stacked. Played that way, cards and life would always appear impossible.[27]

I [WmD] was also interviewed for this *Horizon* program and my segment was shown before this exchange. But my part in the program had been edited so that it seemed my entire argument rested on improbability and did not include specification. The editors then inserted the above exchange after my segment to discredit my use of design inferences in challenging Darwinian evolution.

In fact, Miller's comments in the BBC program do nothing to discredit the design inference. Suppose Miller and his four friends were playing poker and one of them dealt to himself four aces and to the others four kings, four queens, four jacks, and four tens respectively. One doubts that Miller would be satisfied if his friend were to say, "My goodness, how improbable it is that each of us got four of a kind. And what good fortune that I happened to get the winning hand. Nonetheless, those are the hands I dealt, and those are the hands we got." Of course, what sets these hands apart is not merely their improbability. Rather, what sets them apart is their improbability *combined with* their conformity to the specification "everyone got four of a kind." Given this combination of small probability and specification, everyone, including Miller, would conclude that his friend was cheating, thereby drawing a design inference.

Miller might have tried to set the record straight, but he didn't until it was too late. Thus, after the program had aired and righting the record no longer mattered, Miller wrote a letter to the anti-ID website PandasThumb.org in which he denied that his card shuffling example was meant to respond to or refute my work on the design inference. As he put it, "All I was addressing was a general argument one hears from many ID supporters in which one takes something like a particular amino acid sequence, and then calculates the probability of the exact same sequence arising again through mere chance."[28]

But his card-game illustration patently leaves out specification, and no ID researcher has ever suggested that "X is improbable; therefore, X is designed" constitutes a valid argument. Instead, since the publication of the first edition of this book in 1998, design researchers have repeatedly emphasized the indispensability of both small probability and specification in eliminating chance and inferring design.

1.4 Chance as a Side Effect of Intelligence

In many situations, chance operates through impersonal forces. Patterns of sunspots, quakes of the earth, and radioactive decays all fall within such an impersonal conception of chance. Yet chance can also be a side effect or by-product of intelligence. Patterns of crime rates, standardized test scores, and racial distributions in the workplace all fall within such a personal conception of chance. Where chance is a side effect of intelligence, design inferences can still apply and be used to eliminate chance, though chance in this case will, just as with design, be the result of intelligent activity. Interestingly, chance can serve as a foil to design even if both chance and design could arise from the same intelligent agent.

A particularly clear example of chance as a side effect of intelligence is the distribution of letters in written communication. We'll focus on English, but similar considerations apply to other languages.[29] Because of the words in use and the conventions for writing them, letters and combinations of letters occur in English with certain relative frequencies. The letter "e" appears most frequently, approximately 13 percent of the time. The letter "t" appears next most frequently, approximately 9 percent of the time. With letter combinations, "u" always follows "q" except for a few transliterated foreign words (such as Iraq and Qatar). Such relative frequencies correspond to well-defined and stable probabilities.

Such frequencies occur reliably even though the texts in which they occur result from intelligent activity. Moreover, any significant deviation from these frequencies will itself result from intelligent activity. These frequencies raise clear probabilistic expectations.

To understand how such deviations of letter frequencies from probabilistic expectations can result from intelligence, consider Ernest Vincent Wright's 50,000-word novel *Gadsby*.[30] This novel admitted no occurrence of the letter "e." Here is the novel's opening paragraph:

> If youth, throughout all history, had had a champion to stand up for it; to show a doubting world that a child can think; and, possibly, do it practically; you wouldn't constantly run across folks today who claim that "a child don't know anything." A child's brain starts functioning at birth; and has, amongst its many infant convolutions, thousands of dormant atoms, into which God has put a mystic possibility for noticing an adult's act, and figuring out its purport.[31]

How do we know that the omission of the letter "e" from *Gadsby* was by design? As it is, the author plainly admitted in the novel's introduction to intentionally omitting that letter: "The entire manuscript of this story was written with the E type-bar of the typewriter tied down; thus making it impossible for that letter to be printed. This was done so that none of that vowel might slip in, accidentally; and many did try to do so!"[32]

But what if Wright had not been forthcoming about intentionally omitting the letter "e" from his manuscript? What if he had died before his precise motives and methods for constructing this novel had been made plain? In that case, a design inference could nonetheless be drawn. Design inferences, in the absence of direct causal evidence of intelligent activity, argue from effects back to causes based on markers of intelligence. The key marker for intelligence in a design inference, as urged in this book, is specification combined with small probability.

That combination is clearly in evidence in the case of *Gadsby's* missing e's. The absence of the letter "e" constitutes a specification. Moreover, the event specified has small probability. To see this, consider that a novel of 50,000 words can expect to see on average 5 letters per word, or at least 250,000 letters total (omitting spaces and punctuation). Given that the letter "e" occurs, on average, 13 percent of the time, a novel of such length could be expected to contain more than 30,000 occurrences of the letter. By any reasonable probability assumptions, the complete absence of the letter "e" under these circumstances will have minuscule probability.

Probabilities associated with letter frequencies in English writing have precise numerical values and thus allow for a clear design inference to be drawn in the case of a novel like *Gadsby*. At times, however, we are confronted with intelligent activity for which it's clear what we mean by chance and where a violation of chance seems obvious, and yet where we have no clearly given probability values. Take plagiarism. Person A writes a text and publishes it. Person B writes on the same topic and happens to include, word for word, a long paragraph identical with one in A's text.

Chance, in this case, would constitute independent reinvention of the same paragraph. The exact match between the two paragraphs is a highly suspicious coincidence. But is it a specified event of small probability, where together these are enough to trigger a design inference? A's paragraph certainly constitutes a prespecification of B's. But how do we determine probability of reinvention here? The authors, after all, are writing on the same topic, so similarities in the texts they generate would be bound to exist. But even though we might expect the texts to be similar, we would not expect them to be identical. Common sense suggests that the probability of identity will be very small even if we are unable to assign precise numerical values to it, except perhaps by some broad appeal to experience. Design inferences like this, intuitively drawn yet fiercely compelling, abound. In the case of plagiarism, the consequences of such a design inference can be severe, including failing grades, tarnished reputation, and loss of employment.

Or consider the work of actuaries. They readily appreciate that the aggregate behavior of intelligent agents can follow well-defined probability distributions.[33] As long as events are reasonably congruent with those probability distributions, their occurrence may legitimately be ascribed to chance. Thus, actuaries will calculate probabilities associated with human behaviors in the form of accident statistics. These probabilities are then used to assess risk in setting home, auto, and life insurance rates. Actuarial data that characterizes the aggregate behavior of intelligent agents can thus lead to a chance backdrop against which design may be inferred. For instance, in the film *Coma*, far too many people were "accidentally" dying in a given operating room. That was the tip-off. A subsequent investigation revealed that the vastly disproportionate death rate was due to an organ-harvesting scheme.

The question may now be asked why the side effects of intelligent activity, such as the letter frequencies of English texts, follow well-defined probability distributions. The reason seems to be that intelligences, when they act, typically operate within a settled infrastructure based on certain standards, policies, conventions, predilections, aversions, etc. These infrastructures characterize the environments (physical, cultural, digital, or some combination) in which intelligent agents operate as well as the types of activities they can engage in. Such infrastructure-informed environments, when in place, then induce certain routine as well as anomalous behaviors from intelligent agents that reside in them, and their behaviors in turn induce reliable probabilistic side effects. Change the infrastructure, and the probabilities of the side effects will change as well.

For instance, letter frequencies of English texts would be less inclined to follow well-established probabilities unless English spelling conventions were largely fixed and in widespread use, as is the case. In the absence of widely used spelling conventions, such probabilities would be less stable (perhaps varying from locale to locale) and might even change dramatically. In Elizabethan times, for instance, William Shakespeare's name was spelled in over 80 different ways, including Shaxper, which has two fewer occurrences of the letter "e" than the standard spelling of his name in use today.[34] Or imagine that Queen Elizabeth, the monarch through most of Shakespeare's life, had decreed that the letter "e" was overused and should be rooted out as much as possible, action she modeled by changing the spelling of her own name to "Elizabath." In that case, English orthography and even pronunciation might have taken a sharp turn so that the letter "e" would occur far less in English writing than it does now.

Or consider car accidents. Any probability distribution that characterizes car accidents will depend on the quality and maintenance of roads on which cars operate, on the types of signage and warning lights used to regulate traffic, on laws for driving on the same side of the road, on incentives for obeying speed limits, on the speed limits assigned to certain types of roads with certain types of traffic patterns, on the penalties for drunk driving, and on many other infrastructure conventions. Change any of these, and the statistics for car accidents will change as well. Intelligent agents implement such infrastructures by design, and they do so to advance purposes and intentions that ride atop these infrastructures (e.g., the intention of going on a vacation by car presupposes an infrastructure that allows for cars on roads in the first place). At the same time, those infrastructures will have probabilistic consequences, often unintended and unpredictable when taken individually, but more stable when taken collectively.

The takeaway of this section for the rest of this book is that it doesn't matter to a design inference whether the form of chance that it eliminates is impersonal or personal, whether it is a consequence of brute natural forces or a side effect of intelligent activity. When intelligent agents act, their behavior can have probabilistic properties that are stable, inducing well-defined and well-recognized probability distributions. Specified events of small probability that violate such probabilistic expectations can then become the basis for a design inference, where the design inferred stands in stark relief to a backdrop of chance that itself is the result of intentional activity.

The Metaphysics of Chance

In this section, we distinguished chance as an unguided natural phenomenon from chance as an epiphenomenon (or side effect) of intelligence. Both could be true in a given case and also complement each other. So far, we've not pressed the metaphysical foundations of chance. But what if chance in its entirety were a side effect of intelligence? That is a distinct possibility, and one explored in my [WmD's] article "Randomness by Design" and book *Being as Communion*.[35] The metaphysics of chance is pulled in two directions. From a naturalistic or materialistic viewpoint, all intelligence or design is ultimately an outgrowth of blind, purposeless physical processes and thus of what is commonly meant by chance. In this way, chance becomes metaphysically prior to intelligence. But we can also flip this priority on its head. If the ultimate reality is not blind, brute nature but rather intelligence, then chance would ultimately be an outgrowth of intelligence.

A third option exists, a hybrid that keeps the other two alive and in tension. Yet this third option is less parsimonious in that it tries to split the difference, taking a dualistic view in which blind nature and purposive intelligence are both metaphysically fundamental and irreducible to each other in accounting for chance. Of the two pure options (i.e., it's all blind nature or it's all purposive intelligence), we would submit that making intelligence prior to chance rather than the other way round makes better sense, answering a conundrum that seems unanswerable if chance is made fundamental and thus taken to be prior to intelligence.

Here's the issue. Probability allows for uncertainty at the level of individual chance events, yet it establishes order when these events are taken collectively. As such, the outcome of a single coin toss might be entirely unpredictable, but when multiple coin tosses are performed, they result in stable patterns. For example, as a coin is tossed repeatedly, the ratio of heads approaches ½. This stable pattern observed in coin tossing is supported both theoretically, as various probabilistic laws of "large numbers" confirm, and practically, since people who flip coins numerous times generally observe a nearly equal distribution of heads and tails.

Although experience has dulled our sense of wonder about chance, it should seem eerie that chance events, when viewed collectively, display stable and predictable probabilistic patterns. Materialism offers no independently ascertainable "fact of the matter" to explain why this should be so. Why should coins demonstrate predictable probabilistic behavior (or substitute quanta of light passing through a polarizing filter if coin tossing seems too deterministic)? What about a solid, uniform disk with distinct sides causes it to result in heads approximately half the time and tails the other half when tossed repeatedly by a nonlinear dynamical system such as a human being? The geometric symmetry of the coin doesn't begin to answer this question. Nor do the properties of matter. Nor do the laws of probability, which guarantee, as we saw in the last section, that in the long run chance can and will behave anomalously, completely at variance with expected probabilistic behavior in the short run.[36]

The point of this postscript, therefore, is to suggest that chance is better understood as a side effect of intelligence rather than as a probabilistic property of matter (a fuller justification of this position can be found in "Randomness by Design" and *Being as Communion*). We would submit that treating chance as a side effect of intelligence resolves a profound problem at the foundations of probability, namely, how short-run probabilistic data can reliably reflect underlying probability distributions and their long-run expectations. This problem vanishes when chance becomes a consequence of intelligence activity. In that case, there is no underlying probability distribution whose expectations must be matched. There is simply the activity of intelligent agents having consistent probabilistic side effects. An environmental infrastructure characteristic of these agents will then determine the observed probabilities—probabilities that can change as the intelligences change their environmental infrastructure.

Even so, it matters less whether in some broad metaphysical sense all of chance is a side effect of intelligence. What does matter is that we possess solid examples that at least some of what we call chance is a side effect of intelligence, and that chance even under these circumstances can form the backdrop for a design inference.

1.5 From Chance Elimination to Design

Let's now return to the Law of Small Probability. According to this law, specified events of small probability do not happen by chance. This law therefore gets us some distance toward inferring design. Nonetheless, the Law of Small Probability is not the design inference. When the Law of Small Probability eliminates chance, it eliminates a specific chance hypothesis. By itself, a given application of the Law of Small Probability therefore falls under that branch of statistics known as hypothesis testing.

With such testing, when a given chance hypothesis gets eliminated, it is typically because an alternative chance hypothesis has displaced it—chance thus gets replaced with chance.[37] By contrast, a successful design inference depends on identifying all the relevant chance hypotheses and then sweeping the field clear of them. The design inference, in inferring design, eliminates chance entirely, whereas statistical hypothesis testing, in eliminating one chance hypothesis, leaves the door open to others.[38]

To appreciate the difference between statistical hypothesis testing and the design inference, imagine that a die is thrown 6 million times. Statistical hypothesis testing considers two hypotheses: H_0, the null hypothesis, which asserts that the die is fair (i.e., each face has probability 1/6); and H_1, the alternate hypothesis, which asserts that the die is in some way loaded or skewed. Suppose now that the die is thrown 6 million times and that each face appears *exactly* 1 million times. Even if the die is fair, getting *exactly* 1 million appearances of each face will seem fishy. Yet the standard statistical method for testing whether the die is fair, namely, a chi-square goodness of fit test, fails to reject the null hypothesis.[39] Indeed, statistical hypothesis testing will advise accepting the null hypothesis H_0: of all possible 6 million throws of the die, having each number appear exactly 1 million times is the most likely outcome.

To accept the null hypothesis in these circumstances, however, is clearly absurd. As with Ronald Fisher's analysis of Gregor Mendel's pea-pod experiments, the fit between data and theory is too close to be explained by chance. If the die is fair and operating by chance, our best single guess—or what is known as the *mathematical expectation*—is that each face of the die will appear a million times. But this mathematical expectation differs sharply from our practical expectation. Practically, we expect to see each face of the die appear *roughly* a million times, but not *exactly* a million times. The probability of an exact fit with mathematical expectation is around 10^{-20}, which in any practical application constitutes a small probability. Moreover, since in any practical application the mathematical expectation will constitute a speci-fication, the Law of Small Probability will advise rejecting the null hypothesis H_0, contrary to statistical hypothesis testing. Thus, whereas statistical hypothesis testing eliminates chance because the divergence from mathematical expectation is too great, the design inference, in this case, eliminates chance because the fit with mathematical expectation is too close.

Bayesian methods, about which we'll say more later, provide one (though not the only) way to reconstruct design inferences. Bayesianism, by giving values, even if only tacitly, to prior probabilities for any hypotheses whatsoever, vastly extends the range of hypotheses to be considered in the previous dice tossing example. In particular, for Bayesians, design hypotheses are assigned probabilities and in turn confer probabilities. How does Bayesianism give credence to cheating (and thus design) hypo-theses in cases where the fit with expectation is too close? Bayesianism is a mechanism for updating probabilities in light of new information. Thus, when an event has high probability given a design hypothesis and low probability given a chance hypothesis, the probability of the design hypothesis given the event (called a posterior probability) goes up and the evidence for the design hypothesis is increased.

What then, on Bayesian grounds, tells us to infer design when we are informed that in 6 million tosses of a die, each face appeared exactly 1 million times? On the one hand, it is that such a precise fit with mathematical expectation has extremely small probability on the chance hypothesis. On the other hand, it is that the fit is readily explained by a design hypothesis in which an experimenter fudged the data to make the outcome appear much closer to mathematical expectation than it actually was. At any rate, Bayesianism, in drawing a design inference here, still requires first showing that the probability of exactly fitting expectation is extremely small.

In the spirit of Abraham de Moivre's epigraph to this book, we may therefore think of design and chance as competing modes of explanation for which design prevails once chance is exhausted. In eliminating chance, the design inference eliminates not just a single chance hypothesis, but all relevant chance hypotheses. Note that by chance here we can also include what's commonly called necessity (recall the title of Jacques Monod's book *Chance and Necessity*). Necessity, in saying that things have to happen in one and only one way, becomes a special case of chance in which the probability of outcomes is either 0 or 1. We assimilate necessity to chance in this way in Section 5.3.

How, then, do we explain an event once a design inference has swept the field clear of relevant chance hypotheses? In such circumstances, the term *design* serves double duty. It constitutes a *logical category* for the conclusion of a certain type of inference that eliminates chance (i.e., the design inference). And it constitutes a *causal category* for an intelligent cause or agent responsible for the match between event and specification that triggered the design inference. Moreover, design applies in both cases even if no agent has been explicitly identified.

Note that we are not claiming that design inferences, in the logical form to be laid out in this book (see Chapter 5), are the only way to legitimately ascribe design as the cause of some event, object, or structure. We ascribe design all the time when we observe a known intelligent agent do something. Also, when a known intelligent agent confesses to doing something, we ascribe design to it even if it exhibits no improbability and thus cannot, apart from the agent's testimony, draw us into a design inference (such as finding a single coin lying on a table in a position easily explained by chance, but a friend admitting to intentionally positioning it that way).

Design inferences as laid out in this book also face an unavoidable limitation: we can never avoid the problem of false negatives. Without relevant background knowledge about the probabilities and specifications involved in a given situation, we may simply miss a design inference even though it could legitimately be drawn if we were more aware and knowledgeable (or, as we might say, had "better intelligence").[40] But false negatives pose no challenge to the design inference—we miss design all the time for lack of knowledge. The challenge, rather, in developing the design inference is to ensure that it avoids false positives (i.e., misattributions of design to things that are in fact the result of chance).

The central claim of this book is that design as a logical category, when ascribed by successfully drawing a design inference, correlates reliably with design as a causal category. Where there's smoke, there's fire. Where there's a signature, there's a signatory. Where there's a true sign, there's the thing signified by the sign. The design inference provides a reliable marker of design. Design inferences thus detect intelligent causes. Exactly why the design inference provides a reliable marker of intelligent agency will be taken up in Section 5.5.

A design inference limits explanatory options. It constrains causal possibilities. It does not identify causal particulars. To identify a particular cause, we need to investigate the particulars of the situation in which design is inferred, which may include finding new evidence, making unexpected observations, and drawing on background knowledge across a range of disciplines. The bottom line is that in moving from inferring design to identifying a particular intelligent agent, we will need more details.

As the output of a small-probability chance-elimination argument, a design inference detects the work of an intelligent or agent cause, but is not in the business of telling causal stories. The term *design*, as used in the phrase *design inference*, therefore focuses on a form of probabilistic reasoning with causal implications but without offering a full-orbed causal analysis. The design inference belongs, in the first instance, within the logical foundations of probability and statistics, and then, by delivering a marker of intelligent or agent causation, becomes a reliable instrument for detecting design.

CHAPTER 2

A SAMPLER OF DESIGN INFERENCES

DRAWING DESIGN INFERENCES IS NOT AN OBSCURE OR RARE occurrence—it happens daily. We distinguish between a neatly folded pile of clothes and a random heap, between accidental physical contact and a deliberate nudge, between doodling and a work of art. Furthermore, we make important decisions based on this distinction. This chapter examines a variety of areas where we apply the design inference. In each case, what triggers a design inference is a specified event of small probability.

2.1 Intellectual Property Protection

Governments regulate intellectual property through patents, trademarks, and copyrights. Patents protect inventions, trademarks protect branding, copyrights protect text, images, music, video, and the like. People avail themselves of patent, trademark, and copyright laws to assert the priority of their work, and thus to keep copycats from unfairly obtaining a market share. The laws are such that if Alice creates some artifact X and files X with the relevant government regulator at time t_1, and Bob claims to have created the same artifact X at any time t_2 subsequent to time t_1, Bob is liable to penalties.[1]

Is this fair? What if Bob created X independently of Alice? In science and engineering, often when the time is ripe for an idea or invention, it is independently created. Werner Heisenberg and Erwin Schrödinger, for instance, invented mathematically equivalent formalisms for quantum mechanics at roughly the same time in the 1920s. But most artifacts exhibit clear marks of their creators that render independent reinvention highly improbable. Thus, whenever the probability of Bob recreating X independently of Alice is minuscule, the presumption is that Bob copied X from Alice rather than that Bob came up with X independently. The presumption of plagiarism in such cases is rock solid, as witnessed by the success of Turnitin and other plagiarism-checking companies.

Interestingly, in this age of web content and Google search, copyright filings are often unnecessary to assert intellectual property. Google, for instance, when it indexes pages on a website, assigns a canonical status to the first appearance of a page's content on the web. If other sites simply repost this material, Google will keep track that this reposted material is not original and downgrade it on its search engine results pages (SERPs), causing the original posting to appear higher in its search. Google will even penalize pages and websites that repost material without proper attribution, lowering their position in the SERPs, thereby limiting traffic and driving away business. Posting duplicate content without attribution is a sure way to get dinged by the search engines.

Sometimes manufacturers strengthen their grasp on intellectual property by introducing "traps" into their artifacts (digital or otherwise) so that anyone who copies the artifacts is likely to get caught red-handed. Copycats typically introduce variations into the things they copy so that the match between original and copy is not exact. They do this to have plausible deniability for their theft of intellectual property, hoping the variations they introduced are sufficient to convince others of independent invention. Thus, after Alice has produced X, the copycat Bob, instead of reproducing X exactly, will produce a variation on X—call it X'. Bob hopes X' will be sufficiently different from X to circumvent the protection offered by patent, trademarks, copyrights, and the like. Traps block this move.

Consider, for instance, a type of trap known as the *fictitious entry*.[2] Makers of reference works, such as dictionaries, encyclopedias, and maps, insert such entries to expose, and thereby discourage, the copying of their work. Thus, a dictionary might introduce never-before-seen words, such as *esquivalience*.[3] An encyclopedia might include the biography of someone who never lived, such as *Lillian Virginia Mountweazel*.[4] A map might identify a nonexistent town, such as *Beatosu* (supposedly in Ohio).[5] What's so effective about fictitious entries is that no amount of paraphrasing or embellishing them can circumvent the copyright trap. Merely to cite esquivalience, Mountweazel, or Beatosu as legitimate terms is to incriminate yourself.

A variation on this theme occurs when two parties, say Alice and Bob, have the power to produce exactly the same artifact X, but where producing X requires so much effort that it is easier to copy X once X has already been produced than to produce X from scratch. For instance, before the advent of computers, logarithmic tables had to be computed by hand. Although there is nothing mysterious about calculating logarithms, doing so without high-speed computing is tedious and time-consuming. Once the calculation has been accurately performed, however, there is no need to repeat it.

What, then, was the problem confronting the makers of logarithmic tables? After expending so much effort to compute logarithms by hand, if they were to publish their results without safeguards, nothing would prevent a plagiarist from copying the logarithms directly, and then simply claiming to have calculated the logarithms independently. To solve this problem, makers of logarithmic tables introduced occasional—but deliberate—errors into their tables, errors which they carefully noted to themselves. Thus, in a table of logarithms that was accurate to six or eight decimal places, errors in the seventh and eight decimal places would occasionally be introduced.[6]

These errors then served to trap plagiarists, for even though plagiarists could always claim to have computed the logarithms correctly by mechanically following a certain algorithm, they could not reasonably claim to have committed the same errors. As Aristotle remarked in his *Nichomachean Ethics*, "It is possible to fail in many ways,... while to succeed is possible only in one way."[7] Thus, when two manufacturers of logarithmic tables record identical logarithms that are correct, both receive the benefit of the doubt that they have actually done the work of computing the logarithms. But when both record the same errors, it is legitimate to conclude that whoever published last plagiarized.

In the same spirit, there's a joke about two college students who sat next to each other while taking a final exam. In the opinion of the professor teaching the course, the one student was the brightest in the class, the other the least bright. Yet when the professor got back their exams, both students were found to have gotten every question but the last perfectly correct. When it came to the last question, however, the brightest student wrote, "I don't know the answer to this question." The other student wrote, "I don't know the answer to this question either." If there was any doubt about who was cheating, the incriminating use of the word "either" removed it.

Traps to assert intellectual property continue to be used in our day, especially with digital technologies. Coders commonly introduce "Easter eggs" into their code. These can be messages, graphics, or sound effects produced by a program, often as a funny way to claim credit. Typically, they are meant to be found. But Easter eggs can be so well hidden that they serve as traps in the code, providing convincing evidence of copying. Easter eggs embedded in computer code are a special case of digital data embedding technologies (DDETs). These also include watermarking and steganography.

Watermarking introduces a digital signature into digital data that can be very difficult to remove even by concerted attempts to vary the data (as in photoshopping a jpeg image). Steganography introduces a second layer of meaning that is intended to be invisible unless you know what you're looking for. Watermarking and steganography are older terms. Digital fingerprinting is a more recent term. For instance, the media library Pixabay, which must keep track of millions of stock images, videos, and music clips, urges its content creators to use digital fingerprints. Thus, for videos and music posted on YouTube, Pixabay urges content creators to use a "digital fingerprinting system… to easily identify and manage copyrighted content on YouTube."[8]

In general, digital data embedding technologies, or DDETs, provide a way for digital content creators to apprehend copycats, even when the copycats are trying to be clever by introducing variations to avoid exact copies. DDETs can also track the use, both legitimate and illegitimate, of digital assets. They can even be used remotely to shut off access to such assets. But all these other use cases depend on DDETs helping content creators to detect the copying of digital assets. The underlying detection method here invariably takes the form of a design inference because it's just too difficult for chance to replicate the embedded patterns.

Traps can effectively assert intellectual property and priority even when they are not deliberately inserted. For instance, idiosyncrasies in the construction of objects may serve no functional purpose (in some cases they may be purely ornamental or even accidental). And yet these idiosyncrasies may be preserved by reverse-engineers intent on making sure that no important function gets missed. Such features can serve as traps even if the reverse engineering is good about hiding other aspects of the redesign.

The former Soviet Union was notorious for reverse engineering artifacts from the West. Go online, for instance, to compare images of the Hasselblad 1600 F (Swedish) versus the Salyut (Soviet) cameras from the two decades after World War II.[9] The similarities are unmistakable. And even though the Soviet Union made no secret about copying the Hasselblad, idiosyncrasies of design are preserved in the Soviet camera, such as details of the popup mechanism for the viewfinder. On a much larger scale, the Soviets under Stalin employed hundreds of factories and research institutes to duplicate Boeing's B-29 Superfortress with their own Tupolev Tu-4 clone. The Tu-4 has even been called a "carbon copy bomber."[10]

Obviously, when artifacts, whether physical or digital, are exactly or almost exactly copied, the copying—and its violation of intellectual property—is unmistakable. But unmistakable as well can be the copying of subtle features where the copycat or reverse engineer may not even be aware of the detail being copied. In both brazen and stealthy copying, what in the end makes it unmistakable is the improbability that two items could coincide so precisely if they were created independently. The improbability, or small probability, here renders it implausible that chance rather than design was responsible for the coincidence between the two.

2.2 Forensic Science

Forensic scientists, detectives, lawyers, and insurance fraud investigators would be out of business without the design inference. Something as common as a forensic scientist placing someone at the scene of a crime by matching fingerprints requires a design inference. Indeed, there is no logical or genetic impossibility preventing two individuals from sharing the same fingerprints. Rather, our best understanding of fingerprints and the way they are distributed in the human population is that they are, with very high probability, unique to individuals. And so, whenever the fingerprints of an individual match those found at the scene of a crime, we conclude that the individual was indeed at the scene of the crime.

Forensic scientists are continually adding to their stock of patterns and types of evidence by which they can detect bad actors. When the first edition of this book was being written in the mid-1990s, DNA testing and its use in a court of law was novel. The Human Genome Project wasn't completed until 2003. So the following 1994 headline was newsworthy at the time: "DNA Tests Becoming Elementary in Solving Crimes." The article went on to describe the type of reasoning employed by forensic scientists in DNA testing. As the following excerpt makes clear, all the key features of the design inference described earlier in this chapter are present in DNA testing:

TRENTON — A state police DNA testing program is expected to be ready in the fall, and prosecutors and police are eagerly looking forward to taking full advantage of a technology that has dramatically boosted the success rate of rape prosecutions across the country...

Mercer County Prosecutor Maryann Bielamowicz called the effect of DNA testing on rape cases "definitely a revolution. It's the most exciting development in my career in our ability to prosecute."

She remembered a recent case of a young man arrested for a series of three sexual assaults. The suspect had little prior criminal history, but the crimes were brutal knifepoint attacks in which the assailant broke in through a window, then tied up and terrorized his victims.

"Based on a DNA test in one of those assaults he pleaded guilty to all three. He got 60 years. He'll have to serve 27½ before parole. That's pretty good evidence," she said.

All three women identified the young man. But what really intimidated the suspect into agreeing to such a rotten deal were the enormous odds—one in several million—that someone other than he left semen containing the particular genetic markers found in the DNA test. Similar numbers are intimidating many others into foregoing trials, said the prosecutor.[11]

At the time this article appeared, the accuracy and usefulness of DNA testing was still a matter for debate. As a *New York Times* article concerned with the then-ongoing O. J. Simpson case put it, "There is wide disagreement among scientific experts about the accuracy and usefulness of DNA testing and they emphasize that only those tests performed under the best of circumstances are valuable."[12] In the intervening years, DNA testing has become much more widespread and convincing. DNA-based paternity tests, for instance, are now routinely accepted to establish paternity.[13] At interest here, however, is not the nuts and bolts of DNA testing or even the precise improbabilities of a DNA match happening by chance, but the logic that underlies DNA testing, a logic that hinges on eliminating chance through small probabilities and therewith drawing a design inference.

Not just forensic science, but detective work in general is inconceivable without the design inference. Indeed, the mystery genre would be dead in the water without it. The literary critic David Lehman proposed what he called "retrospective prophecy" as a key clarifying concept for understanding detective fiction. It is a form of design inference:

> If mind-reading, backward-reasoning investigators of crimes—sleuths like Dupin or Sherlock Holmes—resemble prophets, it's in the visionary rather than the vatic sense. It's not that they see into the future; on the contrary, they're not even looking that way. But reflecting on the clues left behind by the past, *they see patterns where the rest of us see only random signs.* They reveal and make intelligible what otherwise would be dark...[14]

The design inference uncovers patterns (i.e., specifications) that are not reasonably attainable by chance. Detectives, in drawing design inferences, therefore "see patterns where the rest of us see only random signs."

Examples of design inferences abound in the detective genre. The design inference is decisive when, in the 1944 film *Double Indemnity*, Barton Keyes (the insurance investigator, played by Edward G. Robinson) determines that the husband of Phyllis Dietrichson (the femme fatale, played by Barbara Stanwyck) did not die an accidental death by falling off a train. Instead, Keyes concludes that Phyllis and her accomplice/lover Walter Neff (played by Fred MacMurray) murdered her husband to collect on a life insurance policy.

Why hadn't Phyllis' deceased husband made use of his life insurance policy earlier to pay off on a previously sustained injury, for the policy did have such a provision? Why should he die just two weeks after taking out the policy? Why did he happen to die on a train, thereby requiring the insurance company to pay double the usual indemnity (hence the title of the movie)? How could he have broken his neck falling off a train when at the time of the fall, the train could not have been moving faster than 15 mph? And who would seriously consider committing suicide by jumping off a train moving only 15 mph? Too many pieces coalescing too neatly made foul play the much more plausible explanation than accidental death or suicide. Thus, at one point Keyes exclaims, "The pieces all fit together like a watch!" Consistent with this design inference, it is eventually revealed that the victim's wife and her lover murdered the husband for the insurance money.

Whenever there is a mystery, it is a design inference that elicits the crucial insight needed to solve the mystery. The dawning recognition that a trusted companion has all along been deceiving you (Alfred Hitchcock's *Notorious*); the suspicion that someone is alive after all, even though the most obvious indicators point to the person having died (Graham Greene's *The Third Man*); and the realization that a string of seemingly accidental deaths happened under improbably similar circumstances (Michael Crichton's *Coma*) all depend on design inferences. At the heart of these inferences is a convergence of small probabilities and specifications, a convergence that cannot properly be explained by appealing to chance.

2.3 Data Falsification in Science

The eminent statistician Ronald Aylmer Fisher uncovered a classic case of data falsification when, as noted in the previous chapter, he analyzed Gregor Mendel's data on peas. Fisher inferred that "Mendel's data were fudged," as one statistics text puts it, because the data matched Mendel's theory too closely.[15] Interestingly, the coincidence that elicited this charge of data falsification was a specified event whose probability was, as we saw in Section 1.2, roughly 4 in 100,000, or 1 in 25,000. By everyday standards, this probability will seem small enough, but it is huge compared to many of the probabilities we will be encountering. In any case, Fisher saw this probability as small enough to draw a design inference, concluding that Mendel's experiment was compromised and charging Mendel's gardening assistant with deception.

For a more recent example of data falsification in science, consider the case of UCSD heart researcher Robert A. Slutsky. Slutsky was publishing fast and furiously. At his peak, he was publishing one new paper every ten days. Intent on increasing the number of publications in his curriculum vitae, he decided to lift a two-by-two table of summary statistics from one of his articles and insert it—unchanged—into another article.[16] Data falsification was clearly implicated because of the vast improbability that data from two separate experiments should produce the same summary table of statistics. When forced to face a review board, Slutsky resigned his academic position rather than try to explain how this coincidence could have occurred without any fault on his part. The incriminating two-by-two table that appeared in both articles consisted of four blocks each containing a three-digit number. Given therefore a total of twelve digits in these blocks, the odds would have been roughly 1 in a trillion ($= 10^{12}$) that this same table might have appeared by chance twice in his research.

Why did Slutsky resign rather than defend a 1 in 10^{12} improbability? Why not simply attribute the coincidence to chance? There were three reasons. First, at the review board Slutsky would have had to produce the experimental protocols for the two experiments that supposedly gave rise to the identical two-by-two tables. If he was guilty of data falsification, these protocols would have incriminated him. Second, even if the protocols were lost, the sheer improbability of producing so unlikely a match between the two papers would have been enough to impugn the researcher's honesty. Once a specification is in place (the two-by-two table in one paper here specifying the table in the other) and the probabilities become too small, the burden of proof, at least within the scientific community, shifts to the experimenter suspected of data falsification. In lay terms, Slutsky was self-plagiarizing. And third, Slutsky knew that this case of fraud was merely the tip of the iceberg. He had been committing other acts of research fraud right along, and these were now destined all to come into the open.

Moving from medicine to physics, consider the case of Jan Hendrik Schön, which parallels the Slutsky case almost point for point. On May 23, 2002, the *New York Times* reported on the work of "J. Hendrik Schön, 31, a Bell Labs physicist in Murray Hill, N.J., who has produced an extraordinary body of work in the last two and a half years, including seven articles each in *Science* and *Nature*, two of the most prestigious journals." Despite this track record, his career was on the line. The *New York Times* reported further that Schön published "graphs that were nearly identical even though they appeared in different scientific papers and repre- sented data from different devices. In some graphs, even the tiny squiggles that should arise from purely random fluctuations matched exactly." (The identical graphs that did in Schön parallel the identical two-by-two tables that did in Slutsky.) Bell Labs therefore appointed an independent panel to determine whether Schön was guilty of "improperly manipulating data in research papers published in prestigious scientific journals." The hammer fell in September 2002 when the panel concluded that Schön had indeed falsified his data, whereupon Bell Labs fired him.[17]

Exactly how a design inference was drawn in the Schön case is illuminating. In determining whether Schön's numbers were made up fraudulently, the panel noted, if only tacitly, that the first published graph provided a pattern independently identified of the second and thus constituted the type of pattern that, in the presence of improbability, could negate chance to underwrite design (i.e., the pattern was a specification). And indeed, the match between the two graphs in Schön's articles was highly improbable assuming the graphs arose from random processes (which is how they would have had to arise if, as Schön claimed, they resulted from indepen- dent experiments). As with the matching two-by-two tables in the Slutsky example, the match between the two graphs of supposedly random fluctuations would have been too improbable to occur by chance. With specification and improbability both evident, a design inference followed.

But, as noted earlier, a design inference, by itself, does not implicate any particular intelligence. So how do we know that Schön was guilty? A design inference shows that Schön's data were cooked. It cannot, without further evidence, show that Schön was the chef. To do that required a more detailed causal analysis—an analysis performed by Bell Labs' independent panel. From that analysis, the panel concluded that Schön was indeed guilty of data falsification. Not only was he the first author on the problematic articles, but he alone among his co-authors had access to the experimental devices that produced the disturbingly coincident outcomes. Moreover, it was Schön's responsibility to keep the experimental protocols for these research papers. Yet the protocols mysteriously vanished when the panel requested them for review. The circumstantial evidence connected with this case not only underwrote a design inference but established Schön as the designer responsible.[18]

As a final example of where data falsification becomes an issue facing science, consider efforts to debunk parapsychology. Parapsychological experiments attempt to show that parapsychological phenomena are real by producing a specified event of small probability. Persuaded that they've produced such an event, parapsychological researchers then explain it in terms of a quasi-design-like theoretical construct called psi (i.e., a non-chance factor or faculty supposedly responsible for such events).

For instance, shuffle some cards and then have a human subject guess their order. Subjects rarely, if ever, guess the correct order with 100 percent accuracy. But to the degree that a subject guesses correctly, the improbability of this coincidence (which will then constitute a specified event of small probability) is regarded as evidence for psi. In attributing such coincidences to psi, the parapsychologist will draw a design inference. The debunker's task, conversely, will then be to block the parapsychologist's design inference. In practice, this will mean one of two things: either showing that sloppy experimental method was used that somehow signaled to the subject the order of the cards and thereby enabled the subject, perhaps inadvertently, to overcome chance; or else showing that the experimenter acted fraudulently, whether by making up the data or by otherwise massaging the data to provide evidence for psi. Note that the debunker is as much engaged in drawing a design inference as the parapsychologist—

it's just that one implicates the parapsychologist in fraud, the other implicates psi.[19]

The takeaway from this section is that science needs the design inference to keep itself honest. In the years since the first edition of this book was published, reports of fraud in science have continued to accumulate. The publish-or-perish mentality that incentivizes inflating the number of one's publications regardless of quality has only gotten worse. That mentality moves easily from a haste-makes-waste sloppiness to self-serving fudginess to full-orbed fraudulence. Data falsification and other forms of scientific fraud, such as plagiarism, are far too common in science. What keeps scientific fraud in check is our ability to detect it, and it's the design inference that does the detecting.

We've now seen that the design inference makes design readily detectable in everyday life. Moreover, we've just seen that its ability to make design detectable in the data of science is central to keeping scientists honest. The grand ambition of this book is to show that the design inference makes design part of the very fabric of science.

2.4 Financial Fraud—The Madoff Scandal

In late 2008, Bernard L. Madoff's decades-long Ponzi scheme finally unraveled before the eyes of a shocked public. Ponzi schemes are named after the swindler Charles Ponzi, who took this form of fraud to new heights in the early twentieth century. But Bernie Madoff surpassed him and everyone else in the history of Ponzi schemes (and that includes Sam Bankman-Fried, whose FTX debacle is in the news as these words are being written). Madoff defrauded clients for decades, and at the end the total came to $65 billion.

The mechanics of Madoff's fraud may not have been particularly ingenious, but they were effective. He continued the fraud for about forty years, and for much of that time he evaded scrutiny. But by the early 2000s, and especially toward the end, he felt the walls closing in on him. As he told one of his interviewers,

"Jim, I was so desperate and delusional at the end, I was hoping there was a nuclear attack on Wall Street or some world catastrophe that wiped out all financial records so I could get out of the Ponzi scheme. How's that for insanity. I was tired of the fraud. I just wanted it to be over."[20]

One thing that helped keep the Madoff fraud alive all those years was his towering reputation on Wall Street. At the same time as his Ponzi scheme, Madoff also had a legitimate financial business, a brokerage house that handled close to 10 percent of the equities trading on Wall Street. He was on the board of many Wall Street firms, including the Nasdaq, which for several years he chaired. The Securities and Exchange Commission (SEC), which was supposed to regulate the investment advisory side of his business that housed his Ponzi scheme, would often look to Madoff for guidance. This was regulatory capture in the extreme.

Madoff had a knack for enlisting new clients and therewith new moneys. In a Ponzi scheme, you rob Peter to pay Paul. As long as you keep getting new clients (i.e., new Peters) and as long as what they keep putting in exceeds what you need to pay out to existing clients (i.e., old Pauls), the Ponzi scheme can keep going. In a Ponzi scheme, one typically promises an incredible return on investment. Charles Ponzi, for instance, promised to double the money placed with him in ninety days.[21] Madoff, by contrast, took a much more measured approach, delivering with astounding regularity between 12 and 16 percent annual returns.[22]

Madoff's returns were not the best in the business. Other hedge funds had higher annual returns. But what endeared Madoff to his investors was the steadiness of his returns. There was no apparent risk. No volatility, as they say. With all legitimate investments, even the best ones, while the net trendline should be up, there will still be evident downturns. Two steps forward, one step back. Three steps forward, two steps back. On balance, investors are making money. But in their monthly statements, they can regularly expect to see losses, even if they mostly see gains. Madoff almost eliminated losses.

As the work of psychologists Amos Tversky and Daniel Kahneman has demonstrated, humans are intensely loss averse.[23]

We hate losses much more than we enjoy comparable gains. Whether consciously or unconsciously, Madoff capitalized on this aspect of human psychology, ensuring that monthly returns were overwhelmingly positive, so that losses were very few, and minimal when they did occur. As a consequence, Madoff was regarded among his clients as a genius, who always knew how to time the market, getting in when the going was good, and getting out when it turned south.

In reality, Madoff's investment advisory business was a pure Ponzi scheme in the sense that no actual investing ever occurred. In other words, when Peter paid money to Madoff, Madoff did not take that money and invest it in any securities, not even Treasuries, which would have been secure and would have at least fetched some return. Madoff's approach was in line with Charles Ponzi's original scheme. Ponzi promised to provide his clients with fabulous returns by exploiting an arbitrage opportunity between United States and foreign postal coupons. But Ponzi never did any arbitrage; and Madoff never did any investing.

That said, Madoff took things much further than Ponzi ever did by giving the appearance of doing something to make money for his clients. This was pure stagecraft, and it gave Madoff's scheme a longevity that Ponzi's lacked. Ponzi's main scheme began in January 1920, and it was completely debunked by August of that year. Moreover, when examiners inspected Ponzi's records, they only found file cards with names and dollar amounts. They did not find among Ponzi's records even the pretense that he was exploiting any arbitrage opportunities.

Madoff, like Ponzi, did no trades. He simply moved all moneys to and from clients through a J. P. Morgan bank account. But Madoff, unlike Ponzi, went to elaborate lengths to look as though he was trading. Thus, he would provide his clients with monthly statements detailing how their investments were (supposedly) faring. Just as it's easy to win at sports betting if you already know what the outcome of a sports event is (recall the movie *The Sting*), so it's easy to make yourself look like a financial genius if you can simply look at the past performance of securities and pretend you bought them before they went up and sold them before they went down. Madoff's "split strike conversion strategy" added some slight complications to this picture, introducing the selling of a call option and the purchase of a put option to control for risk. But that's the picture in essence.

Madoff's fraud was a Potemkin village. It looked real enough if you didn't look closely. But everything he was doing could easily have been uncovered with a bit of probing. In the early 2000s, it seemed that such probing was about to begin in earnest with two articles examining Madoff's investment advisory business, one by Erin Arvedlund for *Barron's*, the other by Michael Ocrant for *MAR/Hedge*.[24] Falling short of accusing Madoff of outright fraud, both articles suggested that Madoff was not being forthcoming about what he was actually doing to be as successful as he seemed in the investment business. The articles could easily have prompted the level of scrutiny needed to bring Madoff down. But then 911 happened, which diverted attention from Madoff for several more years.

Yet one person understood early on that Madoff was a fraud. That was Harry Markopolos, a quantitative analyst with Rampart Investments. Rampart was intrigued with Madoff's returns and asked Markopolos to reverse-engineer his investment strategy. Within four minutes of looking at Madoff's returns, Markopolos concluded it was a Ponzi scheme. Subsequently, Markopolos found many red flags to further confirm that Madoff was a fraud, not least that there were simply not enough options being traded to carry out Madoff's stated trading strategy.

But leaving aside all the red flags that further investigation of Madoff's fraud revealed, Markopolos saw through Madoff's fraud by means of a simple and straightforward design inference. The overwhelming preponderance of positive over negative monthly returns was the giveaway for Markopolos. As Madoff biographer Diana Henriques puts it, "[Markopolos] noticed that Madoff had lost money in only three of the eighty-seven months between January 1993 and March 2000, while the S&P 500 had been down in twenty-eight of those months."[25] For Markopolos, there were three possible explanations: (1) Madoff was exceedingly lucky; (2) Madoff was using a novel investment strategy unlike the one he claimed to be using; or (3) Madoff was running a Ponzi scheme.

Markopolos dismissed (2) because it seemed utterly implausible that Madoff had simply invented a strategy out of whole cloth that produced virtually risk-free returns, a performance record without precedent among legitimate investment advisors. To reach his conclusion that Madoff was running a Ponzi scheme, or option (3), Markopolos also had to dismiss that Madoff just happened to get super lucky, or option (1). We've seen that only so much luck is permitted in scientific explanations, and the same holds for financial explanations.

So, how did Markopolos, who was the principal whistleblower against Madoff in the years 2000 to 2008, rule out luck? Why was he so intent during that time about warning off investors from Madoff? According to Markopolos, there was no way for Madoff to make all the trades he would have needed to make for his business to be legitimate. Markopolos communicated his concerns at length to the SEC. But in all that Markopolos wrote to alert people to Madoff's Ponzi scheme, he never laid out the precise statistical calculations to justify his charge against Madoff. As a "quant," Markopolos saw instantly that the numbers were too good to be true. And indeed, for the mathematically adept, like Markopolos, drawing a design inference in such cases, even if tacitly, can be personally quite compelling.

Still, it is an interesting exercise to rationally reconstruct the design inference by which Markopolos eliminated chance and inferred design, ultimately charging Madoff with carrying out a Ponzi scheme. Specification is clearly not in question here: the almost complete lack of volatility in the monthly returns constitutes a specification. What about improbability? Markopolos noted the presence of only three negative returns among 87 monthly returns between January 1993 and March 2000, whereas the S&P 500 had 28.

This suggests a back-of-the-envelope probability calculation. If we take the S&P 500 monthly performance of 28 negative returns among 87 total returns as the baseline, that suggests a binomial distribution with p equal to the probability of a negative return equaling 28/87, or roughly .322. Given this baseline, how likely would it be to get only 3 or fewer negative returns among 87 returns? A simple calculation using the binomial distribution for p = .322, n = 87, and x = 3 shows that the probability of getting 3 or fewer negative returns is less than 1 in 40 billion.

Granted, this probability calculation makes various simplifying assumptions, not all of which will be entirely on the mark. Monthly returns, for instance, will not be probabilistically independent in the full sense required by the binomial distribution. And yet, even as a first approximation, this calculation is not going to be way off the mark. Markopolos drew his (tacit) design inference by simply eyeballing Madoff's returns for four minutes. Given that Markopolos focused on the paucity of Madoff's negative returns, some such calculation must have been in the back of Markopolos' mind. Clearly, in ascribing Madoff's returns to a Ponzi scheme, Markopolos was not just drawing a generic design inference, but was giving it a definite form: Madoff was the designer, and the design was a Ponzi scheme.

After Markopolos drew his design inference, he had much additional work to do to sort out the particulars of how Madoff was perpetrating his fraud. But Markopolos made the right call from the start, and the right call stemmed from drawing a design inference. The rest was details—important details, to be sure, but details nonetheless. That's how it is with design inferences in general. They tell us that we're dealing with a designing intelligence. Once a design inference is legitimately drawn, then the task is to determine the particulars of how the design was implemented. On both these points, Markopolos nailed it.

A postscript is worth adding here. Markopolos, when testifying before Congress in the aftermath of the Madoff scandal, spared no invective against the SEC for ignoring his warnings.[26] Yet before the scandal broke, the SEC could have provided numerous rationalizations for why Madoff was an upstanding guy who did right by his investors. He was knowledgeable. He had been in the business for decades. He had never been in trouble with the law. The SEC itself depended on his wisdom. Etc. As it is, the design inference proved a much more reliable indicator of Madoff's dishonesty than his reputation.

2.5 Randomness

In the 1960s, mathematicians and engineers investigated what makes a sequence of numbers, notably strings of 0s and 1s, random. Independently, and virtually at the same time, Gregory Chaitin, Ray Solomonoff, and Andrei Kolmogorov answered this question by formulating what came to be called algorithmic information theory (AIT). If we flip a fair coin and note the occurrences of heads and tails in order, denoting heads by 1 and tails by 0, then a sequence of 100 coin flips looks as follows:

1100001101011000110111111
1010001100011011001110111
0001100100001011110111011
0011111010010100101011110   (R)

This is in fact a sequence obtained by flipping a penny 100 times. But instead, one can imagine flipping a penny 100 times and obtaining the following sequence:

1111111111111111111111111
1111111111111111111111111
1111111111111111111111111
1111111111111111111111111   (N)

Now the problem facing Chaitin was this (we'll tell the story in terms of Chaitin because we know him personally, he's still alive, he's pushed algorithmic information theory farther than anyone, and he invented it while still a teenager): given probability theory and its usual way of computing probabilities for coin tosses, Chaitin was unable to distinguish these sequences in terms of their degree of randomness. Sequences (R) and (N) have been labeled suggestively, R for "random," N for "nonrandom." Chaitin wanted to say that (R) was "more random" than (N). But given the usual way of computing probabilities, Chaitin could only say that each of these sequences had the same small probability of occurring, namely 1 in 2^{100}, or approximately 1 in 10^{30}. Indeed, every sequence of 100 coin tosses has exactly this same small probability of occurring.

To get around this difficulty, Chaitin introduced some concepts from recursion theory, a subfield of mathematical logic concerned with computation and till then considered quite far removed from probability theory. What Chaitin said was that a string of 0s and 1s becomes increasingly random as the shortest computer program to generate the string increases in length.[27] For our purposes, we can think of a computer program as a short-hand description of a sequence of coin tosses. Thus, the sequence (N) is not very random because it has a very short description, such as,

repeat '1' a hundred times

Note that we are interested in the shortest descriptions since any sequence can always be described in terms of itself. Thus (N) has the longer description

copy   '1111111111111111111111111
1111111111111111111111111
1111111111111111111111111
1111111111111111111111111'

But this description holds no interest since there is one so much shorter.

The sequence

1111111111111111111111111
1111111111111111111111111
0000000000000000000000000
0000000000000000000000000   (H)

is slightly more random than (N) since it requires a longer description, for example, repeat '1' fifty times, then repeat '0' fifty times

So too the sequence

$$1010101010101010101010101$$
$$0101010101010101010101010$$
$$1010101010101010101010101$$
$$0101010101010101010101010 \qquad \text{(A)}$$

has a short description,

repeat '10' fifty times

The sequence (R), on the other hand, has no short and neat description (at least none that has yet been discovered). For this reason, Chaitin regarded it as more random than the sequences (N), (H), and (A). Since one can always describe a sequence in terms of itself, (R) has the description

$$\text{copy} \qquad \text{'}1100001101011000110111111$$
$$1010001100011011001110111$$
$$0001100100001011110111011$$
$$0011111010010100101011110\text{'}$$

Because (R) was constructed by flipping a coin, it is very likely that this is the shortest description of (R). It is a combinatorial fact that the vast majority of sequences of 0s and 1s have as their shortest description just the sequence itself, i.e., most sequences are random in Chaitin's computational sense. Within algorithmic information theory, the language of statistical mechanics is used to describe this fact, denoting the random sequences as high-entropy sequences, and the nonrandom sequences as low-entropy sequences.[28] It follows that the collection of nonrandom sequences has small probability among the totality of sequences, so that observing a nonrandom sequence is reason to look for explanations other than chance.

To illustrate Chaitin's ideas, imagine someone informs you she just flipped a coin 100 times. If she hands you sequence (R), you examine it and try to discover a short description. After repeated attempts you find that you cannot describe the sequence more efficiently than the sequence describes itself. Hence you conclude it is a genuinely random sequence—in other words, that it is a sequence she might well have gotten by flipping a fair coin. Of course, you might be wrong—you might simply have missed some simple and short description. But until you have such a description in hand, you will incline toward regarding the sequence as random.

Next, suppose this same individual hands you the sequence (R) on a slip of paper and then disappears. A week later she reappears and says, "Guess what? Remember that sequence I handed you a week ago? Last night I was flipping this penny. And would you believe it, I got the same sequence as on the slip of paper." You examine the coin and observe it is a genuine United States government mint penny that is evenly balanced and has distinguishable sides. Moreover, she insists that each time she flipped the penny, she gave it a good jolt (these were not phony flips).

What do you conclude now? As before, you will be unable to find any shorter description than the sequence itself—it is a random sequence. Nevertheless, you will be justified rejecting her story. The problem is that the timing is all off. When she handed you the sequence a week earlier, she specified a highly improbable event. When she returned and claimed subsequently to have reproduced the sequence, she in effect claimed to prophesy a vastly improbable chance event. Prophecy of vastly improbable chance events is highly suspect. Indeed, anyone with such a gift should be a billionaire many times over, either in Las Vegas or on Wall Street.

Finally, suppose this individual comes to you and says, "Would you believe it? I just flipped this penny 100 times, and it came up heads each time!" As before, the coin she shows you is a genuine penny, and she is emphatic that hers were not phony flips. Rather than being specified in advance, this time the pattern of coin flips is specified by means of its low computational complexity. The sequence (N) has, in the language of algorithmic information theory, about the lowest entropy possible. There are very few sequences with descriptions as short as "repeat '1' 100 times." Once again, you would be ill-advised to trust her story. The problem is not that low-entropy sequences like (N) are highly improbable—indeed, any sequence of the same length is as improbable. Rather, the problem is that there are too many other sequences for which no short description can be found.

Our coin-flipping friend, who claims to have flipped 100 heads in a row with a fair coin, and without phony flips, is in the same position as a lottery manager whose relatives all win the jackpot or an election commissioner whose own political party repeatedly gets first position on the voting ballots (which is an advantage in elections). In each case, public opinion rightly draws a design inference, eliminating chance and attributing fraud. Granted, the evidence is always circumstantial. Moreover, our legal system has yet to formulate a coherent and comprehensive theory of how such evidence should be handled (for discrimination cases, it's clear; for election fraud, it's less clear). Nevertheless, the inference that chance was offset by an act of intentional meddling (design) can in such cases be compelling.[29]

2.6 Cryptography

Cryptography raises an interesting prospect for the design inference. Cryptography studies how to secure communications by transforming text so that it becomes unreadable except to those who possess a cryptographic key. Cryptosystems lie at the heart of cryptography. A given cryptosystem is a method for transforming plaintext into ciphertext and back again by means of a cryptographic key. Central to any cryptosystem is not just a specific cryptographic key that unlocks the cryptosystem but also an ambient keyspace from which the key is chosen. The keyspace must be so large that it will be highly unlikely for current (and, it is hoped, future) computational technologies to find the key by random or brute-force methods. Cryptanalysis is then the study of how to break cryptosystems by finding their keys.

The security of a cryptosystem is always a matter of prob-ability. For any cryptosystem, it is necessary to ask, What is the probability that an outsider with no knowledge of the key will figure it out and thereafter be able to read its encrypted messages? This probability will depend on several factors. It will depend on intercepted ciphertext transmissions. It will depend on intercepted plaintext transmissions that can be matched with intercepted ciphertext transmissions. But all such interceptions should mean nothing for unearthing the cryptographic key provided the key-space is big enough (a needle in a haystack) and the computational resources to search it for the given key are totally inadequate.

When all is said and done, the security of a cryptosystem is therefore encapsulated in a positive probability p. This is the probability that our enemies, even at their most resourceful and computationally extravagant—albeit short of stealing or gaining direct knowledge of the key—will discover it and thereby break the cryptosystem. This probability is often expressed as a waiting time (see Appendix B.5) or as a computational complexity measure. Thus, for an event of probability p, we can ask how long, in terms of independent attempts to achieve the event in question (in this case, breaking the cryptosystem), it would take for it to be reason-ably likely to occur (for instance, to have better than even odds of being broken). Alternatively, we can ask how computationally difficult it is to achieve the event, associating an improbability of finding the key with that difficulty.

In his history of cryptography, Simon Singh offers an example of this waiting-time approach to breaking (or at least partially breaking) a cryptosystem. There he quotes William Crowell, Deputy Director of the National Security Agency from 1994 to 1997: "If all the personal computers in the world—approximately 260 million computers—were to be put to work on a single PGP [i.e., Pretty Good Privacy] encrypted message, it would take on average an estimated 12 million times the age of the universe to break a single message."[30] Crowell said this about thirty years ago, so with Moore's Law ensuring a doubling of computer power every eighteen months, that would yield present computers with speeds a million times faster. And presumably, there are many more personal computers present now than the 260 million back then. Let's say the increase is 100-fold. In that case, with all personal computers operating on nothing but breaking a single PGP-encrypted message, the waiting time to break it would still take over a billion years. A fortiori, finding the underlying PGP key capable of encrypting all messages would require an even longer waiting time.

The computational complexity approach to the probability p of breaking a cryptosystem looks at how much computational power is available to break it, and how varying that power helps or hurts the effort to break it. Consider the following example from Stephen Budiansky's history of the National Security Agency's efforts to break Soviet ciphers. Citing the Baker Panel's 1958 report on foreign intelligence communications, Budiansky writes:

> The panel pointed out that in the case of the most secure Soviet machine systems, an exhaustive key search was beyond the bounds set by the laws of physics. "There is not nearly enough energy in the universe to power the computer" that could test every setting of such a rotor machine [such as were being used by the Soviets], which had an effective cryptanalytic keyspace on the order of 10^{44}. Even the "more modest undertaking" of recovering the setting of an individual message enciphered on such a machine whose internal configuration has already been recovered, which would involve testing about 10^{16} possibilities, would cost \$2,000,000,000,000,000,000,000 per message for the electricity required to power any known or projected computing devices. (In 1998 a \$250,000 machine built with 1,856 custom-made chips successfully carried out an exhaustive key search on the 56-bit key DES encryption system—a keyspace slightly greater than 10^{16}—in two days. But a 128-bit key, with a keyspace of the order 10^{38}, can be shown to resist an exhaustive search even by the most theoretically energy-efficient computer that the laws of physics permit.)[31]

In the computational complexity approach, one asks how much computational power is required to make resolving the problem reasonably probable.

Suppose, now, we have a cryptosystem with an impressively large keyspace. Suppose this space contains the key k so that with this key people can read our mail. Because the keyspace is so large, successfully finding k without any special knowledge or intelligence will thus have a very small probability, call it p. Suppose further that we have incontrovertible evidence that our cryptosystem is compromised, which is to say that our enemies have discovered our key and are indeed reading our mail. In that case, what can we infer about the event responsible for disclosing our key k to our enemies? Is this event the result of chance or design?

Certainly, when cryptosystems are broken so that their keys are disclosed to unauthorized personnel who should not have access to them, design is a natural conclusion. Design in the form of stealing cryptographic keys has, after all, a long history. For instance, the Walker spy network of the 1970s and 80s, in which John Walker and his associates systematically plundered the U.S. Navy's cryptographic keys and delivered them to the Soviet KGB, ranks as perhaps the worst intelligence disaster in U.S. history.[32]

But it's one thing to have smoking gun evidence of design, such as catching John Walker's accomplices red-handed stealing cryptographic keys and encrypted messages off Navy aircraft carriers. It's another thing to draw a design inference when such smoking gun evidence is lacking. In such cases, we confront three possibilities: chance, design, and a hybrid of the two. The first, chance, is that the key was simply randomly guessed. In other words, our enemies just got lucky. In querying the keyspace, they were repeatedly playing a lottery so that even with all those repetitions factored in, the probability of winning the lottery was so small that we should never have expected them to find the key in this way. Notwithstanding, it just happened as a chance fluke.

Obviously, this explanation doesn't cut it. The key k is clearly prespecified in that it is the one key that encrypts and decrypts our messages. So, if finding this key remains improbable even after factoring in every available attempt by our enemies to find it through a brute random search (in line with the waiting-times or computational-complexity protections to our keyspace as described above by Singh and Budiansky), then a design inference seems warranted.

Design, acting as a foil to chance, is therefore the second possibility to be considered. We always have to consider a pure chance hypothesis H in which discovery of the key k happens as a small probability event—call that probability p. Given that this event of discovering the key does not just have small probability but is specified, we also have to take seriously that a design inference may be in play and therefore consider a design hypothesis D. If H accurately captures the security of the keyspace and if the probability p that H assigns to randomly finding the key k is extremely small, then we will be entitled to draw a design inference, thereby inferring D.

But to infer design under these circumstances, the chance hypothesis H must accurately characterize how unlikely it is that the key could be revealed by purely chance methods. This raises an intriguing third possibility, namely, a hybrid between chance and design where H fails accurately to characterize the security of the cryptosystem. We thus imagine that hardworking cryptanalysts have cut the keyspace down to size so that instead of a chance hypothesis H and a probability p characterizing the probability of finding the key k, it is now a modified chance hypothesis H' that allows k to be revealed with a probability q that is much larger than p.

In a sense, finding the key k is still due to chance, but it is a different chance hypothesis H', according to which the discovery of k becomes so probable that it can no longer be regarded as sufficiently improbable to trigger a design inference. At the same time, what allows H' to substitute for H is the hard design work of cryptanalysts. Hence, a chance-design hybrid is a third option alongside pure chance and pure design. Whether it should be taken seriously hinges on how convincingly the security of the cryptosystem can be demonstrated.

In a hybrid scenario, the problem ultimately is that the cryptosystem is not as secure as advertised. The history of cryptography is filled with such examples. Take the German Enigma cryptosystem of World War II. As the Allies demonstrated, it proved unsecure, even by the standards of the time. The Allies at Bletchley Park, notably through the efforts of Alan Turing, were able to exploit weaknesses in the Enigma cryptosystem by which they could drastically reduce the difficulty of finding its keys.

Suppose, then, that a cryptosystem is known to be highly secure irrespective of the computational resources available now or later (the one-time pad, if used just once, constitutes such a known, provably secure cryptosystem—see Appendix B.1). In that case, the probability p for breaking it will be so small that all the computational resources that might ever be available to us still do not make breaking it likely. A pure chance hypothesis—in which it is supposed that our enemies figured out the key by blindly searching the keyspace—does not cut any ice if the cryptosystem is known to be highly secure.

What should we then conclude if we become convinced that such a cryptosystem has been broken? In that case, we'll conclude that a specified event of small probability has happened and that we are justified in drawing a design inference. Our next task will then, presumably, be to alert the counterintelligence services to find the traitor responsible for compromising the cryptosystem. But if, instead, we become convinced that our cryptosystem has a fatal flaw that vastly increases the probability of the key being found, we'll conclude that a hybrid chance-design hypothesis could legitimately explain how the cryptosystem was compromised.

To sum up, highly secure cryptosystems are not broken by chance. Nor is ingenious hard work enough to break them. Highly secure cryptosystems are broken by cheating, and thus by design. Luck is limited and genius, in the absence of clairvoyance, can only go so far.

2.7 SETI—The Search for Extraterrestrial Intelligence

What are the prospects for successfully finding clear signs of extraterrestrial intelligence? With the dawn of radios, this question has been the subject of intense speculation since around 1900. Perhaps aliens, and not just humans, are responsible for some of the radio signals we can detect on Earth. Nikola Tesla, as early as 1896, raised the possibility of radio-based SETI (the Search for Extraterrestrial Intelligence), contending that wireless electrical transmission could establish communications with Mars and its presumed Martian inhabitants. In fact, he came to believe that "intelligently controlled signals" were actually moving back and forth between Earth and Mars.[33]

The journal *Nature*, however, places the advent of SETI as a scientific research program at September 19, 1959, when *Nature* published an article titled "Searching for Interstellar Communications."[34] Since that time, radio observatories have monitored myriads of radio channels hoping to detect radio transmissions from space that reliably indicate ETIs (Extraterrestrial Intelligences). Since unlike the ETIs on *Star Trek*, genuine ETIs are presumed not to communicate in English, or any other human language for that matter, the problem of determining when a radio transmission is the product of an intelligence falls, in a very broad sense, under the cryptographic framework described in the last section. Indeed, the SETI researcher's task is to eavesdrop on interplanetary communications, trying to determine whether a given radio signal was transmitted by an intelligent agent. If it was intelligently generated and ascertainably so, the next task would be to uncover the signal's meaning. But at the very least, the point would be to rule out that it was randomly generated.

Of course, SETI researchers don't think of themselves as engaged in breaking a cryptosystem to discover signs of intelligence from outer space. Typically, it is assumed that extraterrestrial intelligences are so intent on making their presence known that they will do something terribly obvious, like transmit a sequence of prime numbers, as in the film *Contact*, based on a novel of the same name by Carl Sagan. Thus, at the key moment in the film, what convinces the SETI researchers that they have encountered a bona fide extraterrestrial intelligence is a signal from outer space, represented as a sequence of beats separated by pauses, that spells out the prime numbers from 2 to 101. The SETI researcher who first perceives this pattern remarks, "This isn't noise, this has structure." Clearly, this researcher was drawing a design inference.

For the SETI program to have even a chance of being successful, the following claims must hold: (1) ETIs must at some point in the history of the universe have existed; (2) ETIs have been sufficiently advanced technologically to signal their presence by means of radio signals; (3) ETIs have signaled their presence by means of radio transmissions strong enough to reach Earth; (4) we happen to be living at just the right time and place in cosmic history to receive those transmissions; and (5) we could discover salient patterns in those transmissions emblematic of intelligence.[35]

These claims require such a happy convergence of circumstances that SETI researchers avoid the further complication of asking whether ETIs are communicating enigmatically by means of encryption, and thus making it difficult to discern their intelligence against the backdrop of cosmic radio noise. Any "cryptosystem" the ETIs are employing is therefore assumed to be minimal and unintended. The ETIs, we assume, want to make their presence known. If, therefore, we need to do any cryptanalysis to detect them, it is solely because the means by which ETIs communicate are so foreign to ours.

If the SETI program ever proves successful (something it has yet to do, even with the additional twenty-five years since the first edition of this book was published), its success will require drawing a successful design inference, matching radio transmissions it has monitored with patterns it properly recognizes as clear and reliable indicators of intelligence. As it monitors myriads of radio channels, SETI attempts to match patterns it has independently identified, which is to say it attempts to match specifications. Insofar as SETI fails to specify the patterns employed by ETIs, SETI will fail to detect them, their presence slipping past the SETI researchers' detection sieve. Regardless of whether one thinks SETI constitutes an ill-fated research program, it raises important questions about the nature of intelligence, the possibility of detecting intelligences other than human, and the role of design inferences in detecting human as well as non-human intelligences.

Before moving to the related topic of directed panspermia in the next section, we want to respond to an objection by Seth Shostak, of the SETI Institute, against the ability of design inferences to detect alien intelligence. Shostak has argued that design inferences are irrelevant to SETI because the signals SETI researchers use to try to detect intelligence are not complex, such as the long sequence of prime numbers in *Contact*, but in fact simple. According to him, SETI looks for very simple signals to detect design, namely, radio signals with narrow bandwidth transmissions. This simplicity in the radio signals monitored by SETI researchers is thus supposed to undercut the logic of the design inference detailed in the first edition of this book and subsequently.[36]

But in fact, the design inference applies to the very signals that Shostak's SETI Institute is looking for to provide compelling evidence of intelligent life beyond Earth. True, as narrow bandwidth transmissions, the signals are simple to describe. But to decisively implicate intelligence, they need to be difficult for purely physical processes to reproduce by chance—otherwise they would be common and thus not diagnostic of intelligence. But that's just another way of saying the signals must be improbable.

And so, we see in Shostak's supposed counterexample the very convergence of small probability and specification (in the form of narrow bandwidth patterns) that is characteristic of design inferences in general. It's the same reason we detect design in the monolith from Stanley Kubrick's *2001: A Space Odyssey*.[37] That structure, a rectangular cuboid, is easily described and thus specified. And yet it is hard for purely undirected physical forces to produce such regular geometric solids, thus rendering them improbable. It's the same reason we don't ascribe to chance a thousand coin tosses all coming up heads. Such a sequence is easily described and therefore specified; and it's also improbable.

As a final point, it is worth noting that at SETI's modern inception, Frank Drake and Carl Sagan were not thinking merely in terms of narrow bandwidth transmissions when, on November 16, 1974, they sent the "Arecibo Interstellar Radio Message" to the globular cluster M13. This message consisted of 1,679 bits of data meant to be translated into graphics, characters, and spaces.[38] That many bits correspond to that many coin tosses, and thus to an exceedingly small probability (1 in $2^{1,679}$ or roughly 1 in 10^{505}). Moreover, by any lights, that message is specified. So, as senders rather than receivers of SETI signals, humans have traded not just in narrow bandwidth transmissions, but in the type of specified improbability that is characteristic of human communication and that is the basis of design inferences.

2.8 Directed Panspermia

The search for extraterrestrial intelligence (SETI) focuses on electro-magnetic signals transmitted through space. These signals are hoped to reach planet Earth from distant sources and display improbable patterns that warrant a design inference. Yet any intelligence capable of communicating with Earth through such signals is presumably embodied and could, with the right interstellar or intergalactic transportation technology, reach Earth directly. SETI presupposes that alien intelligences are, for whatever reason, keeping their distance from Earth. But what if alien intelligences could arrive here on Earth more in the style of Steven Spielberg's *ET*? That, of course, raises physicist Enrico Fermi's famous question: If such alien intelligences actually exist, where are they? Shouldn't they have made themselves evident by now?[39] Directed panspermia says that they have.

In its modern formulation, directed panspermia traces to Francis Crick and Leslie Orgel. Undirected panspermia, or panspermia as such, is the idea that life on Earth was seeded here as living forms hitched rides on naturally occurring fast-moving interplanetary objects, such as meteors, comets, or asteroids. Such an object is assumed to get close enough to a planet that it is able to pick up some seeds or spores if the planet has them. Then, later, when the object gets close enough to another planet without life, it deposits those seeds or spores there. According to Crick and Orgel, panspermia in this sense, though taken seriously in times past, is no longer tenable. Thus, they begin their seminal 1973 article on directed panspermia as follows:

> It now seems unlikely that extraterrestrial living organisms could have reached the earth either as spores driven by the radiation pressure from another star or as living organisms imbedded in a meteorite [this is panspermia in its original form]. As an alternative to these nineteenth-century mechanisms, we have considered Directed Panspermia, the theory that organisms were deliberately transmitted to the earth by intelligent beings on another planet.[40]

Why would serious biologists like Crick and Orgel even consider such a science-fiction idea? In pondering the origin of life on Earth, Crick and Orgel concluded that many chemical obstacles stood in the way of life forming naturally on the prebiotic Earth. For instance, they worried about too much oxygen on the early earth: oxidation would make it difficult for first life to form here. What if, instead, life emerged by purely natural means in some other corner of the universe more hospitable to its formation, and only then got transported to Earth, where it could flourish? Unfortunately, hitching rides on things like comets does not make for a hospitable trip to Earth, so it would be better for life to get here on climate-controlled rocket ships, where it would stand a better chance of surviving the transit.

This, in broad strokes, is the idea behind directed panspermia, a theory that Crick developed further in his 1981 book *Life Itself*. Crick was well aware of the objections that could be raised against this theory, not least about how it might be difficult to falsify and also that it seemed merely to defer the ultimate question of life's origin, kicking the can down the road to where life's origin would be more probable, but without actually providing a causal explanation for life's ultimate origin. Crick admitted the validity of such objections but stressed that directed panspermia addressed a legitimate historical question about life on Earth, namely, whether life on Earth arose by intelligent space aliens seeding it. For Crick, this question was worth examining even if it left other questions unanswered.

Whatever its ultimate merits, directed panspermia raises an interesting example of a design inference, or at least, of a small-probability chance-elimination argument. Unless life's formation by purely natural means on planet Earth was highly improbable, there would be no reason to introduce directed panspermia. Occam's razor, for instance, would say that if life is no less likely to have originated on Earth than on some other planet, then there's no need to look to other planets for its origin, to say nothing of the trans-portation technology needed to move it from planet to planet. No, it's precisely life's vast improbability on Earth that prompts speculations about directed panspermia.

Even though directed panspermia thus presupposes a small-probability chance-elimination argument, as formulated by Crick and Orgel, it falls short of warranting a full-fledged design infer-ence. The living forms used to seed planet Earth as well as the living forms piloting the rocket ships to planet Earth are all, for Crick and Orgel, the result of purely physical processes that, in the end, require no design or intelligence. Design, in Crick and Orgel's theory of directed panspermia, arises only in the transportation technology that moves seeds or spores across space and then deposits them on Earth. A weakness of their theory is that we lack any evidence of such transportation technology. In consequence, we have no way to ascribe specified improbability to such technology and therewith find in directed panspermia warrant for a full-fledged design inference.

Nonetheless, directed panspermia as developed by Crick and Orgel admits a very natural extension that is conducive to a design inference. What if life on Earth were not merely seeded by alien intelligences but also, to some degree, designed by them? Perhaps the life that these alien intelligences decided to seed on Earth was originally the result of purely natural forces. But what if, through intensive (genetic) engineering, that life also showed clear marks of intelligence? Here on Earth, we've seen genetic engineers like

Craig Venter, through his company Synthetic Genomics, introduce watermarks into DNA. He has even claimed to produce novel synthetic life forms. Venter has clearly imprinted design into these living forms. The design is evident, and, indeed, Venter wants it to be evident so that he can claim it as intellectual property. In the end, what makes Venter's genomic designs evident are specified events of small probability, which is what is needed to trigger a design inference (recall Section 2.1 on intellectual property).[41]

Ben Stein interviewed Richard Dawkins for the 2008 documentary *Expelled*. In their exchange, they discussed how directed panspermia might lead to a biological design inference:

Ben Stein: "How did [life] get created?"

Richard Dawkins: "By a very slow process."

Ben Stein: "Well, how did it start?"

Richard Dawkins: "Nobody knows how it got started. We know the kind of event that it must've been. We know the sort of event that must've happened, for the origin of life."

Ben Stein: "What was that?"

Richard Dawkins: "It was the origin of the first self-replicating molecule."

Ben Stein: "Right, and how did that happen?

Richard Dawkins: "I've told you; we don't know."

Ben Stein: "So you have no idea how it started?"

Richard Dawkins: "No, no. Nor has anybody.

Ben Stein: "Nor has anyone else. What do you think is the possibility that intelligent design might turn out to be the answer to some issues in genetics or in evolution?"

Richard Dawkins: "It could come about in the following way. It could be that at some earlier time, somewhere in the universe, a civilization evolved by probably some kind of Darwinian means to a very, very high level of technology and designed a form of life that they seeded onto, perhaps, this planet. Now, that is a possibility, and an intriguing possibility, and I suppose it's possible that you might find evidence for that. If you look at the details of biochemistry, molecular biology, you might find a signature of some sort of designer."

Ben Stein [voiceover, not part of interview]: "Wait a second. Richard Dawkins thought intelligent design might be a legitimate pursuit? And that a designer could well be a higher intelligence from elsewhere in the universe [this would be directed panspermia].

Richard Dawkins: "Well, but that higher intelligence would itself have had to have come about by some explicable or ultimately explicable process. It couldn't have just jumped into existence spontaneously."

Ben Stein [voiceover, not part of interview]: "That's the point. So, Professor Dawkins was not against intelligent design. Just certain types of designers, such as God."[42]

The precise merits of a biological design inference when the designer is an alien intelligence (consistent with directed panspermia) versus a transcendent intelligence (consistent with various forms of theism) need not detain us here. The point is that directed panspermia raises the prospect of a biological design inference in which biological systems exhibit actual rather than merely apparent design. We will return to such questions in Chapter 7, where we consider how design inferences apply to evolutionary biology.

In sum, we have seen the design inference used to detect design in plagiarism, copyright infringement, criminal activity, fake data, financial fraud, nonrandomness, compromised cryptosystems, and extra-terrestrial intelligence. In each case, we find the same basic pattern of reasoning, at the core of which is a specified event with small probability. Given such an event, we reject chance and infer design. None of this is controversial or even surprising. Design inferences are not inferences of last resort. They are not even inferences that elicit special scrutiny. They are common, ubiquitous, even second nature.

David Hume versus Thomas Reid[43]

Because the specter of David Hume (1711–1776) looms large over design inferences, it is best exorcised at this early stage in the discussion. Interestingly, the chief exorcist here is Thomas Reid (1710–1796), a fellow Scotsman and contemporary of Hume's. In his *Dialogues Concerning Natural Religion*, published posthumously in 1779, Hume criticized the classical design argument for the existence of God.[44] There he criticized it on a number of fronts, including that it was a failed argument from analogy. But his line of argument that has had the greatest traction against design inferences is that they can only legitimately apply to designers in our immediate inductive experience. Otherwise, they are failed arguments from induction. Reid, by contrast, argued that induction cannot curtail the types of designers we might infer because induction is entirely the wrong instrument for making sense of design. According to Reid, knowledge of design depends on effect-to-cause reasoning that holds independently of induction.

To see what's at stake in this debate between Hume and Reid, let's recall Hume's problem of induction and how he proposed to redress it. Because Hume was a strict empiricist, he saw our knowledge of the world as based entirely on experience. The problem with induction for Hume is that it cannot be rationally justified. He pointed out that there is no logical guarantee that our future experiences will resemble our past experiences: just because something has always happened in a certain way does not guarantee that it will continue to do so. Bertrand Russell memorably illustrated the problem of induction with a chicken that gets fed every day and so assumes that it will continue to get fed, only to wind up one fateful day with the farmer wringing its neck.[45]

But even though Hume saw no logical justification for induction, he accepted its practical utility for human cognition. Inductive inferences, according to him, are warranted when they are based on consistent experience because they induce a habit or custom of expecting the future to resemble the past even in the absence of a bulletproof logical foundation. Hume's understanding of induction therefore ties right in with his understanding of causality, which sees causality not as a necessary connection between cause and effect but as a contingent connection based on habit or custom, developed through repeated observations of events occurring in sequence.[46]

Operating within this Humean inductive framework, several critics of intelligent design have argued that the design inference cannot apply to any designer outside of our direct experience. Human designers are therefore okay. Extraterrestrial designers are iffy. And non-naturalistic designers are strictly prohibited. To see how these critics use the Humean inductive framework to derail design inferences, especially where these inferences threaten to implicate non-naturalistic designers, consider the following claim by Robert Pennock:

> When archeologists pick out something as an artifact or suggest possible purposes for some unfamiliar object they have excavated, they can do so because they already have some knowledge of the causal processes involved and have some sense of the range of purposes that could be relevant. It gets more difficult to work with the concept when speaking of extraterrestrial intelligence, and harder still when considering the possibility of animal or machine intelligence. But once one tries to move from natural to supernatural agents and powers as creationists desire, "design" loses any connection to reality as we know it or can know it scientifically.[47]

John Wilkins and Wesley Elsberry make essentially the same point. According to them, there are "two kinds of design—the ordinary kind based on a knowledge of the behavior of designers, and a 'rarefied' design, based on an inference from ignorance, both of the possible causes of regularities and of the nature of the designer."[48]

Such restrictions on inferring design conveniently rule out designers unacceptable to naturalism. Indeed, accepting this Humean inductive framework for design reasoning makes impossible any evidence for a transcendent designer. But this very framework, and the prohibitions it places on drawing design inferences, cannot withstand scrutiny. Suppose we were to receive a radio signal from outer space containing a lengthy sequence of prime numbers, as in the film *Contact*. This signal would clearly constitute evidence of design. Moreover, if we could rule out that this signal originated on Earth, we would be convinced that we had encountered an alien intelligence. SETI researchers would be celebrating.

But what exactly would we know about the intelligence responsible for that signal? Suppose all we had was this signal representing a sequence of primes. Would we know anything about the intelligence's purposes and motives for sending the primes? Would we know anything about the technology it employed? Would we know anything about its physical makeup? Would we even know that it was physical? Our evidence for design in this case would be convincing but entirely circumstantial. We would be confronted with an effect, know that it had an intelligent cause, and yet be unable to learn any particulars about that cause (except what the intelligence might deem to communicate).

Or consider a more extreme example, a magic penny that acts as an oracle when we toss it (recall Section 1.3). We assign 0 to tails and 1 to heads, and use Unicode as the symbol convention for interpreting strings of 0s and 1s. The penny, for all we know, operates by pure chance (suppose it is hooked up to a quantum mechanical device so that no known deterministic process is affecting how the coin lands). And yet the penny, interpreted in Unicode, answers all of humanity's pressing questions. It is, in effect, the ultimate ChatGPT. The output of this penny is therefore obviously designed. Moreover, it surpasses all current or even imagined human design. Yet a science committed to the Humean inductive framework is barred here from inferring design, having nothing more to offer than a shrug of the shoulders and a vapid assurance that someday we may be able to figure out the underlying regularities of nature that may be operating here.

Note that it does no good to say that the magic penny constitutes a far-fetched thought experiment and therefore need not be taken seriously. The point at issue cannot be so easily dismissed.

Its mere possibility creates serious doubts about the Humean inductive framework. The fact is that we are well able to infer design without knowing the characteristics of the designer, or what the designer is likely to do, or by what means the designer could do it. Humeans in their weaker moments admit as much.

Take Elliott Sober. Before he will permit intelligent design into biology, he wants to know the characteristics of the designer, the independent evidence for the existence of that designer, and what sorts of biological systems we should expect from such a designer. According to Sober, if the design theorist cannot answer these questions, then intelligent design is untestable and therefore unfruitful for science. Yet, as Sober will also admit, "To infer watchmaker from watch, you needn't know exactly what the watchmaker had in mind; indeed, you don't even have to know that the watch is a device for measuring time. Archaeologists sometimes unearth tools of unknown function, but still reasonably draw the inference that these things are, in fact, *tools*."[49]

Sober adheres to the Humean inductive framework, which views all knowledge of the world as an extrapolation from past experiences. Consequently, design must align with our preconceptions to be explanatory; otherwise, it lacks empirical justification. Sober believes predicting a designer's actions necessitates examining past experiences and understanding what designers have previously done. However, his remarks on watchmakers and watches contradict this stance, as he concedes that we could recognize design in watches and mysterious tools without knowing anything about their makers or functions.

In the Humean inductive framework, designers and natural regularities are epistemically equivalent, with their explanatory power rooted in an extrapolation from past experience. Both designers and natural regularities can exhibit predictable behavior. Yet designers differ from natural regularities in their capacity for innovation. Unlike natural regularities, which are universal and uniform and whose behavior therefore is in principle predictable, designers are able to create unprecedented novelty, which defies predictability. As a result, design cannot be encompassed within the Humean inductive framework. Designers are inventors. The best we can do to predict what an inventor would do is for ourselves to become inventors.

The contradictions inherent in the Humean inductive framework go still deeper. Not only can't Humean induction tame the unpredictability inherent in design; it can't account for how we recognize design in the first place. Sober, for instance, regards the design hypothesis for biology as fruitless and untestable because it fails to predict what biological contrivances a designer might bring about. Yet take a different example, say from archeology, in which a design hypothesis about certain aborigines predicts certain artifacts, say arrowheads. Such a design hypothesis would, on Sober's account, be testable and thus acceptable to science.

But what sort of archeological background knowledge had to go into that design hypothesis to make it a successful predictor of arrowheads? At the very least, we would need past experience with arrowheads. But how did we recognize that the arrowheads in our past experience were designed? Did we see humans actually manufacture those arrowheads? If so, how did we recognize that these humans acted deliberately as designing agents and did not just randomly chip away at random chunks of rock? After all, while carpentry and sculpting entail design, whittling and chipping, though performed by intelligent agents, do not. As is evident from this line of reasoning, the induction needed to recognize design (even in so clear-cut a Humean case as arrowheads) can never get started. Our ability to recognize design must therefore arise independently of induction and therefore independently of the Humean inductive framework.

That was precisely Thomas Reid's point. In making it, he effectively scuttled the application of Humean induction to design. In 1780, only a year following the publication of Hume's *Dialogues Concerning Natural Religion*, where Hume had raised his most trenchant criticisms against design, Reid delivered a set of lectures on natural theology in Glasgow. In those lectures, Reid remarked:

> No man ever saw wisdom [read "design" or "intelligence"],
> and if he does not conclude [i.e., infer design] from the

marks of it, he can form no conclusions respecting anything of his fellow creatures. How should I know that any of this audience has understanding? It is only by the effects of it on their conduct and behavior, and this leads me to suppose that such behavior proceeds only from understanding. But, says Hume, unless you know it by experience, you know nothing of it. If this is the case, I never could know it at all. Hence it appears that whoever maintains that there is no force in the argument from final causes [i.e., design], denies the existence of any intelligent being but himself. He has the same evidence for wisdom and intelligence in God as in a father, a brother, or a friend. He infers it in both from its effects, and these effects he discovers in the one as well as the other.[50]

Reid epitomized his point with the following dictum: "From marks of intelligence and wisdom in effects, a wise and intelligent cause may be inferred."[51] Thus, according to Reid, we attribute design as an inference from signs or marks of intelligence. We do not get into the mind of such designing intelligences and thereby attribute design. Rather, we recognize their intelligence by examining the effects of their actions and determining whether those effects display clear marks of intelligence. In consequence, a follower of Hume who purports to attribute design on the basis of induction has already presupposed the ability to identify design independently of induction. Induction's role in design is logically downstream from the inherent power of the mind to interpret marks of intelligence and therewith infer design.

Consider an anthropologist watching a native islander chipping away at stones. Is the islander an arrowhead maker and therefore a designer? If induction has any insight to offer about design, it must be able to do so for arrowheads, whose construction forms as clearcut a case of design as induction will ever find. Yet, if our anthropologist saw the native banging away at a stone with a second rock ideal for chipping arrowheads, this by itself would not prove that the islander was designing anything. Only if, at the end of the day, the islander produced an actual arrowhead (or some other known and well-defined stone artifact) would the anthropologist know that the object produced was in fact a designed object and that the islander was in fact an arrowhead maker and therefore a designer.

To sum up, we recognize intelligence by its effects, not by directly perceiving it. A human being who continuously mumbles the same nonsense syllable displays no intelligence and provides no justification for attributing design to the utterances. Design reasoning is effect-to-cause reasoning: it begins with effects in the physical world that exhibit clear signs of intelligence and from those signs infers to an intelligent cause. Humean induction is therefore entirely the wrong framework for how to infer design.

Thomas Reid agreed that signs of intelligence can be learned and confirmed by experience, but he also stressed that our capacity to recognize them as convincing indicators of intelligent causation cannot originate in experience. That ability must be hardwired into us as part of basic human rationality. It is, as Alvin Plantinga would put it, part of our "proper function."[52] Hume and his followers exercise that proper function just like everyone else. What's new with the contemporary intelligent design movement is that it brings analytic precision to our understanding of these signs of intelligence. Within the theory of intelligent design, signs of intelligence get cashed out in terms of improbability and specification, which together serve as an analytic tool for inferring design.

CHAPTER 3

SPECIFICATION

3.1 Patterns That Eliminate Chance

IN THE INNER HARBOR OF VICTORIA, BRITISH COLUMBIA LIES AN arrangement of flowers that spells the words "Welcome to Victoria." Everyone who sees these flowers concludes that they were deliberately planted to form that message. But why? If pressed, people will say, correctly, that it is highly unlikely for randomly arranged flowers to spell any message at all.

In the early 2000s, it was discovered that almost all the top prizes offered in McDonald's Monopoly sales promotion were being won by friends, relatives, and associates of a certain Jerome P. Jacobson. Jacobson, it turned out, also happened to oversee security for the contest, working for the company McDonald's hired to run it. This coincidence, of course, was highly suspicious. But why? It seems highly unlikely that, if the contest were fairly run, almost all the big winners would have a connection to Jacobson.

Nicholas Caputo was a county clerk in New Jersey. One of his duties, in the elections under his supervision, was to randomize the order of candidates on ballots. It is a known fact that appearing first on a ballot improves a candidate's chances of winning. In 40 out of 41 elections, Caputo placed the Democrats, his own party, ahead of the Republicans. This outcome earned him a nickname: "The man with the golden arm." The New Jersey Supreme Court regarded this outcome as highly suspicious and urged better randomization procedures in determining ballot orders. But why? The Court deemed Caputo's results as too improbable.

Even so, for each of the three examples just considered, we can find parallel events that are equally improbable but raise no suspicions. The specific pattern of flowers found in the "Welcome to Victoria" sign is highly improbable. But so too is the specific pattern of weeds found in an untended flower bed. The probability that those weeds would be in that exact pattern is also exceedingly small. But nobody thinks the weed garden was deliberately arranged.

Approximately 200 people become millionaires each year by winning the lottery. The probability that those particular people would happen to win is exceedingly small, and for all we know just as improbable as Jacobson's associates winning all of the big McDonald's prizes (non-associates, it turned out, did win lesser prizes). Even so, none of the lottery winners or lottery organizers appears to be going to jail.

Numerous ballots exist around the world where election clerks are tasked with randomly assigning the order of candidates on ballots. For any election clerk overseeing more than 40 elections, the probability that the order would be exactly the one observed is extremely small—as extreme as the case of Nicholas Caputo. Nevertheless, the Caputo case seems unique for its sheer brazenness in favoring the clerk's own political party (which, no doubt, accounts for Caputo being dubbed "the man with the golden arm").

What is the relevant difference in these examples? Why are flowers arranged to form a message worthy of a design inference whereas randomly strewn weeds in a weed garden are not? Why does the consistent winning of big prizes by Jacobson's associates raise suspicions whereas big lottery winnings in general do not? Why are Caputo's ballot orders suspect whereas most other ballot orders are not? Intuitively, the difference centers on whether a suspicious pattern identifies the event in question.

But what distinguishes the suspicious patterns that lead us to infer design from the unremarkable patterns that lead us to think that nothing of consequence needs to be explained? The difference boils down to whether a highly improbable event matches a simple, easily describable pattern. The description, "flowers spelling 'Welcome to Victoria'" is short and simple, and any arrangement of flowers answering to that description is improbable. By contrast, any description that identified the positions and types of weeds in the weed garden would be long and complicated. True, "weed garden" is a very simple description. But it does not describe an improbable arrangement because many possible gardens fit that description.

The collection of winners of big McDonald's prizes could be simply described as "associates of Jacobson." No such simple description exists to describe the other major lottery winners. A description that identified all of the other major lottery winners would be long and complicated. True, one could simply describe them as an "assortment of people," which is short. But that description would not correspond to an improbable event because numerous assortments answer to it, including mostly non-lottery winners.

Caputo's ballots can be described as placing the Democrat candidate first in all but one instance. "Democrats first in 40 out of 41 ballots" describes the event. But most ballot orders, because random, require descriptions that are long and complicated. For instance, "Democrats first on the initial ballot, Republicans first on the next three ballots, Democrats first on the next two ballots, ..." The ellipsis here needs to be explicitly filled in. Any short description of such ballot orders would apply to a wide range of ballots and thus not describe an improbable event.

The key difference, therefore, between the two types of cases is whether a match exists between an improbable event and a simple description of it. When only a long and complicated description of the event exists, no suspicions are aroused even if that event is highly improbable. As a frivolous example of this point, consider a line from the film *The Empire Strikes Back*. Darth Vader tells Luke Skywalker, "No, I am your father," revealing himself to be Luke's father. This is a short description of their relationship, and the relationship is surprising, at least in part because the relationship can be so briefly described.

In contrast, consider the following line uttered by Dark Helmet to Lone Starr in *Spaceballs*, the Mel Brooks parody of *Star Wars*: "I am your father's brother's nephew's cousin's former roommate." The point of the joke is that the relationship is so complicated and contrived, and requires such a long description, that it evokes no suspicion and calls for no special explanation. With everybody on the planet connected by no more than "six degrees of separation," some long description like this is bound to identify anyone.[1]

Probabilistic Complexity and Descriptive Simplicity

Before we get into the nuts and bolts of specification in this chapter, we want to preempt a possible confusion about the roles of complexity and simplicity in design inferences. Complexity and simplicity apply to both the probability of events and the description of patterns, but in design inferences these concepts pull in opposite directions. If an event has small probability, such as when a coin is tossed a thousand times and the outcome has an improbability of 1 in $2^{1,000}$, the exact sequence of coin tosses corresponds to 1,000 bits. An improbable, or small-probability, event is thus complex in the sense of corresponding to a lot of bits. A medium-to-large probability event, on the other hand, such as when a coin is tossed only two times and the outcome has a probability of 1 in 4 (= 2^2), corresponds to very few bits, in this case 2. Such an event is therefore probabilistically simple. In general, a probability of p corresponds to a bit sequence of length $-\log_2 p$, which can be fractional, and in fact take on any nonnegative real value. A given probability therefore corresponds to a precise bit length.

The role of complexity and simplicity, however, looks quite different for specifications as compared to events. Specifications are patterns that identify events and that have short or simple descriptions. But descriptions in general can be represented in terms of bits. For instance, consider a random sequence of 1,000 coin tosses and also a sequence of 1,000 coin tosses each of which landed heads (let 0 represent tails and 1 heads). A truly random sequence cannot be described any more simply than by repeating the sequence exactly (recall Section 2.5). The all-heads sequence, on the other hand, has a simple description: "all heads" (which will require only a few bits when represented in ASCII or Unicode). The reason we can legitimately eliminate chance for 1,000 heads in a row is that the sequence is "probabilistically complex" (it has small probability and therefore its occurrence corresponds to lots of bits) but it is also "descriptively simple" (it can be described with few bits).

Design inferences engage in a balancing act between the complexity of events and the simplicity of descriptions. To draw a design inference for an event, it must have small probability, or be improbable. In other words, its occurrence must correspond to a high complexity of bits. But in addition, to draw a design inference, the event must precisely match a pattern of short description length, the pattern thus constituting a specification. In other words, the event's description must correspond to a low complexity, or simplicity, of bits. High probabilistic complexity combined with low descriptive complexity, or descriptive simplicity, thus becomes the hallmark of design inferences. All the other ways of matching up complexity and simplicity with events and patterns (e.g., probabilistic simplicity combined with descriptive complexity) do not make for a viable design inference.

In Chapter 6, we will develop specified complexity as a precise information measure for inferring design. The "complexity" in "specified complexity" refers to probabilistic complexity, or small probability. The "specified" in "specified complexity" refers to a specification, which is a pattern of short description length, and therefore one of descriptive simplicity. Specified complexity will therefore constitute a rigorous formal way of combining the dual requirements of design inferences, which is to say probabilistic complexity and descriptive simplicity.

3.2 Minimum Description Length

We call patterns that have a simple description *specifications*. But what makes a description simple? To answer this question with mathematical precision requires measuring the length of descriptions and then designating the simple descriptions as those that have short length. This approach turns specification into a complexity-theoretic notion, making descriptions more or less complex depending on their length. The underlying metric that measures description-length is therefore a complexity measure.[2]

The mathematical, linguistic, and computer science literature is replete with complexity measures that measure description length, although the specific terminology to characterize such measures varies with field of inquiry. For instance, the abbreviation MDL, or minimum description length, has currency in the statistical literature dealing with model selection and Bayesian hypothesis testing.[3] The abbreviation AIT, or algorithmic information theory, also has wide currency, where the focus is on compressibility of computer programs, so that highly compressible programs are the ones with shorter descriptions.[4]

Measures of description length focus on those descriptions that are shortest or minimal. So, for a given item whose description length is of interest, such measures will give the length of the shortest description that describes it. The precise calculation of such minimum description lengths, however, can be difficult. At the heart of algorithmic information theory, for instance, is a complexity measure that gives the length of the shortest (most compressible) program that computes a given string of characters. Nonetheless, it is a result of computer science that finding these shortest programs, and thus determining their length, is non-computable.[5]

Noncomputability here means that no algorithm exists that can find these shortest programs or calculate their lengths. In consequence, there's typically no guarantee that a short description that someone can find is in fact the shortest. Rather than precisely calculate minimum description lengths, in practice we must content ourselves with estimating them. In forming such estimates, we often can do no better than simply to identify the shortest description we can think of and take its length as our best estimate of the minimum description length.

Measures of description length attempt to solve an optimization problem, finding the minimum length among all descriptions of an item. In practice, however, we just need descriptions that are short enough to convince us that we are dealing with a specification, not descriptions that are optimally short. If we are able to come up with a short description, we know that the shortest description is at least as short. Thus, having a short description is typically enough to determine that we actually have a specification.

A rigorous characterization of specifications and design inferences depends on complexity-theoretic measures that calculate description length. These measures, when identified and applied, will then gauge what counts as a "short" or "long" description (or somewhere in between). Nonetheless, we all draw design inferences in practical life based on an intuitive understanding of the simplicity/complexity needed to describe a pattern. When we infer design intuitively, we judge whether an event is sufficiently improbable and also whether its description is sufficiently simple. If so on both counts, we recognize, even if only in some broad pre-theoretic sense, that we are dealing with a specified event of small probability. Such a confluence of specified event and small probability then leads us to infer design.

To sum up, to be specified is to have the property of being simply described, which, when given mathematical precision, comes down to having a short description according to a complexity measure that calculates description length. The rest of this chapter lays out the basics of specifications. Given these basics, readers with sufficient technical knowledge should, setting aside one key sticking point, be in a position to reconstruct the full formal theory of specification. The sticking point centers on the choice of language for measuring description length, and this is why, in an upcoming chapter on specified complexity (Chapter 6), we return to the concept of specification, supplying the technical details lacking in this chapter.

To make specification a rigorous and usable concept for characterizing the types of patterns that trigger design inferences, we need to show that the choice of language for describing these patterns and the means of calculating their description lengths do not undermine the validity of design inferences. Specifications are language relative. They depend on the language (information-theoretic or linguistic conventions) used to formulate the descriptions, which in turn determine the types of complexity measures available to measure description length.

This language relativity or dependency is a feature and not a bug of specifications, and it is therefore a feature as well of design inferences. Yet, as we will see later in this book, this feature in no way undermines design inferences. It simply underscores the role that context and background knowledge play in design inferences. It also makes clear why design-inferring agents are not all going to draw the same design inferences. Language dependence, and the way it constrains context and background knowledge, simply means that we may not know enough to draw certain design inferences, not that the design inferences we do draw are invalid. The role of language dependency in specifications will be explained at length in the upcoming chapter on specified complexity (Chapter 6).

3.3 Events and Their Corresponding Patterns

Ordinarily, when we infer design, we first witness an event. Initially, as we become aware of the event, we may fail to see in it any suspicious or unusual pattern. Later, however, we may notice that this event exhibits a salient pattern and that for any event to match that pattern by chance is highly improbable. If, in that case, we also discover that this pattern has a simple description (i.e., that it is a specification), a design inference is triggered. Let's call the event E and the corresponding pattern V. The event E is something that happens in the world. The pattern V is a linguistic or information-theoretic entity that identifies an event in the world. As a handy notation, we use an asterisk to distinguish between V as a descriptive entity and V^* as the event identified by V. (See Appendix A.1 for this asterisk notation, which characterizes event-pattern duality.)

How we discover a salient pattern having a simple description admits no precise classification. Typically, the discovery of such patterns depends on background knowledge, context of inquiry, and even psychological factors, such as priming effects. It is a basic feature of human rationality that we are able to discover such patterns. But to date, no algorithm or step-by-step procedure shows how we do it. The issue is not how to discover these patterns, but what to do with them once we have discovered them. If an AI program such as ChatGPT is able to discover such patterns, they will form perfectly good specifications for drawing design inferences.

A brief digression is worth inserting here. Over the years, I [WmD] have debated professional skeptic Michael Shermer before various college audiences. Shermer, to dismiss the use of patterns to infer design, appeals to evolution. According to him, evolution biases us to read more into patterns than may actually be there. Thus, we are supposed to be hardwired to commit false positives, falsely attributing design even when it is absent. Here is how Shermer makes such an argument:

> Perceiving the world as well designed and thus the product of a designer... may be the product of a brain adapted to finding patterns in nature. We are pattern-seeking as well as pattern-finding animals... Finding patterns in nature may have an evolutionary explanation: There is a survival payoff for finding order instead of chaos in the world, and being able to separate threats (to fight or flee) from comforts (to embrace or eat, among other things), which enabled our ancestors to survive and reproduce. We are the descendants of the most successful pattern-seeking members of our species. In other words, we were designed by evolution to perceive design.[6]

But there's a problem with Shermer's argument: To refuse to attribute patterns to design on the grounds that our brains are hardwired by evolution to overinterpret patterns as designed begs the question. That's because some patterns are indeed rightly interpreted as signaling design. We all recognize a valid distinction between patterns that convincingly demonstrate design (as when seeing arrowheads, Egyptian pyramids, or Stonehenge) and patterns that result from unreliable psychological factors, such as an overactive imagination (as when seeing human faces in clouds, burnt toast, or soap films—examples of the sort that Shermer uses to discredit intelligent design).

Shermer's claim that we are hardwired by evolution to be pattern-seeking, pattern-finding animals does nothing to draw this distinction. Specification, on the other hand, provides a rigorous basis for distinguishing patterns that rightly signal design from those that don't. Ironically, Shermer's argument here assumes that he can separate errant design inferences from solid ones. That ability is precisely what we argue for and seek to advance by rigorously characterizing the two key theoretical underpinnings of design inferences, namely, small probability and specification.

Back to the main discussion. To say that a pattern V describes an event E, the event E must fall under the event identified by V. Nevertheless, E need not precisely coincide with the event identified or denoted by V. As noted, we use an asterisk notation to associate a description/pattern with the event it identifies. In the case of the pattern V, the corresponding event is V^*. Accordingly, V can describe E even though $V^* \neq E$. All that's required for V to describe E is that V^* include or subsume E, or in set-theoretic notation, $V^* \supset E$. The occurrence of E thus guarantees the occurrence of V^*. But V^* could include other outcomes that do not intersect with E.

Description, as we understand and use the concept, therefore differs from exact identification. To see the difference, consider the toss of a die. Let E be the event of rolling a six, but let V be the description "an even number was rolled." V therefore (precisely) identifies the event of rolling either a two or four or six, an event of probability ½. But V also describes the event E, which falls under it but has probability 1/6. To say that E falls under V is to say that E (logically) entails the event identified by V, that is, V^*. In other words, if E happens, then the event V^* must also happen. In this example, the roll of a six entails an even roll.

Patterns that trigger a design inference often describe without exactly identifying the event in question. In design inferences, there tends to be a balancing act between coming up with a description that is sufficiently short but where the precise event identified by the description is also sufficiently improbable. Increase the description length (i.e., make it less simple), and you can decrease the probability of the event described. But the cost of the lengthier description may be, as in the *Spaceballs* example, that a design inference is no longer warranted.

Often context sets the salient patterns that become grist for the design-inferential mill. Suppose, for instance, the context is poker. E, let us say, is the event of drawing a royal flush in the suit of hearts. What should V be in this case? V could be the pattern described as "royal flush in the suit of hearts." There are 2,598,960 different possible poker hands, and so the probability of V^* is 1/2,598,960. In this case, the event E coincides precisely with the event described by the pattern V, in other words $V^* = E$.

But what if, instead, we let V be the pattern denoted by "royal flush"? In that case, the description is shorter, but the event associated with V has larger probability. Four poker hands answer to the description "royal flush," and so the probability of V^* in this case would be 4/2,598,960, which equals 1/649,740. V now describes the hand we observed, that is, the event E, but it does not fully describe, or precisely identify, the event E.

To appreciate how improbability and description length play off each other in design inferences, consider next the poker description "any hand." This description is roughly the same length as "royal flush." Yet whereas "royal flush" refers to 4 hands among the total number of 2,598,960 poker hands, "any hand" refers to all poker hands whatsoever. Whereas "royal flush" describes an event of probability 1/649,740, "any hand" thus describes an event of probability 1.

Clearly, if we witnessed a royal flush, we'd be inclined, on the basis of its short description and low probability, to resist attributing it to chance. Now granted, with all the poker that's played worldwide, the probability of 1/649,740 is not small enough to decisively rule out its chance occurrence (in the history of poker, royal flushes have appeared by chance). But clearly, we'd be less inclined to ascribe to chance a hand that answers to "royal flush" than a hand that answers to "any hand."

The difference between "royal flush" and "any hand" illustrates the concept of *specificity*. If E is a royal flush in the suit of hearts, then "royal flush" describes E with high specificity, narrowing E down to one of four possible hands. "Royal flush in the suit of hearts" goes further, and describes E with maximal specificity, narrowing E down to exactly E. "Any hand," by contrast, describes E with low (indeed minimal) specificity. It describes E, but only in the loosest sense, without in any way narrowing the field. The probability of the event identified by the pattern "any hand" is 1, compared to 1/649,740 for "royal flush" and 1/2,598,960 for "royal flush in the suit of hearts."

Patterns thus have high specificity to the degree that they narrow the range of possible events. In practice, this means patterns have high or low specificity to the degree that the events with which they are exactly identified have respectively high or low improbability (or, conversely, have respectively small or large probability). These in turn may be thought of, respectively, as the high- and low-information patterns, corresponding as they do respectively to high and low information events (in line with Appendix A.9–11). In consequence, a pattern V describing an event E will trigger a design inference only if the *specificity* of V is high, which is to say that V both describes E and identifies an event V^* that is improbable in its own right.

3.4 Recognizing Patterns That Signal Design

Finding a salient pattern that triggers a design inference often results from a flash of insight. Such flashes are common in detective work, where a sleuth sees a multiplicity of seemingly unrelated elements all suddenly converge to incriminate a guilty party. These insights arise from creatively seizing on one's background knowledge to discern a suspicious or unusual pattern in an event. The pattern may not be immediately obvious. Yet once it is discerned, its salience cannot be denied. In most instances, these patterns can be succinctly described and therefore constitute specifications.

To illustrate how a pattern can signal a design inference, consider the following sequence of numbers: 11121121111122 1312211etc. Is this sequence the result of a random process? What if we represent this number by separating it into blocks as follows: 1 | 11 | 21 | 1211 | 111221 | 312211 | etc. It now becomes clear that each block gives a recipe for writing the previous block. To write "1" means putting down one one. Hence "11." To write "11" means putting down two ones. Hence "21." To write "21" means putting down one two and one one. Hence "1211." Etc. With this insight, it's clear that the sequence is contrived. Filling in the details for a full-fledged design inference is straightforward.

Here's a particularly insightful example of how specifications work. It was this example that finally clarified for me [WmD] what makes a pattern a specification. Consider the following event E_ψ, an event that to all appearances was obtained by flipping a fair coin 100 times (0 for tails, 1 for heads):

$$0100011011000001010011100$$
$$1011101110000000100100011$$
$$0100010101100111100010011$$
$$0101011110011011110111100 \qquad E_\psi$$

Is E_ψ due to chance or not? A standard ploy of statistics professors in teaching an introductory statistics course is to have half the students in a class each flip a coin 100 times, recording the sequence of heads and tails on a slip of paper, and then have each student in the other half, as a purely mental act, attempt to mimic a sequence of 100 coin tosses, also recording the sequence of heads and tails on a slip of paper. When the students then hand in their slips of paper, it is the professor's job to sort the papers into two piles, those generated by flipping a fair coin, and those concocted in the students' heads. To the amazement of the students, the statistics professor is typically able to sort the slips of paper with 100 percent accuracy.

There is no mystery or magic here, nor is the statistics professor drawing on any deep mathematical knowledge. Rather, the professor simply looks for a repetition of six or seven heads or tails in a row to distinguish the truly random from the pseudo-random sequences (truly random sequences being those derived from flipping a fair coin, pseudo-random sequences being those concocted in the students' heads). In a hundred actual coin flips, a person is likely to see six or seven such repetitions. The proof of this fact is easy to understand: In 100 coin tosses, on average half will repeat the previous toss, implying about 50 two-repetitions. Of these, on average half will repeat the previous toss, implying about 25 three-repetitions. Continuing this line of reasoning, we find on average 12 four-repetitions, 6 five-repetitions, 3 six-repetitions, and 1 (or 1.5) seven-repetitions.[7]

On the other hand, people concocting pseudo-random sequences with their minds tend to record far fewer repetitions—perhaps three or four maximum. That's because people, as a matter of human psychology, tend to alternate between heads and tails too frequently. Whereas with a truly random sequence of coin tosses, there is a 50 percent chance that one toss will differ from the next, people tend intuitively to expect that one toss will differ from the next around 70 percent of the time. So, when they concoct a sequence of coin tosses that they want to appear random, they tend to switch between heads and tails too frequently. Thus, after three or more repetitions they think, whether consciously or unconsciously, that it's time to alternate from tails to heads or vice versa. But coins that are actually being tossed have no memory of how they came up last, so they don't alternate nearly as much.[8]

How, then, will our statistics professor fare when confronted with E_ψ? Will the professor attribute E_ψ to chance or to the musings of someone trying to mimic chance? According to the professor's crude randomness checker, E_ψ would be considered truly random because it contains a repetition of seven tails in a row. Everything that at first blush would lead us to regard E_ψ as truly random checks out. There are exactly 50 alternations between heads and tails (as opposed to the 70 that would be expected from humans trying to mimic chance). Moreover, the relative frequencies of heads and tails check out: there were 49 heads and 51 tails. Thus, it's not as though the coin supposedly responsible for E_ψ was heavily biased in favor of one side versus the other.

And yet, E_ψ is anything but random. To see this, rewrite the sequence of heads and tails by separating it into blocks as follows: 0 | 1 | 00 | 01 | 10 | 11 | 000 | 001 | 010 | 011 | 100 | 101 | 110 | 111 | 0000 | 0001 | 0010 | etc. It now becomes clear that this sequence was contrived by writing the binary numbers in ascending order, starting with the one-digit binary numbers (0 and 1), moving next to the two-digit binary numbers (00, 01, 10, and 11), and continuing on until 100 digits were recorded. It's therefore intuitively obvious that E_ψ does not describe a truly random event such as could be obtained by tossing a fair coin, but instead is a pseudo-random event concocted by doing a little binary arithmetic (hence the subscripted Greek *psi*, for *pseudo*, in E_ψ).

Obviously, there is a short description of this event, such as "count in order the one-digit binary numbers, then the two-digit ones, and keep going." This event is therefore specified. Moreover, it is also improbable, on the order of 1 in 10^{30}, corresponding to 100 coin tosses, and growing ever more improbable as the corresponding number of coin tosses is increased. We therefore have here the two key elements needed to triangulate on a design inference— specification and small probability. But we know more than simply that this sequence was designed. In fact, it is known who designed it, namely, the mathematician David Gawen Champernowne, who invented it back in 1933.[9] This sequence is known as the Champernowne sequence and it has the property that any n-digit combination of bits appears in this sequence with limiting frequency 1 in 2^n (as would be the case with any truly random infinite sequence of coin tosses).

An important lesson to be drawn from this example is the asymmetry between seeing that a pattern is a specification and not seeing that it is a specification. At the start of the class demonstration, our statistics professor is likely to believe that all the slips with random and pseudo-random coin tosses would be handed in by neophyte statistics students (it is, after all, an introductory statistics class). By simply using the crude randomness checker of looking for a repetition of six or seven heads or tails in a row, the professor would have signed off on the Champernowne sequence as truly random. But if, for whatever reason, the professor suspected that something was amiss, this sequence may have come under closer scrutiny, at which point the professor may have seen the incriminating pattern (specification) of binary digits arranged in ascending order.

Perhaps the professor remembered the Champernowne sequence from a course on number theory. Perhaps the professor, in a flash of insight, saw the binary pattern, and thus essentially reinvented the Champernowne sequence from scratch. The exact process of discovery is irrelevant so long as the discovery is made. Regardless, once an incriminating or suspicious pattern (a specification) that refutes randomness is seen, it cannot be unseen. This example points up the fundamental asymmetry between randomness and nonrandomness, and thus respectively between non-specifications and specifications.

Randomness is always a *provisional* designation. Something can therefore switch from being designated random to being designated nonrandom—as when a salient pattern is discovered in something previously thought to be random. Nonrandomness, however, is always a *settled* designation. Something, once seen as nonrandom, cannot switch back to being seen as random. We regard something as random only so long as we have not discovered a chance-defeating pattern in it. Once we perceive such a pattern, however, the item switches from the provisional designation of random to the settled designation of nonrandom.

The patterns that induce us to make this switch are the specifications. Randomness corresponds to the non-specifications (which may become specifications with further insight by discovering a hitherto unknown short description). Nonrandomness corresponds to the specifications. The statistician Persi Diaconis underscored this asymmetry between randomness and nonrandomness in his closing remarks at a 1988 conference on randomness: "We know what randomness isn't, not what it is."[10]

We've all experienced this asymmetry between randomness and nonrandomness. For instance, we see some image that looks random, but on closer examination we see that it exhibits a pattern that can't be the result of chance. It may look like a chaotic inkblot, but when viewed in the right way it is an image of a well-known object. Or it may look like a swirl of colors, but when viewed in the right way it is a 3D optical illusion—perhaps of dinosaurs running at the viewer. Once you see "what's really there," you can't unsee it, and what at first appeared to be random can no longer be regarded as random. The converse, however, doesn't hold: if an image is perceived to be nonrandom, we know it to be nonrandom, and we can't unknow or unsee its nonrandomness. Short of amnesia or some perceptual defect, it cannot thereafter be regarded as random. Randomness can be unlearned, nonrandomness cannot.

Before leaving this section, we want to address some older terminology used previously to motivate and characterize specifications. When the first edition of this book came out, and for a few years following, it was common to refer to specifications as "detachable patterns," "patterns that satisfy a tractability condition," "independently given patterns," and, in the context of biology, "patterns that satisfy independently given functional/ structural requirements." All this terminology is still usable provided it is understood as an intuitive restatement of patterns that have short descriptions.[11]

For instance, the original idea with detachable patterns was that it should be possible to "detach" or separate a pattern (*qua* specification) from a corresponding event, making it practicable to describe the event apart from its actual occurrence. Detachability here was defined in terms of a conditional independence criterion that attempted to model probabilistically the ability to reconstruct the pattern in question without the event happening and thus without being able to read the pattern off the event. But the actual criterion proved to be superfluous because such an ability to reconstruct follows directly from short description length: events with a short description are few and far between and thus readily identifiable even if the corresponding events don't happen. For example, "1,000 heads in a row" is a readily identifiable description of an event even if no one ever witnesses the event so described (namely, 1,000 heads in a row).

Likewise, reference to "independently given patterns" was meant to suggest the ability to derive such patterns independently from the corresponding events, which is again an automatic consequence of short description length. Moreover, such pattern independence in the context of biology and in reference to functional or structural requirements is merely a special case of the last point, where the description is cashed out in terms of function and/or structure. Finally, the term "tractability" was introduced as a synonym for descriptions of events that require minimal computational effort, which is easily reframed in terms of minimal description length (shorter descriptions require less computational effort, and less computational effort yields shorter descriptions).

Bottom line: All the earlier terminology to characterize specifications is readily subsumed under the present terminology in terms of minimal description length. At the same time, minimal description length makes precise and explicit what was implicit in the older terminology.

3.5 Prespecifications

In practice, we recognize a specification by carefully examining an event and perceiving in it a simply describable pattern. Thus, we first witness an event, and then we try to discover a pattern in it that can be simply described. If we succeed in uncovering such a simple description, the pattern is a specification. Event first, specification second is the usual order in drawing a design inference. Yet this order can be reversed. Thus, we can start with a pattern that is highly improbable to be matched by chance, and then attempt to elicit an event that matches it. Specifications in such cases are called *prespecifications*. Prespecifications, as "ante hoc" specifications, work a bit differently from the typical "post hoc" specifications, and it's worth understanding the difference.

Take, for instance, the following sequence of coin tosses, which came up at the start of Section 2.4, where 0 stands for tails and 1 for heads:

$$1100001101011000110111111$$
$$1010001100011011001110111$$
$$0001100100001011110111011$$
$$0011111010010100101011110 \qquad E_r$$

The event E_r did in fact result from tossing a fair coin 100 times (hence the subscripted r, for *random*). If we now treat this sequence of 0s and 1s not as an event but as a pattern T_r of characters consisting of the numeric symbols 0 and 1, then T_r is "long and complicated," the result of simply reading off the event after the fact and forming it into a pattern. Indeed, any sequence of coin tosses can be transformed into a pattern that simply lists the exact sequence of heads and tails tossed. Such sequences are therefore not short and simple, as compared to descriptions like "all heads" or "all tails" or "the Champernowne sequence." It thus appears that there's no specification here—nothing salient to capture our imagination, no design inference to be had.

Or not. As noted in Section 2.4, if someone, a week after E_r was first tossed, claims likewise to have tossed a coin and witnessed E_r, we would reject chance as a valid explanation for this second occurrence of the event. Why? Even though T_r is long and complicated, we don't need to invoke it as the pattern to describe E_r. We could instead invoke the pattern "same event as last week." If a sequence of coin tosses has already been explicitly identified, then the pattern we use to identify it does not require an unabridged description. An abridged description that merely points to the event is enough. Such an abridged description *pre*specifies E_r, ensuring that any reoccurrence of this event would not occur by chance.

The rationale here is one we learn as school kids, stated in the form of a proverb that lightning does not strike twice in the same place. The first strike prespecifies the second. The truth of this proverb, however, assumes that anywhere lightning strikes is highly improbable and that lightning strikes are probabilistically independent. In fact, those assumptions don't hold. For instance, the Empire State Building is struck about twenty-five times a year by lightning.[12] Still, the proverb, though not quite hitting the mark, valiantly gestures at the truth that specified events of small probability don't happen by chance.

Whereas specifications, by sporting short descriptions, tend to be identified after an event has occurred, prespecifications, whether sporting short or long descriptions, are identified before an event of interest occurs. How prespecifications arise, whether by being read off an event that happened by chance or through human invention, is unimportant. What is important is that they are identified *in advance* of the events whose chance occurrence they will be used to assess. Prespecifications are common in statistics, where they are known as critical or rejection regions. If a sample (event) falls within a rejection region (prespecification), its chance occurrence is ruled out.

The most obvious descriptions of prespecificational patterns tend to be long and complicated. That's because the obvious descriptions require no insight and simply rehearse the pattern in full detail, leaving nothing to the imagination. This was the case with E_r, whose most obvious description, in plain English, would look something like "two ones, four zeros, two ones, zero, one, zero, two ones, three zeros, two ones, zero, seven ones, zero, one, three zeros, two ones, three zeros, two ones, zero, two ones, two zeros, three ones, zero, three ones, three zeros, two ones, two zeros, one, four zeros, one zero, four ones, zero, three ones, zero, two ones, two zeros, five ones, zero, one, two zeros, one, zero, one, two zeros, one, zero, one, zero, four ones, zero." Such a description will therefore plainly fail the key defining condition of specifications, which is to have a short description. In line with our distinction between specifications and fabrications in Section 1.1, it will thus seem that we are dealing with a fabrication and not a specification.

But just because one description is long doesn't mean all of them have to be long. With prespecifications there's always a convenient way to reduce their length. Just as a pronoun can significantly diminish the descriptive length of the noun phrase to which it refers, so a reduced description is always available to refer to a prespecification, ensuring that the event in question is also specified (in the minimum-description sense). Prespecifications have short descriptions because the underlying patterns have already been identified with prior explicit descriptions, and whether these be short or long, they are already available in the descriptive language and can therefore be abbreviated. Prespecifications by their very nature admit abbreviations, and it's the abbreviations that render their descriptions simple.

Why are prespecifications capable of brief, abbreviated descriptions? Prespecifications are identified in advance of the events whose chance occurrence they are intended to assess. Their descriptive complexity is therefore controlled in ways that don't apply to specifications in general. Think of a specification as a book that has yet to be written. The totality of different possible books that could be written is immense and unwieldy. Moreover, there is no way to describe with any degree of specificity a book that exists solely in the realm of possibility. For such a description, the book must leave the realm of possibility and enter the realm of actuality.

The books that actually exist, however, are much more limited than those that are merely possible. For instance, the Library of Congress houses 40 million books. That's a lot of books, but it is easily catalogued using the Library of Congress Classification (LCC) system, which uses only a handful of letters, numbers, spaces, and punctuation to identify every book in its collection. QA279.D455 1998, for instance, consists of fifteen characters (including one space) and is the LCC number that uniquely identifies the first edition of *The Design Inference*. The LCC system can easily identify every book in the Library of Congress with twenty or fewer characters. Such descriptions are therefore short. Prespecifications, in precisely this way, take advantage of existing, fully articulated descriptions.

For an amusing illustration of this point, consider a scene from the film *The Big Short*, which portrayed the 2008 financial crisis. Charlie Geller and Jamie Shipley, two young investors, are describing how they made their initial $30 million (which they then used to short the housing market to make hundreds of millions of dollars). Charlie, the brains of the duo, remarks, "Our investment strategy was simple. People hate to think about bad things happening, so they always underestimate their likelihood." Jamie, the marketing face of the duo, doesn't elaborate but merely remarks, "What he said." When something has already been stated, it doesn't need to be restated, at least not at the same level of detail as the initial statement.[13] Instead, it simply needs a signpost to point to it. That's what Jamie was doing here, offering a signpost rather than a full restatement. Accordingly, prespecifications always have short descriptions, even if the short descriptions are only implicit.

Prespecifications arise in statistical settings where a rejection region is set prior to an experiment (for more on this, see the next section). The rejection region will then be used to decide whether a given chance process was operating. To that end, a researcher then runs the experiment, determining whether the sample taken (i.e., the event) falls within the rejection region. If it does, the researcher will refuse to attribute the sample to the chance process under consideration. Nonetheless, prespecifications also arise in less stylized settings. Even when a pattern is not proposed before an event takes place, multiple coinciding events, or coincidences, can at times be identified after the events have taken place. From the coincidence, once a common pattern becomes evident, a prespecification then becomes evident.

Without prejudging the ultimate cause of a coincidence, we can think of a coincidence as a grouping of events where one event happens in conformity with a pattern (not necessarily a specification), and then another event happens in conformity with that same pattern, with perhaps still more such events happening and all conforming to the pattern. For the coincidence to raise doubts about its chance occurrence, the pattern needs to have high specificity, and thus signify an event of small probability. A coincidence is therefore such a grouping of pattern-conforming events. In a coincidence, the first event becomes a prespecification for those that follow.

We've seen such coincidences in Chapter 2. With intellectual property, for instance, reinvention of the same item (unless it is very simple) cannot reasonably be attributed to chance. The original invention thus becomes a prespecification for the reinvention.

Plagiarism is a case in point (Section 2.1). Or recall in our account of data falsification (Section 2.3) how getting exactly the same distribution of errors twice was "too coincidental" to happen by chance. Particularly memorable here were the cases of scientific fraud perpetrated by Robert Slutsky and Jan Hendrik Schön, who simply recycled the same pattern of errors among their respective papers.

Coincidences that induce prespecifications also arise in coding. *Easter eggs*, for instance, are hidden undocumented features in computer software that may be revealed when certain conditions are met and the program responds accordingly. Alternatively, such features may simply stay hidden unless they are explicitly identified in the source code. Inserting Easter eggs into code helps to guard against copycats who may modify the code but inadvertently still copy enough of the Easter egg to prove that the code was not independently created. In this way, the Easter egg can serve as a prespecification.

Here's an example of a coincidence-induced prespecification from evolutionary biology. John Maynard Smith, in his 1958 *The Theory of Evolution*, concluded that flatworms, annelids, and mollusks, representing three different phyla, must nonetheless have descended from a common ancestor because their common cleavage pattern in early development "seems unlikely to have arisen independently more than once."[14] "Unlikely" is, of course, a synonym for "improbable," and "arisen independently" is a synonym here for chance. Smith is thus saying here that improbability has led him to rule out chance. But clearly, it was not mere improbability that led to this conclusion. In addition, it's that the emergence of flatworms, annelids, and mollusks followed a common cleavage pattern. That cleavage pattern therefore constituted a prespecification, on the basis of which Smith inferred a common evolutionary cause. Granted, he did not here draw a design inference. But he did use a prespecification to eliminate a particular chance hypothesis.

3.6 Specification-Induced Rejection Regions— The Idea

For a rejection region R, if an event E falls within R and if R has small probability under a chance hypothesis H, then the chance hypothesis H is to be rejected for failing adequately to explain the occurrence of E. More formally, given

(1) a rejection region R that's a subset of the possibility space Ω,

(2) the occurrence of an event E within Ω that falls under R (i.e., $E \subset R$),

(3) the question whether E occurred by chance according to the chance hypothesis H,

(4) a positive number α that acts as a cutoff for determining small probabilities, and

(5) $P(R|H) \leq \alpha$,

it then follows that E did not happen (or should not be regarded as having happened) according to the chance hypothesis H. Just what it means for a probability cutoff α to be small enough for a rejection region to eliminate chance is the subject of the next chapter on probabilistic resources. But the point here is simply that given such a cutoff, this is how rejection regions work.

Prespecifications induce straightforward rejection regions. In the previous section, we considered explicit prespecifications taking the form of a pattern V set prior to an observation or sample, as is common in applied statistical research. Provided the corresponding event V^* had small probability and E fell within V^*, the chance occurrence of E would be rejected. V^* in this case was the rejection region R. Prespecifications make working with specifications straightforward. They are common in applied statistical research and were the first types of specifications articulated in statistical theory.

We want next to consider specification-induced rejection regions in which we don't have the advantage of prespecifications (i.e., those given in advance) and instead must deal with specifications discovered only after the fact. In other words, the event E occurs and only then do we discover a pattern V to which E conforms (i.e., $E \subset V^*$). In such cases, the specification-induced rejection regions for eliminating chance take on additional complications.

To see how, let's return to the roulette example of Section 1.2. There we found that the world record for spinning reds in a row with a roulette wheel was 32, which with exactly 32 spins of the roulette wheel is an event, as we noted, of probability 1 in 25 billion. We also noted that roughly 46 billion spins of the roulette wheel would bring up 32 reds in a row with a probability of ½.

Thirty-two reds in a row is certainly a salient pattern, and it has a short description ("thirty-two reds in a row"). It is therefore a specification. But it is not the only relevant pattern here. Occurrences of red and black in roulette are entirely parallel. Indeed, any sequence of one is a mirror image of the other. Thus, to the degree that 32 reds in a row is extreme enough to establish a world record at roulette, so too is 32 blacks in a row. And such a succession of blacks likewise has a short description, namely, "thirty-two blacks in a row." And let's not forget about "thirty-two greens in a row," which would correspond to getting 0 or 00 thirty-two times in a row.

Whereas 32 reds in a row and 32 blacks in a row each have a probability of roughly 1 in 25 billion, 32 greens in a row has a probability of roughly 1 in 80 thousand trillion trillion trillion. Moreover, each of these descriptions ("thirty-two reds in a row" etc.) has roughly the same length (taken informally in terms of ordinary English usage). Suppose, then, we witnessed an event E of spinning 32 reds in a row, and suppose we noted that it corresponds to a specification V so that V^* doesn't just include but is identical with E. In other words, V is just "thirty-two reds in a row." The seemingly low probability of the specification-induced event $V^* (= E)$ would, however, not be enough to count against this event happening by chance. That's because we must also factor in the specifications for 32 blacks in a row and 32 greens in a row.

A useful way to think about what's happening here is in terms of arrows and targets: the greater the number of targets, the greater the probability of some arrow hitting one of them by chance. "Thirty-two reds in a row" paints a target on a wall. The probability of hitting it with an arrow by chance is small. But then there's also "thirty-two blacks in a row" as well as "thirty-two greens in a row," which also paint small targets (the latter smaller than the former). The probability of individually hitting these other targets by chance is therefore also small. But what happens when we pool all these targets together? All these targets need to be pooled together to determine whether an arrow hitting the target corresponding to "thirty-two reds in a row" hit it by chance. That's because hitting the other targets would likewise cause us to question chance. The other targets are parallel to this one, both in terms of smallness of probability (some even smaller) and in terms of shortness of description length.

In this roulette example, patterns with the same short description length identified events of small probability. The description lengths stayed roughly constant, but the corresponding probabilities, though all small, could vary (32 greens in a row was much more improbable than 32 blacks in a row). But what happens when the description length is also allowed to vary? Consider another example: suppose you witness ten brand new Chevy Malibus drive past you on a public road in immediate, uninter-rupted succession. Maybe you are standing on a street corner. Maybe you are driving in traffic. In any case, the question that crosses your mind is: Did this succession of ten brand new Chevy Malibus happen by chance?

Your first reaction might be to think that this event is a publicity stunt by a local Chevy dealership. In that case, the succession would be due to design rather than to chance. But you don't want to jump to that conclusion too quickly. Perhaps it is just a lucky coincidence. But if so, how would you know? Perhaps the coincidence is *so* improbable that no one should expect to observe it as happening by chance. In that case, it's not just unlikely that you would observe this coincidence by chance; it's unlikely that anyone would. How, then, do you determine whether this succession of identical cars could reasonably have resulted by chance?

Obviously, you will need to know how many opportunities exist to observe this event. It's estimated that in 2019 there were 1.4 billion motor vehicles on the road worldwide.[15] That would include trucks, but for simplicity let's assume all of them are cars. Although these cars will appear on many different types of roads, some with traffic so sparse that ten cars in immediate succession would almost never happen, to say nothing of ten cars having the same late make and model, let's give chance a chance by assuming that all these cars are arranged in one giant succession of 1.4 billion cars arranged bumper to bumper. Giving chance a chance, however, can't stop there. Cars are in motion and thus rearrange themselves continually. Let's therefore assume that the cars completely reshuffle themselves every minute, and that we might have the opportunity to see the succession of ten Malibus at any time across a hundred years. In that case, there would be no more than 74 quadrillion (= 7.4×10^{16}) opportunities for ten brand new Chevy Malibus to line up in immediate, uninterrupted succession.

So, how improbable is this event given these 1.4 billion presumed cars subject to repeated reshuffling? To answer this question requires knowing how many makes and models of cars are on the road and their relative proportions (let's leave aside how different makes are distributed geographically, which is also relevant, but introduces needless complications for the purpose of this illustration). If, *per impossibile*, all cars in the world were brand new Chevy Malibus, there would be no coincidence to explain. In that case, all 1.4 billion cars would be identical, and getting ten of them in a row would be an event of probability 1 regardless of reshuffling.

But clearly, nothing like that is the case. Go to Cars.com, and using its car-locater widget you'll find 30 popular makes and over 60 "other" makes of vehicles. Under the make of Chevrolet, there are over 80 models (not counting variations of models—there are five such variations under the model of Malibu). Such numbers help to assess whether the event in question happened by chance. Clearly, the event is specified in that it answers to the short description "ten new Chevy Malibus in a row." For the sake of argument, let's assume that achieving that event by chance is going to be highly improbable given all the other cars on the road and given any reasonable assumptions about their chance distribution.[16]

But there's more work to do in this example to eliminate chance. No doubt, it would be remarkable to see ten new Chevy Malibus drive past you in immediate, uninterrupted succession. But what if you saw ten new *red* Chevy Malibus in a row drive past you? That would be even more striking now that they all also have the same color. Or what about simply ten new Chevies in a row? That would be less striking. But note how the description lengths covary with the probabilities: "ten new red Chevy Malibus in a row" has a longer description length than "ten new Chevy Malibus in a row," but it corresponds to an event of smaller probability than the latter. Conversely, "ten new Chevies in a row" has shorter description length than "ten new Chevy Malibus in a row," but it corresponds to an event of larger probability than the latter.

What we find in examples like this is a tradeoff between description length and probability of the event described. In a chance elimination argument, we want to see short description length combined with small probability. But typically these play off against each other. "Ten new red Chevy Malibus in a row" corresponds to an event of smaller probability than "ten new Chevy Malibus in a row," but its description length is slightly longer. Which event seems less readily ascribable to chance (or, we might say, more worthy of a design inference)? A quick intuitive assessment suggests that the probability decrease outweighs the increase in description length, and so we'd be more inclined to eliminate chance if we saw ten new red Chevy Malibus in a row as opposed to ten of any color.

The lesson here is that probability and descriptive complexity are in tension, so that as one goes up the other tends to go down, and that to eliminate chance both must be suitably low. We see this tension by contrasting "ten new Chevy Malibus in a row" with "ten new Chevies in a row," and even more clearly with simply "ten Chevies in a row." The latter has a shorter description length (lower descriptive complexity) but also much higher probability. Intuitively, it is less worthy of a design inference because the increase in probability so outweighs the decrease in descriptive complexity. Indeed, ten Chevies of any make and model in a row by chance doesn't seem farfetched given the sheer number of Chevies on the road, certainly in the United States.

But there's more: Why focus simply on Chevy Malibus? What if the make and model varied, so that the cars in succession were Honda Accords or Porsche Carreras or whatever? And what if the number of cars in succession varied, so it wasn't just 10 but also 9 or 20 or whatever? Such questions underscore the different ways of specifying a succession of identical cars. Any such succession would have been salient if you witnessed it. Any such succession would constitute a specification if the descriptive complexity were low enough. And any such succession could figure into a chance elimination argument if both the descriptive complexity and the probability were low enough. A full-fledged chance-elimination argument in such circumstances would then factor in all relevant low-probability, low-descriptive-complexity events, balancing them so that where one is more, the other is less.

That, then, is what specification-induced rejection regions do, namely, factor in and balance out all the relevant low-probability, low-descriptive-complexity events. In Chapter 6, we define *specified complexity* as an information measure that does just this, inducing a rejection region with the following property: if the specified complexity of an event is greater than or equal to n bits, then the corresponding rejection region, which includes that event in question, has probability less than or equal to 2^{-n}. This is a powerful result and it provides a conceptually clean way to identify specification-induced rejection regions. But it presupposes some deep results from theoretical computer science, and in practice it's intuitively easier to take a less elegant approach, which we do next.

3.7 Specification-Induced Rejection Regions— The Math

Our task is to lay out a formal approach to specification-induced rejection regions that is both practical and intuitively appealing. We start with the occurrence of an event E and want to find a specification-induced rejection region R with which to rule out that E occurred by chance. We find that E falls under a specification V (i.e., $V^* \supset E$, the asterisk mapping patterns/descriptions to events). As a specification, V has a short description. But V was not given in advance of E's occurrence, and so it is not a prespecification. If V were a prespecification, we would be done, given our earlier work in this chapter for rejection regions built from prespecifications. But since we are assuming that V is a specification that is not also a prespecification, there's more work to do.

Some probabilistic infrastructure now needs to be put into place. We assume that E and V^* belong to (which is to say, are subsets of) a probability/possibility space Ω. Since we are engaging in a chance-elimination argument, we also need a probability distribution P over Ω to characterize the chance occurrence of events such as E and V^*. We assume that P assigns probability over Ω in line with a chance hypothesis H. We could thus write $P(\cdot|H)$ or P_H in place of simply P, but since H is the only chance hypothesis whose elimination we will be considering, we'll simply write P. In full-fledged design inferences, where we need to eliminate an entire range of chance hypotheses, dependence of probability distributions on chance hypotheses will be more explicit.

The specification V will be the basis for a rejection region R that we'll construct, so we'll refer to V as a *basis specification*. R will include the event E whose chance occurrence is in question. To eliminate chance, the probability of R will need to be small enough. Small enough in these situations always means small in light of the probabilistic resources that are available for the event E to happen by chance and fall within R. This will then come down to such probabilistic resources inducing a (local or universal) probability bound α. The details here are left for the next chapter, but the intuition will by now be clear. Long story short, to eliminate the chance hypothesis H, we'll need to confirm that $P(R)$ is less than or equal to α (i.e., $P(R) \leq \alpha$).

The basis specification V, on which a specification-induced rejection region R will be based, in the end is always a description, and descriptions always belong to some language. The basis specification V will therefore be formulated within a given language $\mathscr{L}$. This language is assumed to include all descriptions, and thus specifications, of interest, and these will map onto events in Ω. $\mathscr{L}$ will have a *description length metric*, denoted by $|\cdot|$. Thus, the description length of V will be given by $|V|$.

We assume that $\mathscr{L}$ is finitely generated in the sense that there are only finitely many characters or finitely many words out of which all the expressions of the language are generated, and that any expression of $\mathscr{L}$ is a finite sequence of such characters or words (for specified complexity, all the expressions are bit strings). It follows that for any expression in $\mathscr{L}$, the description length metric will be well-defined and finite. The same holds for the minimum description length metric, to be defined next.

As it is, we also need a minimum description length metric because specifications are defined in terms of short descriptions, and descriptions can always be padded and made longer than they need to be. The *minimum description length metric* on $\mathscr{L}$ will be denoted by $D(\cdot)$. For V, but this holds for any element of $\mathscr{L}$, $D(V)$ will then be defined as equal to the minimum $|W|$ of all descriptions W in $\mathscr{L}$ for which W maps to the same event in Ω as V, i.e., for all W such that $W^* = V^*$. More formally,

$$D(V) = \min \{ \, |W| \mid W \in \mathscr{L} \text{ and } W^* = V^* \, \}.$$

Note that V is necessarily in the set $\{ \, |W| \mid W \in \mathscr{L} \text{ and } W^* = V^* \, \}$, and so $D(V)$ will always be less than or equal to $|V|$. In other words, $D(V) \leq |V|$ regardless of V. This inequality will be important later as we estimate the probability of a specification-induced rejection region R. But the immediate lesson here is that if we have a description that's short enough to count as a specification, then its minimum description length can only shorten the description further, making it even more of a specification.

Chance-elimination arguments, as we've stressed throughout this book, depend on probabilities being sufficiently small and patterns being sufficiently short. We've defined a probability to be sufficiently small if it is less than or equal to a given probability bound α. Similarly, we define a description to be sufficiently short, and therefore to count as a specification, if its description length is less than or equal to a *descriptive complexity bound m*. Here *m* is just a natural number greater than 0 such that if a description has length less than or equal to *m*, then it can be regarded as short enough to count as a specification. How are α and *m* determined? They are determined by the inquirer's needs and interests in controlling for false positives in eliminating chance (i.e., eliminating chance when it shouldn't be eliminated).

Associated with a descriptive complexity bound *m* is a number that counts all the elements of the language $\mathcal{L}$ with description length less than or equal to *m*. Let's denote this number by $L(m)$. Formally, it is defined as

$$L(m) = \text{cardinality of } \{\, W \in \mathcal{L} \mid |W| \le m \,\}.$$

In other words, $L(m)$ is the number of descriptions in $\mathcal{L}$ with length less than or equal to *m*. Because $\mathcal{L}$ is finitely generated, $L(m)$ is always well-defined and finite. It will be convenient simply to let *M* denote $L(m)$ (i.e., $M = L(m)$). In that case, *m* sets a bound on the description length, and *M* sets a corresponding bound on the number of descriptions whose length is less than or equal to *m*.

With all this stage-setting, we are now finally in a position to define the specification-induced rejection region *R*. Provided that $|V| \le m$ and $P(V^*) \le \alpha/M$, the rejection region *R* associated with the basis specification *V* can now be defined as follows:

> R = the union of all events W^* from Ω where W belongs to $\mathcal{L}$ such that $D(W) \le m$ and $P(W^*) \le \alpha/M$.

Because we are stipulating that $|V| \le m$ and $P(V^*) \le \alpha/M$, *V* is clearly among the *W*s that form the *W**s in this union. That's because $D(V) \le |V|$, and therefore $D(V) \le m$. Also, because V^* includes the observed event *E*, *R* also includes it. As is clear from this definition, *R* depends on Ω, $\mathcal{L}$, *V* (and therefore V^*), *m* (and therefore $M = L(m)$), α, *P*, and *H*. Thus, technically, we might write $R(\Omega, \mathcal{L}, V, m, \alpha, H)$, suppressing V^* since it is defined in terms of

V, suppressing M because it is defined in terms of m, and suppressing P because it is defined in terms of H (as $P(\cdot|H)$ or P_H).

In general, in a chance elimination argument, Ω, $\mathcal{L}$, V, m, and α will all be explicitly given and stay the same throughout the argument. The one thing that can vary is the chance hypothesis H. That's because chance elimination arguments can eliminate a range of chance hypotheses, as often happens in design inferences. The chance hypothesis H is therefore the one we need to watch in defining and making use of R. In general, we'll suppress all these terms, even H, and simply refer to R. But it's important to bear in mind this dependence of R as a specification-induced rejection region, especially on underlying chance hypotheses.

The rationale for defining specification-induced rejection regions in this way parallels and extends what we do in statistical significance testing (recall especially Section 3.6). There we observe an extreme outcome in the tail of a distribution, and then consider the combined probability of all outcomes at least as improbable as that outcome observed (i.e., as far or still further out in the distribution tails).[17] With specification-induced rejection regions, we need to go further. Thus, we need to factor in all specifications whose description length is at least as extreme as (i.e., less than or equal to) the descriptive complexity cutoff m and with probability at least as extreme as (i.e., less than or equal to) the probability cutoff α divided by M. "Extreme" here means minimizing both length of descriptions and probability of corresponding events. The basis specification V is pivotal here in ensuring that the rejection region actually includes the event E, whose chance occurrence stands in question.

Calculating the exact probability of R can be difficult to impossible. Nonetheless, there's an easy way to form an upper bound for the probability of R. To calculate such a bound, simply consider how many events there could be with a description length less than or equal to m and with a probability less than or equal to α/M. That number, the way we've set things up, is bounded by M, which counts all the descriptions with descriptive complexity less than or equal to m. This fact leads to a straightforward upper bound probability for R, namely, $P(R) \leq \alpha$. That's because R is a union of at most M events of the form W^*, each of which is bounded by the probability α/M. So, $P(R) \leq M \times \alpha/M = \alpha$. Moreover, because R includes V^* and therefore E, and because α is a probability bound chosen in light of a relevant set of probabilistic resources for eliminating chance, E's chance occurrence (according to the chance hypothesis H) can therefore be ruled out.

There are a lot of moving pieces here, but this approach to specification-induced rejection regions is really not that complicated. Here's an example that shows how all the pieces work together. Imagine a roulette wheel that has 100 evenly situated slots, with one slot red and the rest black. Suppose this roulette wheel is spun 100 times and each time comes up red. That's the event E. We assume all the slots on the roulette wheel are symmetrical and therefore equiprobable. That sets the probability P and the chance hypothesis H. The event E therefore has probability $.01^{100}$, or 1 in 10^{200}.

We suppose our language $\mathscr{L}$ consists of English words strung together. Thus, the specifications we deal with in this example are English phrases and/or sentences. Let's assume that there are $100,000 = 10^5$ distinct possible words in each position (a generous word limit given the size of most people's vocabularies). There are thus 10^5 one-word phrases/sentences, 10^{10} two-word phrases/sentences, 10^{15} three-word phrases/sentences (one of which is "hundred red repeated"), and so on. To control for false positives in mistakenly rejecting chance, we set a descriptive complexity bound of $m = 10$ and a probability bound of $\alpha = 10^{-150}$ (corresponding to our go-to universal probability bound in the next chapter). $L(m) = M$ then comes to almost exactly 10^{50} (it is precisely $10^5 + 10^{10} + 10^{15} + \cdots + 10^{45} + 10^{50}$, but all the other terms are minuscule compared to the last).

We now let $V = $ "hundred reds repeated" be the basis specification for the rejection region R that we are forming. V^*, the event that V identifies, then not only includes but also coincides with the event E, whose chance occurrence we are assessing. R will then need to be defined as

R = the union of all events W^* from Ω where W belongs to $\mathscr{L}$ such that $D(W) \leq m = 10$ and $P(W^*) \leq a/M = 10^{-200}$.

Because $D(V) \leq |V| = 3$ and because $P(V^*) = 10^{-200}$, it follows that V^* belongs to this union and is a subset of R. All the pieces are now in place and have been confirmed. It follows that E's chance occurrence (according to the chance hypothesis H) can therefore be ruled out.

This example illustrates specification-induced rejection regions in general. But note, this example involves a binomial distribution in which the event E is in the tail of that distribution. As we will see in the next section, we can in the right circumstances also form rejection regions that arise from modes and tails of probability distributions. So, an added lesson from this example is that there can be different ways of approaching specification-induced rejection regions, yielding distinct rejection regions all conducive to eliminating chance. The approach outlined in this section is quite general, applying to specifications and not just to prespecifications. We outline still another general approach applying to specifications and not just to prespecifications in the upcoming chapter on specified complexity (Chapter 6).

A lingering question might now remain: Is a specification-induced rejection region itself specified? Specifications always arise in the context of a language $\mathscr{L}$, a minimum description length metric D, and a chance hypothesis H incurring a probability P. There's nothing in all this to require that $\mathscr{L}$ include what might be called a "rejection-region operator" that acts on specifications of $\mathscr{L}$ to produce new specifications of $\mathscr{L}$ that are specification-induced rejection regions. Rather, it makes sense to think of specification-induced rejection regions as not themselves identified by specifications but as a canonical way of associating a basis specification with as broad an event as is needed for a successful chance-elimination argument.

3.8 Modes and Tails of Probability Distributions

Prespecifications are the easiest specifications to deal with. They are given in advance of an event whose chance occurrence stands in question. In such cases, the question is simply whether an observed event conforms to the prespecification. In symbols, the prespecification V is given in advance, the event E happens, and V^* is then seen to include E. If so, the path is clear for a small-probability specification argument. Moreover, the rejection region R in such a small-probability prespecification argument is then simply V^*. Thus, with prespecifications, we don't have to factor in all the ways that E might have been identified with short descriptions, as in the general case of specification-induced rejection regions just covered in the last two sections.

There's another way, however, for rejection regions associated with specifications to avoid factoring in numerous short descriptions. That occurs when the specification has a description that's as short as it could be, thereby drastically reducing the number of short descriptions to be considered. The most widely known instance of this occurs in Ronald Fisher's account of statistical significance testing. There the chance occurrence of an event E with respect to a chance hypothesis H is rejected if E either matches expectation too closely or departs from it too sharply under the probability distribution $P(\cdot|H)$.[18] To minimize the number of short descriptions to be considered, the underlying probability space Ω needs a canonical metric or distance function that defines how far apart points (outcomes) in Ω are. Moreover, this metric needs to induce a canonical probability density with respect to the probability distribution $P(\cdot|H)$.

To see how this all works, think of a normal distribution. We will tend to reject an event E with respect to this distribution when E is too close to the mode (i.e., where the normal distribution is highest) or too far out in the tails (i.e., where the normal distribution is lowest). Rejecting chance when events are too far out in the tails is standard Fisherian significance testing (see Appendix A.5). Rejecting chance when events are too close to the mode (too close to expectation) occurs, as we saw, with data falsification (Section 2.4) where the results are too good to be true. Fisher himself rejected chance in such a case when he analyzed Gregor Mendel's data on peas and concluded that the data were fudged because they were too close to expectation.[19]

In practice, the way this works is that we assign a probability density function f over Ω that captures how the probability distribution $P(\cdot|H)$ distributes probability over Ω. In these situations, Ω will have a canonical metric d that for points x and y in Ω assigns a distance $d(x,y)$. This metric then induces an underlying geometry that will itself be canonical, and that in turn then induces a canonical measure μ over Ω so that $f \cdot d\mu$ represents the probability distribution $P(\cdot|H)$.[20] To say that this distance, geometry, and measure are canonical is simply to say that these are uniquely determined with respect to the language we are using to identify events in Ω.

To recap, Ω has a metric (distance function) that induces a geometry on Ω and that thereby induces a measure μ on Ω, all of which are canonical with respect to our underlying descriptive language. This measure will generally be a uniform probability or a limit of uniform probabilities (such as Lebesgue measure on Euclidean space). The details here require advanced probability theory and real analysis.[21] But the point being made is easily appreciated by anyone who has worked with a normal distribution on the real line.

Given the geometric structure of the real line, the normal distribution density function is well defined and given (when the standard deviation is 1) by $f(x) = (2\pi)^{-1/2} \times \exp(-x^2/2)$. In general, the canonical measure μ on Ω (in this case Lebesgue measure) will induce a canonical probability density function f. The portion of Ω where f is large will then reliably indicate where the probability distribution $P(\cdot|H)$ is concentrating (or focusing) its maximal amount of probability. On the other hand, the portion of Ω where f is small will then reliably indicate where $P(\cdot|H)$ is minimizing (or withdrawing) probability.

Given f, we now form two event-classes in Ω. Specifically, for r and s positive real numbers, we form event T^r and T_s in Ω as follows: $T^r = \{\, x \in \Omega \mid f(x) \geq r \,\}$ and $T_s = \{\, x \in \Omega \mid f(x) \leq s \,\}$. As r gets larger, T^r closes in on the mode of f and as s gets smaller (close to but always bigger than zero), T_s moves further and further out in the tail of f. If for any r, T^r includes E (i.e., $T^r \supset E$), then the density function f specifies T^r and therewith E. Likewise, if for any s, T_s includes E (i.e., $T_s \supset E$), then the density function f specifies T_s and therewith E.

The specification of E here is inherent in the simplicity of description of the underlying mathematics, which matches E with a well known and widely used statistical property, namely, the event E residing in the mode or tail of $P(\cdot \mid H)$. In other words, we would succinctly describe T^r or T_s (in keeping with these being specifications) as respectively "matching the mode of f" and "matching the tail of f." But the point to realize is that as mathematical descriptions based on the geometry of Ω and the density f, these modes and tails are as simple as can be described.

Note that the larger r is taken, the smaller the probability of T^r. In the limit, as r gets arbitrarily large, the probability of T^r, i.e., $P(T^r \mid H)$, will go to zero. Likewise, note that the smaller s is taken (albeit always larger than zero), the smaller the probability of T_s. In the limit, as s gets arbitrarily small (close to zero), the probability of T_s, i.e., $P(T_s \mid H)$, will go to zero. Given a probability cutoff α, what's needed to specify E will be to find the smallest r' or the largest s' such that $T^{r'}$ includes E or $T_{s'}$ includes E and such that $P(T^{r'} \mid H) \leq \alpha$ or $P(T_{s'} \mid H) \leq \alpha$. In this way, $T^{r'}$ or $T_{s'}$ will respectively become small-probability rejection regions and thus underwrite a valid chance-elimination argument. But note, there's no guarantee that either $T^{r'}$ or $T_{s'}$ will actually include E. If either does, then E will be specified by a mode or tail. But whether one of them does will depend on E itself and the density that induces $T^{r'}$ and $T_{s'}$.

The tail (singular) of a distribution is by definition where the density f is small (or close to 0). But in many ordinary instances, we speak in terms of the tails (plural) of a distribution. That happens when the probability space Ω also has a natural ordering, as when Ω is the real line $\mathbb{R}$ (which can support a normal distribution), or a portion of the real line, such as the nonnegative reals $[0, \infty)$ (which can support a continuous Poisson distribution), or the natural numbers $\mathbb{N}$ (which can support a discrete Poisson distribution), or finite subsets of the natural numbers (which on the set $\{0,1,\ldots,n\}$ can support a binomial distribution).

In such situations, there's both a left and right tail: $T_{s-} = \{ x \in \Omega \mid x < c$ and $f(x) \leq s \}$ as the left tail and $T_{s+} = \{ x \in \Omega \mid x > c$ and $f(x) \leq s \}$ as the right tail where c is a natural dividing point between the left tail and the right tail (for instance, 0 for a univariate normal distribution with mean 0). If we now let $T_{s'-}$ denote the largest s where the left tail includes E, or similarly let $T_{s'+}$ be the largest s where the right tail includes E, then $T_{s'-}$ or $T_{s'+}$ can specify E, and thus be used to preclude its chance occurrence. What makes all such specifications work is that we are exploiting inherent mathematical structures of the possibility space Ω, which therefore don't just have short descriptions but optimally short descriptions. They induce rejection regions that at most require factoring in such tails and modes.

The key takeaway here is that these tails and modes are not just induced by specifications, but equate directly with rejection regions. It's not that these rejection regions need to be expanded because there are so many other ways that what was specified could have been specified with equal or less descriptive complexity, as was typical with the specification-induced rejection regions of the last two sections. Because modes and means as specifications capitalize on an underlying canonical geometry, they are descriptively simplest mathematically and therefore equate directly with rejection regions.

One might quibble that insofar as tails and modes constitute optimally short descriptions with respect to an underlying canonical mathematical structure, they should both be factored into a rejection region. Likewise, if we have a right tail in a rejection region, then we should also factor in the left tail, and vice versa. In practice, however, we don't expand rejection regions in this way. A left tail, a right tail, a tail as such, and a mode, as captured respectively by $T_{s'-}$, $T_{s'+}$, $T_{s'}$, and $T^{r'}$ serve just fine in practice as rejection regions. Typically, with an α cutoff, we factor in quite a bit of slack so that even if α is off by a factor of 10, conclusions from small probability arguments still hold (see Section 4.6).

In this vein, Fisher, when he charged Mendel's experiment with fraud, simply looked to the mode and did not feel the need also to add the tail to his rejection region. Likewise, when we witness an event far out in the tail of a probability distribution, we focus on a rejection region that's simply a tail rather than also add into this rejection region a mode. So too with left and right tails, if we find an event far out in the left tail, we feel no compulsion also to include the right tail in the rejection region, or vice versa. Technically speaking, perhaps we should. But in practice, such modes and tails serve as adequate rejection regions.

For a practical application of the ideas in this section, see the example of Nicholas Caputo in Section 5.1.

ⅅⅭⅩⅭⅢⅩⅭⅢ CHAPTER 4

PROBABILISTIC RESOURCES

4.1 Calculating Probabilistic Resources

WHETHER AN EVENT IS PROBABLE OR IMPROBABLE DEPENDS ON THE number of opportunities for the event to occur. The greater the number of opportunities, the more probable is the event. Ten heads in a row may seem improbable, having roughly a one-in-a-thousand probability of occurring with ten tosses of a fair coin. But with thousands of coin tosses, ten heads in a row becomes highly probable, and with enough coin tosses ten heads in a row becomes inevitable. Probabilistic resources factor in the number of opportunities for an event to occur and thereby determine its real probability.

A principle of counseling psychology is that "the presenting problem" is never "the real problem." The two are related, and the presenting problem is the point of entry to the real problem. But the real problem is the one that requires redress. With probabilistic resources, a similar truth holds, namely, "the presenting probability" is never "the real probability." The real probability is determined by factoring in all the relevant probabilistic resources. Only by factoring these in can we realistically assess how probable or improbable an event actually is.

Probabilistic resources are defined, then, as the number of opportunities for an event to occur. To calculate this number, we need to identify the different relevant ways an event could occur and then count them. Many industries depend on calculating the number of opportunities for an event to occur. These industries then use that number to determine the real probability of an event occurring. Even if these industries don't explicitly use the term "probabilistic resources," they tacitly understand the underlying concept. Often, probabilistic resources play a crucial role in assessing risk, everything from the risk of a password getting hacked to the risk of death by playing Russian roulette.

How many attempts do bad actors have to break the passwords on a server? Even with all these attempts factored in, would it still be improbable that any one of them might succeed in cracking even a single password, thereby penetrating the server? Cybersecurity experts need to ask and answer such questions. Or take the game of Russian roulette. With six chambers in a revolver and a bullet in one of them, how many times does the player intend to spin, point, and shoot? The probability of death depends on that number.

Perhaps Russian roulette seems like a far-fetched example, but consider that we humans engage in many less flamboyantly dangerous behaviors that we repeat and where the repetition is bound eventually to catch up with us. (*Would you like another cigarette? Would you like sprinkles on that donut?*) Actuaries and insurance companies understand, even without using the term, how probabilistic resources bear on their business.

Unless otherwise stated, we assume that the separate opportunities for an event to occur, which make up a set of probabilistic resources, are all probabilistically independent (see Appendix A.2). But this assumption is not strictly speaking necessary, and we will relax it on occasion. Consider, for instance, an urn with 1 white ball and 999 black balls, all the balls being geometrically identical, and thus equally probable when sampled. With selections from the urn where a ball is replaced into the urn every time it is selected, the opportunities to witness a white ball are all probabilistically independent. But with selections from the urn where a ball is left outside the urn every time it is selected, the opportunities to witness a white ball are probabilistically dependent.

When the urn is sampled without replacement, fewer probabilistic resources are needed to increase the probability of finding the white ball as compared to sampling with replacement. Given 1,000 selections without replacement, the probability of finding the white ball increases to 1, which is to say we are guaranteed to find it because in 1,000 selections all the balls in the urn will be sampled. On the other hand, given 1,000 selections with replacement, and thus given probabilistic independence of the selections, the probability of finding the white ball is only .632. Of course, we can also make probabilistic resources less efficient than independent sampling to achieve an event. Imagine, for instance, adding 1,001 black balls to the urn every time a black ball is selected. Thus, when we sample a black ball, we replace it and add 1,000 more black balls to the urn. In that case, given 1,000 selections, the probability of finding a white ball goes down to approximately .00746.[1]

Cryptography provides a particularly clear example of how to calculate probabilistic resources. Risk is always at the forefront of cryptography, the risk being whether a cryptosystem can withstand the attempts (opportunities) it might face from actors attempting to break it. The National Research Council (NRC) addressed this risk by posing and then answering the following question: What is the level of security required for cryptosystems, and specifically for their cryptographic keys, to resist cryptanalytic attack regardless of how much faster and bigger computers get on account of technological improvements?

As it is, technological improvements are always subject to the constraints of physics and thus cannot increase indefinitely. Such a limitation on technology imposed by physics is evident in the NRC's analysis of cryptographic security. Through this analysis, the NRC drew a conclusion that holds up well even though it is now a generation old (1996). Indeed, the probabilistic resources calculated by the NRC (i.e., 3×10^{94}) show no signs of ever needing to be increased to ensure the security of cryptosystems at the scale of planet Earth:

Computers can be expected to grow more powerful over time, although there are fundamental limits on computational capability imposed by the structure of the universe (e.g., nothing travels faster than light in a vacuum, and the number of atoms in the universe available to build computers is large but finite). Thus, the minimum [cryptographic] key size needed to protect a message against a very powerful opponent will grow as computers become more powerful, although it is certainly possible to choose a key size that will be adequate for protecting against exhaustive search for all time...

The mass of Earth is about 6×10^{24} kg. A proton mass is 1.6×10^{-27} kg, so that Earth contains about 4×10^{51} protons. Assume one proton per computer, and that each computer can perform one operation in the time that it takes light to cross its diameter (i.e., 10^{-15} meters divided by 3×10^{10} meters per second, or $1/3 \times 10^{-25}$ seconds). Each computer can thus perform 3×10^{25} operations per second. If all of these computers work in parallel, they can perform $4 \times 10^{51} \times 3 \times 10^{25}$ operations per second, or 10^{77} operations per second. The age of the universe is on the order of 10 billion years, or 3×10^{17} seconds. Thus, an earthful [sic] of proton-sized computers can perform 3×10^{94} operations in the age of the universe. With the assumptions made before, this corresponds to a key size of 95 decimal digits, or about 320 bits...

[This calculation] demonstrate[s] that it is clearly possible to specify a key size large enough to guarantee that an attack based on exhaustive search will *never* be feasible, regardless of advances in conventional computational hardware or algorithms.[2]

According to the NRC, 3×10^{94} therefore characterizes the maximal number of computational operations available at the scale of the earth. All this computational firepower is here described as devoted to solving for one particular cryptographic key. This key therefore sets the pattern that must be matched, and it is specified *prior to* all the computational efforts to uncover it. It is therefore a prespecification (in the sense defined in the previous chapter). The NRC's assessment of cryptographic security assumes that crypt-analytic attackers have no special insight or knowledge about the cryptographic key—knowledge that might make the search easier. Thus, any program running on all these proton-sized computers simply attempts to run through all possible keys of a given length, and will stop as soon as (if ever) it finds that key. The proton-sized computers are therefore attempting to do an exhaustive parallel search.

What if all this computational firepower were spent on multiple keys? If there were multiple cryptographic keys and we had no special knowledge of them, all those 3×10^{94} computational operations would need to be allocated across the different keys, which would mean diluting the number of computational opera-tions available per key. The math is straightforward, and shows that the probability of finding even one of those multiple keys with 3×10^{94} computational operations—and regardless of the method for allocating those operations—is no larger than finding a single key with the full complement of 3×10^{94} computational operations.

This NRC assessment of cryptographic security illustrates several important lessons. Consider the key size that the NRC cites, namely, 95 decimal digits, or 320 bits. The NRC's numbers are slightly off. Three-hundred twenty bits coincides with 2^{320} dif-ferent possible bit sequences of that length, which comes to approximately 2×10^{96}, and thus to at least 96 decimal digits (not 95 decimal digits). If each of the 3×10^{94} computational opera-tions identified as available on Earth were devoted toward guessing a cryptographic key 320 bits in length, all 3×10^{94} guesses would still be unlikely to guess the key. In fact, to have an even chance of finding that sized key would require about 67 Earth-sized planets each running 3×10^{94} computational operations.

This NRC assessment of cryptographic security exemplifies probabilistic resources and indicates how they are to be calculated. Yet the role of chance and probability in this example requires some elaboration. All the activity described in this example is computational, devoted to exhaustive search for a particular crypto-graphic key. The keyspace, over which the exhaustive search takes place, consists of a certain number of possible keys. But exhaustive searches are deterministic, simply attempting to plough through the search space in some straightforward, often wooden-headed, way. We apply exhaustive search when we have no knowledge that helps us to home in on the item we're looking for. In consequence, any way of traversing the search space that gets through it efficiently and avoids repetitions is as good as any other.

Thus, while it makes sense to refer in this example to *com-putational resources*, it might seem to make less sense to refer to *probabilistic resources*. Where, then, does chance reside in this example? Ultimately chance here resides in the probabilistic way the cryptographic key that we're looking for was identified in the first place. With cryptographic keys, they are typically selected at random, which is to say by taking a uniform random sample over the keyspace, thereby assigning equal probability to each key. If the key has 320 bits (the minimal number of bits assumed in this example necessary for full cryptographic security) and if all bit sequences of this length are fair game for a workable cryptographic key (that's not always the case, as in public key cryptography, where the keys need also to satisfy certain number-theoretic constraints), that means the key's selection has a probability of approximately 1 in 2×10^{96}.

It follows that the search process, if it operates with no special knowledge of the key (as is typically the case with exhaustive search), will simply be guessing at random. Consequently, the ability of the search to successfully find the key will correspond to a certain probability of success at each search query. Suppose, for instance, the search space has K possible keys, with each key having equiprobability (as is assumed in this example). The key would therefore have been generated via a chance process that gives it a probability of $1/K$. An exhaustive search will, on its first search attempt, have probability $1/K$ of finding it. On its second attempt, if unsuccessful on the first attempt, it will have probability $1/(K-1)$ of finding it. On its third attempt, if unsuccessful on its first two attempts, it will have probability $1/(K-2)$ of finding it. And so on.

Search problems like this are typically so big that exhaustive search is powerless to solve them. They are needle-in-a-haystack problems where the haystack so overwhelms the needle that for all practical purposes it is pointless to search for the needle. In such cases, an exhaustive search can get started, but it cannot finish, being virtually guaranteed to end in defeat. Thus, for any attempted exhaustive search, if K is big enough, no feasible number of steps will exhaust a search space consisting of K items. In such circumstances, K will be much bigger than the number of search queries—call them M—that can, practically speaking, be made in any actual search. To depict this disparity between M and K, mathematicians suggestively write $M \ll K$. In consequence, the probability of locating the item of interest, here the cryptographic key, in M queries is $M \times (1/K) = M/K$. In practice, this probability will be so small as to make finding the item of interest highly unlikely.[3]

To sum up, in the National Research Council's assessment of cryptographic security, cryptographic keys whose key size is at least 320 bits may thus be considered utterly secure, come what may in technological advance. Physics itself ensures that cryptographic keys beyond that size will be immune to crypt-analytic attack.[4] Specifically, fundamental physical limits exist on the number of cryptographic keys that can be searched regardless of how advanced and efficient technology becomes.

4.2 Relative Versus Absolute Probabilistic Resources

The limit imposed by physics on the number of search queries differs starkly from the number of search queries that can, practically speaking, be executed. This latter number corresponds to the probabilistic resources available given current technology. That number will always be far less than what fundamental physics could conceivably allow.

How close is current technology to the theoretical limits that fundamental physics imposes on the cryptanalytic search for cryptographic keys? For the theoretical limits described in the last section, we were asked to imagine that every proton on planet Earth could operate at around 10^{25} computational operations per second over the course of 10 billion years in the service of breaking a single cryptographic key (or multiple keys, but then with those computational operations diluted among them). We concluded that keyspaces with keys of 320 bits were thus effectively intractable given the computational operations (probabilistic resources) available on Earth.

By contrast, how many computational operations are possible given current technology? The fastest computer in the world as of June 2022 is the HPE Cray EX235a supercomputer at Oak Ridge National Laboratories, operating at 1,686 petaflops, or roughly 1.7×10^{18} floating point operations per second.[5] Breaking cryptographic keys tends to be time-sensitive, so let's imagine that we have at most 10 years (= 315 million seconds) to break them. That time period is more than generous—actionable intelligence typically requires cryptanalytic breakthroughs in hours or days, not years.

But let's not just be generous—let's be profligate. Imagine therefore a billion such Crays all operating at full throttle and all fully devoted to solving for just one cryptographic key over the course of 10 years. In that case, there would be at most $3.15 \times 10^8 \times 10^9 \times 1.7 \times 10^{18} = 5.36 \times 10^{35}$, or approximately 5×10^{35} computational operations that all these Crays could perform in an exhaustive search for a cryptographic key. Let's now set M equal to 5×10^{35}. A keyspace with 10^{36} (= $2 \times M$) searchable keys would, with M computational operations, therefore stand only an even probability of successfully finding the right key.

The disparity between M (= 5×10^{35}) and K (= 3×10^{94}) is stark and illustrates the difference, respectively, between *relative* and *absolute* probabilistic resources. M, as calculated here, provides an upper bound on the number of probabilistic resources that might be available for finding a given cryptographic key *relative to* current technology. Go back a decade, and the fastest supercomputer in the world was still at Oak Ridge and still a Cray, but it was the Cray XK7. Its top speed was 18 petaflops, about 100 times slower than today's fastest supercomputer.

The probabilistic resources relative to today's technology are therefore bounded by 5×10^{35}, but relative to the technology available a decade ago, the probabilistic resources would have been bounded by 5×10^{33}. Obviously, the assumption that a billion such machines could be used over a ten-year span to break just one cryptographic key is supremely generous and wholly unrealistic. Supercomputers are built for use on multiple problems. Moreover, coordinating a billion such computers to work on uncovering a single cryptographic key would itself be a huge challenge in both hardware and software engineering.

The point of these musings is to distinguish relative from absolute probabilistic resources. Relative probabilistic resources often reflect the practical limits on the opportunities to bring about a result by chance. Relative probabilistic resources may push the limits of current technological plausibility. Absolute probabilistic resources, by contrast, reflect the absolute limits on the opportunities to bring about a result by chance given the constraints of fundamental physics. Absolute probabilistic resources of 3×10^{94} are vastly more than the current (albeit generous) relative probabilistic resources of 5×10^{35}.

Indeed, we have no convincing grounds to think that technological advance can get anywhere close to approaching this 3×10^{94} absolute limit, and it's safe to say that Moore's Law will never take us there. The only possible exception might be quantum computation. But it has proven extremely difficult to scale up quantum computation to real-world problems. Moreover, the cases, such as Peter Shor's factorization algorithm, where full-scale quantum computing could provide a decisive advantage over conventional computing do not generalize, being few and far between.[6]

For relative probabilistic resources, practical considerations determine how many opportunities are deemed relevant to an event's chance occurrence. With Visa credit cards, for instance, we showed in Section 1.2 that a single random guess of a Visa card's numerical data had a one in a billion billion (= 10^{18}) probability of success. Thus, so long as the (relative) probabilistic resources for randomly guessing a Visa card's numerical data were well below 10^{18}, which is true in practice, individual Visa cards would be secure from random hacking.[7] And indeed, hackers find this range of probabilistic resources intractable and thus look to other means to discover valid Visa card numbers, such as phishing expeditions. Depending on how stringent the need is to rule out chance, relative probabilistic resources can therefore vary quite a bit.

4.3 Variations in Absolute Probabilistic Resources

Like relative probabilistic resources, absolute probabilistic resources can also vary in size. Such variations arise from the different ways physics can be enlisted to limit the extent of physical reality and the performance of physical processes. Such variability in absolute probabilistic resources may sound strange given that if something is absolute, we tend to think it should be immune to variability. But, as it turns out, absolute probabilistic resources, depending as they do on physical limits, can be calculated in different ways by focusing on different types of physical limits, some of which are more stringent than others.

Consider again the NRC's assessment of cryptographic security given in Section 4.1. The NRC came up with 3×10^{94} as the total number of computational operations at the scale of the earth. A factor in calculating this number was the mass of the earth, which came to 6×10^{24} kg. But why focus simply on the mass of the earth? Why not instead go with the mass of the sun, which is 333,000 times the mass of the earth? In that case, the mass to be factored in would come to 2×10^{30} kg.

But why stop there? Why not go with the mass of the Milky Way galaxy, which comes to 3×10^{42} kg? This line of reasoning can be continued but has an obvious stopping point, namely, the total mass of the observable universe, which current estimates place at 1.5×10^{53} kg (this number is for ordinary matter, and thus excludes dark matter or other forms of exotic matter, whose computational powers are unknown). By swapping out this last number for the mass of the earth in the NRC's calculation, the maximal number of computational operations for finding a cryptographic key at the scale of the entire universe then comes to 7.5×10^{122}, or roughly 10^{123} (which corresponds to about 409 bits).

Could there ever be any reason to go bigger than 10^{123} as an upper bound on absolute probabilistic resources? In fact, there could be. In all the cryptographic calculations given so far, the speed of computation depended on the time it takes for light to cross the diameter of a proton, which comes to $1/3 \times 10^{-25}$ seconds. But the smallest physically meaningful unit of length is not the diameter of a proton (10^{-15} meters), but the *Planck length* (1.6×10^{-35} meters), which is 20 orders of magnitude smaller.[8] Thus, in place of a proton-based top computing speed of $1/3 \times 10^{-25}$ seconds per operation, the absolute limit for computation could be based on the time it takes for light to cross the Planck length. This time period, known as the *Planck time*, comes to 5.4×10^{-44} seconds. Compared to the time it takes for light to cross the diameter of a proton, the Planck time would raise the absolute limit on computational speed by almost 20 orders of magnitude.

At the scale of the observable universe rather than merely the earth, substituting the reciprocal of the Planck time for the reciprocal of the proton-light-crossing time in the NRC's calculation thus raises the absolute probabilistic resources from 7.5×10^{122} to 4.6×10^{140} (the latter corresponds to about 467 bits). Moreover, if we focus not just on protons, of which there are about 10^{80} in the observable universe, but on the total number of elementary particles (baryons, photons, neutrinos, etc.), that raises the number of particles potentially capable of computation by roughly another ten orders of magnitude, and thus raises absolute probabilistic resources to 10^{150}. This number received special treatment in the first edition of *The Design Inference*, where it was the basis for the universal probability bound calculated there, namely, 1 in 10^{150}. It retains that special status in this second edition. We'll return to the relation between absolute probabilistic resources and universal probability bounds later in this chapter.[9]

At the scale of the observable universe, probabilistic resources anywhere in the range from 10^{123} to 10^{150} seem totally secure, impervious to any new and undreamt-of physical processes that might significantly increase their number. Yet there's another approach that increases these numbers still further, developed by quantum computational theorist Seth Lloyd. This approach looks at the universe as a giant quantum computer. Calculating the number of elementary quantum logic operations that the universe can perform based on entropic and quantum-field theoretic considerations, Lloyd comes up with a larger number of absolute probabilistic resources, though he doesn't use the language of probabilistic resources, referring instead to "the ultimate physical limits of computation" and "the computational capacity of the universe."

Lloyd calculates two sets of numbers. In a 2000 *Nature* article, he imagines what he calls "the ultimate laptop," which consists of one kilogram of matter residing in one liter (i.e., a volume of 1,000 cubic centimeters). He calculates, based on quantum and entropic considerations, that such an ultimate laptop could perform 5.4258×10^{50} elementary logical operations per second acting on roughly 10^{31} bits.[10] This amounts to roughly 10^{81} bit operations per second for Lloyd's ultimate laptop.

Two years later, in a 2002 article for *Physical Review Letters*, Lloyd calculated that the observable universe can, throughout its history up to the present, have performed 10^{120} elementary logical operations acting on 10^{90} bits, or even as many as 10^{120} bits if gravitational degrees of freedom are factored in. If we go with the larger number of bits here, this means that the $10^{120} \times 10^{120} = 10^{240}$ is the maximal number of bit operations that the observable universe could have performed in its 13 or so billion years of existence from the big bang to the present. If we don't venture outside the observable universe (such as by appealing to a multiverse), then 10^{240} sets the maximal limit on probabilistic resources consistent with the current scientific literature (Lloyd's numbers have not been improved upon in the intervening twenty years).

So, do we really need to go as high as 10^{240} probabilistic resources in trying to understand and control for the number of opportunities for an event to occur by chance in the observable universe? 10^{240} seems like overkill. Even 3×10^{94}, which in the NRC's assessment of cryptographic security determined the size of the keyspace needed for complete safety from cryptanalytic attacks at the scale of the earth, seems excessive. We will argue in Section 4.5 that, in practice, we can make do with probabilistic resources that are at once smaller than this and yet likely to stay ahead of technology and physics, come what may.

Nonetheless, even though in practical applications we will never encounter probabilistic resources in the range between 10^{94} and 10^{240}, it's worth noting that possibility/search spaces of this size are readily encountered:

(1) A coin tossed 1,000 times yields 2^{1000}, or roughly 10^{300}, possible bit strings, and thus represents a space larger than 10^{240}. There are roughly 10^{19} ways of filling out an NCAA "March Madness" bracket, and thus over 10^{240} ways of separately filling out thirteen such brackets (a number you'll reach if you and twelve friends fill out brackets).[11] The lottery with the least chance of winning is Mega Millions. It is played twice weekly, and for any drawing there exist slightly more than 3×10^8 possible outcomes. Separate drawings over the course of 15 weeks thus exceed 10^{240} possible outcomes.[12]

(2) QR codes (short for *quick response codes*) have become ubiquitous. They are two-dimensional grids or matrices, and they come in various sizes. The largest is a 177 by 177 matrix, consisting of 31,329 positions that can each be on or off, black or white. QR codes can therefore represent over 30,000 bits. Even with error correction, which limits the number of effective bits for encoding and communication, the number of

possibilities capable of being represented by QR codes vastly exceeds 10^{240}.[13]

(3) Proteins are polymers consisting of 20 possible amino acids at each position. The number 20^{185} is roughly 10^{240}, so the space of possible proteins made up of 185 amino acids likewise represents a space larger than 10^{240}. An average sized protein, in any of life's domains (Archaea, Eubacteria, or Eukarya), is considerably larger than 185 amino acids. Eukaryotes like us (whose cells have nuclei) have well above 400 amino acids on average per protein.[14]

(4) DNA is a double helical arrangement consisting of 4 possible nucleotide base pairs at each position. The number 4^{400} exceeds 10^{240}, so the space of possible DNA coding regions made up of 400 nucleotide base pairs (which at most could code for a protein with 133 amino acids—one nucleotide triplet or codon for each amino acid) likewise represents a space larger than 10^{240}. DNA spaces, corresponding to the complete genomes of organisms, are of course much larger. The number of base pairs in the DNA of any human cell is roughly 3 billion, or 3×10^9, and yields a possibility space whose size is roughly $10^{(10^9)}$, which is superexponential and, in this case, represents a 1 followed by a billion zeroes!

4.4 Avoiding Superexponentiality by Not Miscounting

Superexponentials, or what are also called compound exponentials, merit some comment in any discussion about probabilistic resources. Superexponentials are numbers with exponents that are themselves exponentials or, alternatively, numbers whose logarithms are exponentials. Superexponentials arise both in fundamental science and in real-world applications. We just saw $10^{(10^9)}$ for the human genome space, a modest superexponential as superexponentials go. Roger Penrose has argued that the present universe required overcoming an entropic constraint of one part in $10^{(10^{123})}$.[15] That's a much bigger superexponential.

In an article by Haug, Marks, and Dembski, we argue that superexponentials are unavoidable in artificial intelligence (AI), posing a potentially insuperable obstacle to the success of AI in solving a variety of problems. We argue that as the complexities of AI systems increase linearly, the contingencies with which these systems must deal increase exponentially, and so the possible ways of the systems dealing with those contingencies (their performance designs) will increase superexponentially.[16]

The number of probabilistic resources in any practical situation need never become superexponential. If it does, we are making a mistake in counting. To see how probabilistic resources can avoid becoming superexponential, we must understand an important rule for properly counting probabilistic resources. The rule is this: probabilistic resources must not count events that overlap with other events that have already been counted.

A probabilistic resource is an opportunity to repeat an event whose chance occurrence stands in question. Just as we want with currency to avoid double spending, so with events we want to avoid double counting them. Double counting events occurs when events overlap with other events. If we don't control for such overlap, we can greatly overestimate probabilistic resources, even to the point of superexponentiality. Of course, we must always also guard against underestimating probabilistic resources.

Consider, for instance, the tossing of a fair coin 500,000 times, and let the event E denote tossing 500,000 heads in a row. Next, consider a sequence of 1,000,000 actual coin tosses. How many probabilistic resources do these 1,000,000 coin tosses contribute toward explaining the chance occurrence of E? If we think of probabilistic resources as separate arrows that we get to shoot at a target, then these 1,000,000 coin tosses seem naturally to represent two arrows that are shot at random to try to achieve E, conceived of as a target. Thus, the first 500,000 coin tosses represent the first arrow, and the second 500,000 coin tosses represent the second arrow. This answer is acceptable—1,000,000 coin tosses, according to this method of counting, yields two opportunities to bring about, by chance, 500,000 coin tosses.

Nonetheless, this is not the only acceptable answer. In a million tosses, we might want to allow that any uninterrupted sequence of 500,000 coin tosses should count as an opportunity to achieve E, and that would include a sequence of 500,000 tosses that straddles the first 500,000 coin tosses and the second. There are 499,999 ways of straddling the first and the second 500,000 coin tosses with 500,000 heads in a row, and thus 500,001 total different ways of getting 500,000 heads in a row if we're counting coin tosses one after another.

The arithmetic here is straightforward. There are two ways of getting 500,000 heads in a row that don't straddle the first and second half-million tosses. And then there are 499,999 that do. In this method of counting probabilistic resources, we restart the count for 500,000 heads in a row whenever a tail is observed. In short, by counting opportunities to bring about 500,000 heads in a row as soon as a tail is observed, the number of possible opportunities for 500,000 heads in a row is 500,001. There could be other ways of counting probabilistic resources in this example. For instance, we might want to reshuffle the order of the 1,000,000 coin tosses, now considering 500,000 heads in a row within the reshuffled order. Any such alternative way of counting probabilistic resources would be acceptable provided it doesn't double count overlapping events.

To see what could go wrong in counting probabilistic resources in this example, consider that from elementary combinatorics, there are roughly $10^{300,000} = 10^{(3 \times 10^5)}$ ways of individuating, or identifying, 500,000 coin tosses among 1,000,000 coin tosses.[17] This number is a small superexponential, as superexponentials go, but it is still a superexponential. It is therefore well beyond any of the probabilistic resources at the scale of the observable universe that we calculated earlier.

But note, each such way of individuating 500,000 coin tosses constitutes an event. If we therefore blithely count up all these events and treat them as probabilistic resources, we can easily explain the chance occurrence of 500,000 heads in a row. In 1,000,000 coin tosses, there's an even chance that at least 500,000 heads will occur by chance. Given 500,000 heads distributed in some way among these 1,000,000 coin tosses, one of these ways of individuating 500,000 among 1,000,000 coin tosses could then be seen as coinciding with 500,000 heads in a row.

This example of overestimating probabilistic resources may seem so obviously wrong-headed as not to be worth addressing. After all, we're inquiring into how many chances there are to get 500,000 heads *in a row*, so it seems obvious that we are not going to count results where 1,000,000 coin tosses have at least 500,000 heads sprinkled through the million total and where those half a million heads are broken up by tails. But it's all a question of how the tosses are individuated. What if one looked at the first, third, fifth, etc. tosses and they were all heads? What if one looked at the second, fourth, sixth, etc. tosses and they were all heads?

A million coin tosses supplies a superexponential number of events. We need therefore to rule out the vast majority of these events from serving as probabilistic resources. Otherwise, probabilistic resources will be useless. The point is to prevent double counting from running amuck. The way to do this is to insist that probabilistic resources for an event consist of nonoverlapping opportunities to repeat the event. Events can consist of sub-events (which are themselves events), and such sub-events, when held in common by other events, will mean that the events overlap. In set-theoretic terms, events E and F don't overlap if their intersection is empty: $E \cap F = \emptyset$. They overlap if their intersection is nonempty: $E \cap F \neq \emptyset$. Such overlap invalidates the counting of probabilistic resources.

As is clear in the preceding coin-tossing example, for any two distinct subsequences of 500,000 coin tosses within the sequence of 1,000,000 coin tosses, unless the subsequences are mutually exclusive and exhaustive (such as even-numbered versus odd-numbered coin tosses), they will overlap, sharing some individual coin tosses, perhaps only 1, perhaps 499,999, or some number in between. It's such overlap that our nonoverlap rule precludes.

In most circumstances, there won't be a temptation to double count probabilistic resources. Ordinarily, there will be an obvious natural way to individuate probabilistic resources so that each probabilistic resource won't overlap with others, just as separate arrows fired at a target don't overlap (i.e., don't share any common sub-events). Moreover, we are often incentivized to avoid over-estimating probabilistic resources.

For instance, gambling strategies that allow overlaps and therefore double counting invariably fail, and fail quickly. Thus, if we allow for overlapping events and calculate a superexponential number of probabilistic resources based on a mere 1,000,000 coin tosses, as in the example just considered, we might think that factoring in those probabilistic resources could render 500,000 heads in a row to be quite probable. But any betting scheme based on that conclusion would fail.

There's one other way that superexponentiality can arise and threaten to derail improbabilities that would otherwise make for a successful small-probability design-inferential argument. The problem here consists in failing to distinguish the total number of possibilities from the realizable number of possibilities. More simply, the problem is to mistakenly count *mere possibilities* in place of *realizable possibilities*.

Consider 1,000,000 DNA base pairs. There are $4^{1,000,000}$, or roughly $10^{(6 \times 10^5)}$ possible DNA sequences consisting of 1,000,000 base pairs. Yet the actual number of such sequences that could be realized at any one time if all the atoms in the universe were dedicated to forming such sequences would be far less. Indeed, it is far less than 10^{80}, which is the number of protons in the universe. It will have to be far less than 10^{80} because many protons will be involved in even a single DNA base pair, to say nothing of a sequence of 1,000,000 base pairs.

The prohibition against not counting every conceivable possibility in assessing probabilistic resources may seem too obvious to mention, but it is a mistake that we've seen committed. For instance, Howard Landman, in criticizing an earlier formulation of specified complexity by WmD, argues that Seth Lloyd's 10^{90} bits on which the known physical universe can expend 10^{120}

operations entail $2^{(10^{90})}$ "accessible states." He then argues that this number (a sizable superexponential) is the proper bound on probabilistic resources at the scale of the whole universe.[18]

Granted, all those $2^{(10^{90})}$ states are in principle accessible, comprising the total number of (or "mere") possibilities that might be achieved by the universe acting as Lloyd's giant quantum computer. But in fact, each of those 10^{120} operations changes the 10^{90} bits from one state to another, and thus yields a totality of probabilistic resources as described in the last section, limiting them to $10^{120} \times 10^{90}$, which equals 10^{210} (or, if we allow for gravitational degrees of freedom, this number will go up to 10^{240}). Clearly, these numbers, which describe the realizable number of possibilities here, are vastly smaller than $2^{(10^{90})}$.

4.5 Minimizing Absolute Probabilistic Resources

Let's return to absolute probabilistic resources, but this time to consider how small they might reasonably be taken. So far, we have seen absolute probabilistic resources range between 10^{94} on the low end and 10^{240} on the high end. These are sizable numbers. But in fact, in our reasoning with small probabilities, we can make do with considerably fewer probabilistic resources, even when understood in absolute terms.

Other things being equal, we should prefer fewer to more probabilistic resources. Fewer probabilistic resources mean that we don't have to run through as many possible opportunities to show that something is or is not the result of chance. We need 10^{94} probabilistic resources to render probable 312 heads in a row, but we need 10^{240} probabilistic resources to render probable 797 heads in a row. The latter is more difficult to achieve probabilistically than the former, but the former can still be difficult to achieve by chance.

In practical life, we prefer not to set the bar too high for probabilistic resources. The greater the number of probabilistic resources, the more difficult it is to eliminate chance, and thus the more likely it is that chance is retained even when a design inference may be staring us in the face. The greater the number of probabilistic resources, the greater the number of false negatives, which is to say ascriptions to chance that should be ascriptions to design.

If, for instance, 10^{50} set the number of absolute probabilistic resources (as implicit in Emile Borel's universal probability bound discussed in Section 1.2), then improbabilities at the level of 166 heads in a row would decisively contravene chance provided the events in question are also prespecified. A diminished set of probabilistic resources that are nonetheless absolute would thus make it easier to eliminate chance while at the same time making it harder to commit false negatives.

How much, then, can absolute probabilistic resources be decreased below the 3×10^{94} lower limit calculated so far? Let's consider two alternative approaches, both at the scale of the earth. In the first approach, we imagine ourselves as earth-bound scientists trying our best to witness a prespecified chance event of improbability p given M absolute probabilistic resources for repeating the event in question by chance, but where $M \times p$ is still so small that we should never expect to see it happen by chance. The first approach assumes the vantage of all possible human observers. In effect, it takes an empiricist approach to absolute probabilistic resources. The second approach we will consider takes a biological approach.

The empiricist approach starts by noting that no more than 100 billion humans are estimated to have existed throughout Earth's history. Let's imagine that each of these humans has 100 years (= 3.15 billion seconds) to conduct chance-based experiments. Moreover, assume each has a metric ton (= 1,000 kg) super-duper computer at their disposal, which is to say a computer that consists of 6.25×10^{29} protons (i.e., a metric ton of protons), each operating at 3×10^{25} operations per second.

These are massively parallel machines with each operation occurring in the time it takes for light to cross the diameter of a proton. Multiplying these numbers together yields $10^{11} \times 3.15 \times$

$10^9 \times 6.25 \times 10^{29} \times 3 \times 10^{25}$, which comes to just under 6×10^{75}. These are the number of computational operations these 100 billion super-duper computer users could marshal in an exhaustive search for a chance-based outcome. This empiricist approach thus yields absolute probabilistic resources of around 10^{76}, which is 18 orders of magnitude less than our previous low-end estimate for absolute probabilistic resources of 3×10^{94}.

The second approach to absolute probabilistic resources looks to biology. It estimates the total number of possible biological replication events over the course of natural history on Earth. The biomass on planet Earth is estimated to be 500 gigatons = $500 \times 10^9 \times 10^3$ kg = 500 trillion kg = 5×10^{14} kg.[19] Given pollution and extinctions, let's double this number and assume that the total biomass throughout the earth's 4 billion or so years since life on Earth first emerged has, moment by moment, been 1 quadrillion kilograms, in other words, 10^{15} kg. The mass of a typical bacterium is 1 picogram, or 10^{-12} grams, or 10^{-15} kg. Some bacterial cells are smaller than this average (e.g., *Mycoplasma genitalium*); other cells are much larger (e.g., the eukaryotic cells that make up animals).

In consequence, if the entire biomass of planet Earth consisted of 1-picogram bacteria, there would at any given time be no more than 10^{30} separate organisms on planet Earth. Moreover, the fastest reproduction time for organisms is around 4 minutes (for bacterial cells). But let's be generous and assume organisms can reproduce every second.[20] Over how many seconds could Earth's organisms reproduce? In the roughly 4 billion years that life has existed on Earth, there have been fewer than 10^{18} seconds. Accordingly, an extremely generous upper limit on the number of organisms that could have existed throughout the natural history of planet Earth comes to 10^{48} (= $10^{30} \times 10^{18}$).[21]

Because evolutionary events take place as heritable changes are transmitted from parent to offspring, anyone looking for a particular prespecified evolutionary event of probability 1 in 10^{48} would thus have only around an even probability of witnessing it among all these organisms. It follows that for any evolutionary event of probability 1 in 10^{50}, there would be less than a 1 in 100 chance of witnessing it among all these 10^{48} organisms. We therefore have grounds to take the absolute probabilistic resources for biology to be 10^{50}. This number seems secure as an upper bound on probabilistic resources for chance events across all biological systems on planet Earth.

The number 10^{50} seems like all the absolute probabilistic resources that earthbound scientists will ever need for their scientific inquiries. Because this 10^{50} number sets a generous upper bound on all biological replications that have ever occurred on planet Earth, this bound applies especially to all chance-based evolutionary events. Design inferences are most contentious within evolutionary biology (see Chapter 7), and yet this number of probabilistic resources is all that's needed for drawing design inferences in relation to the evolution of life on Earth. But if it's adequate for inferring design in biology, where wouldn't it be adequate? Indeed, we know of no design inference in any realistic context that requires more than 10^{50} probabilistic resources. If a scientific theory needs this many probabilistic resources, and thus allows itself the luxury of ascribing to chance prespecified events whose improbability is 1 in 10^{50}, it is conceding way too much to chance.

To sum up, we've seen absolute probabilistic resources range between 10^{50} at the low end and 10^{240} at the high end. What's remarkable is that these numbers, calculated using such widely varying physical and biological characteristics, are nonetheless in a reasonably constrained range, passing sanity checks on both ends and avoiding obvious counterexamples. A 10^{20} absolute probabilistic resource, for instance, would be way too low, accessible to exhaustive search by current supercomputers. In contrast, 10^{50} is secure not just by current computational standards but also by a wide margin, namely, by 20 or so orders of magnitude.

With Moore's Law ending in the next few decades (estimates range from 2025 to 2055 for the end of performance doubling of semiconductors every 18 or so months), we have no convincing reason to think that computers will be able to achieve 10^{50} operations any time soon, if ever. Sure, there may be some unexpected technological breakthroughs. But any such breakthrough would still be bounded by our upper-end absolute probabilistic resource numbers between 10^{94} and 10^{240}. Our current understanding of science and technology suggests that 10^{50} will serve as an adequate absolute probabilistic resource across subjects for at least this century and perhaps in perpetuity, not just in, say, biology, where computer speeds seem less important, but also in subjects where computer processing speed is pivotal, such as cryptography.

Yet for the rest of this book, we will go with the absolute probabilistic resources advanced in the first edition of this book, namely, 10^{150}. We'll treat this as the default for absolute probabilistic resources. This level of probabilistic resources received additional justification earlier in this chapter in Section 4.3. We take 10^{150} to be a rock-solid estimate for absolute probabilistic resources at the scale of the universe. It corresponds to 498 bits. Not only does this figure seem utterly secure against advances in technology, but it seems to capture the full range of specifications and events that might ever arise in any chance elimination argument based on specifications and small probabilities.

Think of all the specifications and events that might play into any such chance elimination argument as forming a giant list. Specifically, imagine the cosmic history of all specifications and events that might ever arise in any small-probability specification chance-elimination argument conducted across the observed physical universe. Any item on that list will require the activity of at least one proton over the time it takes light to cross its diameter. That seems like an utterly reasonable and absolutely minimal constraint on the specifications and events on which our approach to chance elimination depends. Most such specifications will require multiple protons over time periods much greater than the time it takes for light to cross the diameter of a proton.

Some of these specifications will induce rejection regions, and these rejection regions will likewise require proton activity and make it into the cosmic history. The takeaway here is that only so many specifications, only so many specification-induced rejection regions, and only so many ways of matching them with events can in fact be realized and make it into this cosmic history. The figure

10^{150} sets a hard limit on the number of list items in this cosmic history. Moreover, any universal probability bound based on these probabilistic resources will make matching a specification-induced rejection region with an event so improbable that we should never expect to see it by chance even if the universe spent itself on nothing else than trying to match that rejection region.

4.6 Universal and Local Probability Bounds

We now round out this chapter with a deeper dive into probability bounds. A simple rule of thumb connects probabilistic resources to their corresponding probability bounds: For probabilistic resources of size M, take the corresponding probability bound to be $1/M$. We then say that if the probabilistic resources are absolute, the corresponding probability bound is *universal*. Otherwise, we say the probabilistic resources are relative and the corresponding probability bound is *local*.[22] The underlying idea here is that events of probability less than such a probability bound (whether universal or local) are so unlikely that even with all the probabilistic resources that could realistically be factored in, the events would still be improbable to the point of implausibility. We've seen this connection between probabilistic resources and probability bounds already, back in section 1.2, with Borel's universal probability bound of 1 in 10^{50} corresponding to probabilistic resources of 10^{50}. We now generalize this connection.

Why does this simple relationship between probabilistic resources and probability bounds, in which one is the reciprocal of the other, work to rule out chance? In identifying probabilistic resources, whether absolute or relative, we always err on the side of excess or generosity, overestimating the number of opportunities that would, practically speaking, be available. This is a conservative approach, and we've applied it consistently to the probabilistic resources discussed earlier in this chapter. For instance, in assigning probabilistic resources based on computational firepower, we've assumed that computations can be performed in the time it takes light to cross the diameter of a proton (or even a Planck

length). No computations currently available or realistically imagined operate anywhere close to this speed. Or, in assigning probabilistic resources based on the earth's biology, we've assumed that
organisms could potentially reproduce every second. No organisms
reproduce anywhere close to this fast.

By consistently erring on the side of excess or generosity, and
thereby significantly overestimating the probabilistic resources that
might realistically be in operation, the number of probabilistic
resources calculated ends up being far more than the opportunities
that might actually arise for the event in question to occur by
chance. In consequence, a $1/M$ probability ensures that even if all
available opportunities of probability $1/M$ or less are factored into
the chance occurrence of an event, the number of those
opportunities, call it N, will be orders of magnitude less than M (in
quasi-rigorous notation, $N \ll M$). The event will therefore be
unlikely to occur, not just with probability less than ½ but with
probability orders of magnitude less than ½. Events of probability
$p = 1/M$ should therefore be regarded as highly improbable even in
light of whatever probabilistic resources may realistically be
factored in.

A probability of ½ constitutes an important cutoff in assessing
the chance occurrence of events in light of a given set of
probabilistic resources. For instance, this cutoff plays a pivotal role
in betting strategies: events of probability less than ½ are not worth
betting on if the point is to guess simply whether the event will
happen (assuming the payoff is even). Events whose probabilities
are less than ½, even when all relevant probabilistic resources are
factored in, are less likely than not to happen, and therefore should
not be expected to happen. Thus, if we increase the estimated
probabilistic resources by a factor of 100, as we did by raising 10^{48}
to 10^{50} with biologically based absolute probabilistic resources, it
becomes that much more unlikely for an event of probability 1 in
10^{50} to occur by chance—two orders of magnitude more. An event
of probability less than ½ for 10^{50} probabilistic resources becomes
an event of probability less than $1/200$ for 10^{48} probabilistic
resources.

The relevant math here is elementary and straightforward. Note that it does not assume that the events are probabilistically independent (Appendix A.2). It does assume that each opportunity yields an event of probability less than or equal to $p = 1/M$. Moreover, it assumes that the number of opportunities to match an event in question is N and that N is much less than the probabilistic resources M (orders of magnitude less because we always vastly overestimate M in relation to probabilistic resources that in fact are feasible). Given $p = 1/M$, we then find that N events each of probability no more than p cannot, when considered together, yield a probability of more than $N \times p = N/M$. N/M will therefore be orders of magnitude less than $\frac{1}{2}$ because M vastly overestimates the probabilistic resources that may actually be in play and thus will be much bigger than N.

More formally, the probabilistic resources in such a situation allow that there could be N events, denoted by $E_1, E_2, \ldots, E_N$, each of probability no more than $p = 1/M$ that match the event in question. The probability of matching the event in question with any of these events will thus be the probability of the union (or disjunction) of these events—namely, $P(E_1 \cup E_2 \cup \ldots \cup E_N)$, which by the rules of probability will be less than or equal to $P(E_1) + P(E_2) + \ldots + P(E_N)$, which in turn will be less than or equal to $N \times p = N/M$ because all the terms in this sum are, by assumption, less than or equal to p. Moreover, we are assuming that $N/M \ll \frac{1}{2}$.

Probability bounds, whether universal or local, are thus straightforwardly calculated on the basis of probabilistic resources. As noted, absolute probabilistic resources determine universal probability bounds; relative probabilistic resources determine local probability bounds. We all have experience of local probability bounds and the relative probabilistic resources on which they depend. Take, for instance, logging into an online account. Let's say the username offers no security (suppose usernames are common knowledge), so all security is in the password. Let's say the password is unknown and has only eight decimal digits (thus covering a mere 10^8 or 100 million possibilities). Local probability bounds then gauge the security of these passwords.

Is such a password secure? Certainly not for any probabilistic resources numbering 10^8 or more. But what if three mistaken attempts at logging in, whether by a single user or a combination of users, shut down the account and require it to be manually restarted? In that case, the relevant probabilistic resources have the conspicuously small number of 3. If there are only 3 opportunities across users to guess an eight-decimal password before the account is disabled, and assuming the password was selected randomly, a simple calculation shows that the chance of logging into the account by chance is less than 1 in 10 million (1 in 10^7). That's a level of security many people would regard as acceptable (if not for one's bank account, then perhaps for a magazine subscription that's behind a paywall). Such a level of security would correspond to a local probability bound.

With relative probabilistic resources, and thus with their corresponding local probability bounds, we always need to make clear who are the relevant stakeholders and what their interests and needs are in ruling out chance. A subscriber to an online magazine behind a paywall, where users are blocked once they mistakenly punch in the wrong password three times, may be entirely happy with probabilistic resources at the level of 10^8, unworried about the account being hacked by random guessing. But the owner of the magazine, especially if the magazine has millions of subscribers, will want to ensure that none of these subscribers has their passwords randomly hacked. The owner will thus require a level of security far more stringent than needed by individual subscribers.

Such a higher level of security, as desired by the owner, will factor in considerably more relative probabilistic resources than 10^8, and thus induce a considerably smaller (and thus more stringent) local probability bound. The owner might therefore require a password that's at least 12 characters long and includes numbers, special characters, and letters both capitalized and lower case. Such a move vastly increases the number of probabilistic resources needed to randomly hack a password. Prudently implemented, this move will ensure security against random hacking sitewide. (The usual provisos, of course, apply, such as users need to avoid over-obvious passwords and also to guard against passwords being stolen or compromised.)

All these insights that apply to local probability bounds now apply to universal probability bounds in relation to their corresponding absolute probabilistic resources. As always, we choose our probabilistic resources generously. Thus, the probabilistic resources will vastly exceed the number of opportunities realistically available for an event in question to come about by chance. As noted and justified in the last section, we agree upon 10^{150} as our go-to set of probabilistic resources at the scale of the observable universe, and thus as our preferred set of absolute probabilistic resources. Given our reciprocal rule for relating probabilistic resources and probability bounds, this means that our go-to universal probability bound comes to 1 in 10^{150}. Events at this scale of improbability are improbable across the entire known physical universe.

CHAPTER 5

THE LOGIC OF THE DESIGN INFERENCE

MOST DESIGN INFERENCES ARE DRAWN INTUITIVELY. OUR ASSESS-ments of specifiability (whether an event answers to a short description) and improbability (whether an event falls within a small probability rejection region) tend to happen quickly without precise numerical calculations. Thus, when you walk into a children's room that was previously in a chaotic state and find it organized and neatly arranged, you immediately conclude that someone (other than the children who made the mess in the first place) carefully and deliberately cleaned it up. You do not stop to evaluate the exact improbability or specifiability of a neatly arranged room.

Likewise, when you come across a chess board with all the pieces correctly placed at their starting positions, you do not need to calculate the probability of the particular arrangement of pieces or the length of a description for that placement. Rather, you know intuitively that that arrangement is extremely unlikely to have occurred by chance and that the arrangement is simply described ("the starting chess position"). Implicitly, you do evaluate the improbability and specifiability of the arrangement, but you make no explicit calculations.

If you find a long list of words written in alphabetical order, you conclude that somebody deliberately alphabetized the list. You do not have to stop and calculate the probability of the particular arrangement occurring by chance. Nor do you have to estimate the length of the description of that pattern. You know, intuitively, that the probability is very low and the description is short and simple. Making precise calculations is unnecessary and might even cause others to doubt your sanity.

Such intuitive judgments of improbability and specifiability are valuable for drawing design inferences in everyday life. Even so, they are not always sufficient. People may disagree about whether an event is indeed improbable and specified, in which case a design inference may be disputed. To demonstrate that the inference is valid in the face of such disagreement, there must be at least a rough estimate of probability and of description length. For example, when Nicholas Caputo was accused of fraudulently selecting ballot orders (see Section 5.1), it was insufficient that the result he claimed to obtain by chance intuitively suggested specification and small probability. Rather, because he was brought to trial, the specification needed to be made explicit and the probability had to be calculated.

With specification and probabilistic resources under our belts, we are now in a position to go beyond mere intuition and, instead, lay out the logic of the design inference. In this chapter, we there-fore bring formal rigor to the design inference. We'll start with a detailed example, namely, the case of Nicholas Caputo, which has been briefly mentioned already. Then we'll consider a general schema for chance elimination arguments, namely, the Generic Chance Elimination Argument. Next, we'll turn to the actual design inference, which deploys the Generic Chance Elimination Argu-ment to sweep the field clear of all relevant chance hypotheses, thereby justifying a conclusion of design. And finally in this chapter, we'll consider some implications and applications of this design-inferential logic.

Although the previous chapters have identified all the key pieces for formulating and drawing design inferences, one key piece remains to be fully explained—namely, the role of language and description lengths in defining specification. A thorough discussion of this topic, especially a precise characterization of how language and minimum description lengths figure into the concept of specification, is deferred until the next chapter. There we combine specification and small probability into a unified information measure called *algorithmic specified complexity*, or simply *specified complexity*. Because of the technical challenges this material presents, we thought it better, pedagogically, to lay out the overall logic of the design inference here, and then address these more technical matters afterward.

5.1 The Man with the Golden Arm

The design inference appears widely in human inquiry, even if it doesn't go by that name and even if it is used without being fully explained or formalized (the goal of this book being, of course, to provide such an explication). So far, we've given some quick overviews of different design inferences (notably in Chapter 2). In addition, we've laid out the two key ingredients needed to make a design inference work, namely, the low description length patterns that constitute the specifications (Chapter 3) and the probabilistic resources that gauge what it means for a probability to be "small" (Chapter 4). As a first detailed example of the design inference, let us consider the following case of alleged election fraud:

> TRENTON, July 22 — The New Jersey Supreme Court today caught up with the "man with the golden arm," Nicholas Caputo, the Essex County Clerk and a Democrat who has conducted drawings for decades that have given Democrats the top ballot line in the county 40 out of 41 times.
>
> Mary V. Mochary, the Republican Senate candidate, and county Republican officials filed a suit after Mr. Caputo pulled the Democrat's name again last year.
>
> The election is over—Mrs. Mochary lost—and the point is moot. But the court noted that the chances of picking the same name 40 out of 41 times were less than 1 in 50 billion. It said that "confronted with these odds, few

persons of reason will accept the explanation of blind chance."

And, while the court said it was not accusing Mr. Caputo of anything, it said it believed that election officials have a duty to strengthen public confidence in the election process after such a string of "coincidences."

The court suggested—but did not order—changes in the way Mr. Caputo conducts the drawings to stem "further loss of public confidence in the integrity of the electoral process."

...

Justice Robert L. Clifford, while concurring with the 6-to-0 ruling, said the guidelines should have been ordered instead of suggested.[1]

The Republican party claimed that Caputo had consistently rigged the ballot line in the New Jersey county where he was clerk. As is widely known, first position on a ballot increases one's chances of winning an election (other things being equal, voters are more likely to vote for the first person on a ballot than the rest). Since, except for one instance, Caputo positioned the Democrats first on the ballot line, the Republicans argued that in selecting the order of ballots, Caputo had intentionally favored his own Democratic party. In short, the Republicans claimed Caputo cheated. The advantage of cheating, of course, is that it leaves nothing to chance. Short of catching cheaters red-handed or gaining their confession, we need a design inference to catch them (see Section 2.3 on data falsification in science).

The question, then, before the New Jersey Supreme Court was, Did Caputo actually rig the order of ballots, or was it without malice and forethought that Caputo assigned the Democrats first place 40 out of 41 times? Caputo denied wrongdoing. At the same time, he conducted the drawing of ballots so that witnesses were unable to observe how he actually did draw the ballots (this was brought out in a portion of the article omitted in the preceding quote). So, if he did cheat, he at least had the presence of mind to avoid getting caught red-handed. Nor did he ever admit to cheating.

Thus, determining whether Caputo did in fact rig the order of ballots becomes a matter of evaluating the circumstantial evidence connected with this case. How then is this evidence to be evaluated?

In trying to explain the remarkable coincidence of Nicholas Caputo selecting the Democrats 40 out of 41 times to head the ballot line, the court faced two explanatory options. The first option was chance: In selecting the order of political parties on the state ballot, Caputo employed a reliable random process that did not favor one political party over another. The fact that the Democrats came out on top 40 out of 41 times was simply a fluke. The other option was design: Caputo, acting as a fully conscious intelligent agent and intending to aid his own political party, rigged the ballot line selections to keep the Democrats coming out on top. That is, Caputo cheated.

Caputo told the *New York Times* that he placed capsules designating the various political parties running in New Jersey into a container, swished them around, and then chose one at random.[2] In other words, Caputo claimed to have used an urn model. Since urn models are among the most reliable randomization techniques available, the court had no reason to suspect that Caputo's randomization procedure was at fault. The key question, therefore, was whether Caputo actually put this procedure into practice when he made the ballot line selections, or merely claimed to while in fact rigging the ballot order to favor his party.

Having noted that the probability of picking the same political party at least 40 out of 41 times was less than 1 in 50 billion, the court concluded that "confronted with these odds, few persons of reason will accept the explanation of blind chance." Now this certainly seems right. Nevertheless, more needs to be said. As we've stressed since the start of this book, extreme improbability is by itself not enough to preclude an event from happening by chance. Anyone dealt a bridge hand participates in an exceedingly improbable event. Anyone who plays darts will find that the precise position where the darts land represents an exceedingly improbable configuration. In fact, just about anything that happens is exceedingly improbable once we factor in all the other ways what actually happened might have happened. The problem, then, does not reside simply in an event being improbable.

All the same, in the absence of a verified causal story detailing what happened, improbability remains a crucial ingredient in eliminating chance. Suppose that Caputo actually was cheating right from the beginning of his career as Essex County Clerk. Suppose, further, that the one exception in Caputo's career as "the man with the golden arm" (i.e., the one case where Caputo placed the Democrats second on the ballot line) did not occur till after his third time selecting ballot lines. Thus, for the first three ballot line selections of Caputo's career, the Democrats all came out on top, and they came out on top precisely because Caputo intended (designed) it that way. Simply on the basis of three ballot line selections, and without direct evidence of Caputo's cheating, however, an outside observer would be in no position to determine whether Caputo was cheating or selecting the ballots honestly.

With only three ballot line selections, the probabilities are too large to reliably eliminate chance. The probability of randomly selecting the Democrats to come out on top given that their only competition is the Republicans is in this case 1 in 8 (here p equals 0.125; compare this with the p-value computed by the court of 0.00000000002, or 1 in 50 billion). Because three-Democrats-in-a-row could easily happen by chance, we would be acting in bad faith if we did not give Caputo the benefit of the doubt in the face of such large probabilities. Small probabilities are therefore a necessary condition for eliminating chance, but they are not a sufficient condition.

What, then, besides small probability do we need for evidence that Caputo cheated? As stressed throughout the previous chapters, the event in question also needs to conform to a pattern. But not just any pattern will do. Some patterns successfully eliminate chance while others do not. Suppose an archer stands 50 meters from a large wall with bow and arrow in hand. The wall is so large that the archer cannot help but hit it. Now suppose every time the archer shoots an arrow at the wall, a target is painted around the arrow so that the arrow is positioned squarely in the bull's-eye. What can be concluded from this scenario? Absolutely nothing about the archer's ability as an archer. For all we can tell, all the archer's shots at the wall may be purely random.

But suppose instead that a fixed target is placed on the wall, and the archer then shoots at it. Suppose the archer shoots one hundred arrows, and each time hits a perfect bull's-eye. What can be concluded from this second scenario? Such a performance would be highly improbable if it happened by chance. Or, in the words of the New Jersey Supreme Court, "confronted with these odds, few persons of reason will accept the explanation of blind chance." Indeed, confronted with this second scenario, we would infer that here is a world-class archer. The difference between the first and second scenario is, of course, that only the pattern in the second is a specification, and thus able in the presence of small probability to eliminate chance.

Let's now consider more closely how specification and small probability play out in the case of Nicholas Caputo. By selecting the Democrats to head the ballot 40 out of 41 times, Caputo, according to the New Jersey Supreme Court, participated in an event of probability less than 1 in 50 billion ($p = 0.00000000002$). But how exactly was that improbability calculated? To reconstruct the court's reasoning, consider the possibility space Ω consisting of all sequences of Ds and Rs that are 41 characters in length ("D" for Democrat first on the ballot line, "R" for Republican first on the ballot line). A typical member of Ω will now look as follows:

DRRDRDRRDDRDRDDRDRRDRRDRRRDRRRDRDDDRDRDD

In this sequence of Ds and Rs, the Democrats came out on top only 20 times, and the Republicans 21 times. As it is, this sequence was formed by tossing a fair coin, assigning D to heads and R to tails. There are 2^{41}, or approximately 2.2 trillion, such sequences of Ds and Rs. Given that any such sequence of ballot line selections is equally probable, as it would have to be if the electoral process were fair, it follows that any such sequence of 41 Ds and Rs has probability 1 in 2^{41}. If this last sequence of Ds and Rs had been Caputo's actual ballot lines, there would, obviously, have been no court trial to determine whether he cheated. So, what was it about his actual ballot lines that inspired litigation going all the way up to the New Jersey Supreme Court?

The *New York Times* did not say in which election Caputo gave the top ballot line to the Republicans, only that there was just one. For definiteness, let's suppose that it was the 23rd election, resulting in the following sequence of Ds and Rs:

DDDDDDDDDDDDDDDDDDDDDDRDDDDDDDDDDDDDDDDDD

That is, Caputo chose the Democrats to head the ballot line the first 22 times, the Republicans the 23rd time, and then the Democrats again for the remaining 18 times. We'll call this event E.

The probability of E, like that of all the other sequences consisting of 41 Ds and Rs, is 1 in 2^{41}, or approximately 1 in 2.2 trillion. How, then, did the New Jersey Supreme Court arrive at a 1 in 50 billion probability, which considerably exceeds 1 in 2.2 trillion? The answer can be gathered from the following statement in the *New York Times* article: "The court noted that the chances of picking the same name 40 out of 41 times were less than 1 in 50 billion." Clearly, the court was counting the number of times the Democrats appeared first on the ballot line. In other words, it was calculating the probability of the broader event F that included E and that consisted of all sequences with 40 or more Ds.

F consists of 42 different outcomes, namely, one outcome consisting of all Ds, and 41 outcomes consisting of all Ds except for one R. All these outcomes are mutually exclusive, and so the probability of F is calculated by adding 2^{-41} to itself 42 times, which is just 42×2^{-41}. This last number comes to just under 1 in 50 billion. It is this event F and this probability for F that led the court to reject that Caputo's ballot line selections were the result of chance, deeming blind chance as an unacceptable explanation of such a coincidence.

That explains how the court calculated the probability in this case. But in a design inference, there also needs to be a specification. On the role of specification, the court's reasoning was less forthcoming. It's clear that the court regarded F as exhibiting a suspicious pattern. The court thus cited a string of "coincidences" the odds of which were such that "few persons of reason [would] accept the explanation of blind chance." But without a clear concept of specification, the court did the best that it could.

A number of ways exist to see that F (and thus E, which is included in F) is specified. One is to note that F has the short description "at least 40 Ds," although this will then require forming a specification-induced rejection region that includes but doesn't necessarily coincide with F. Specifically, we'll need a basis specification V such that $V^* = F$, with V corresponding to a short description such as "at least 40 Ds," and then expand V^* to a full-fledged rejection region R that includes F but also more than F (see Sections 3.6 and 3.7). That rejection region R would, presumably, have a probability much bigger than 1 in 50 billion. This is not the approach the court took, even tacitly.

Instead, the preferred approach, and the one that the court implicitly used, is to note that F, consisting as it does of 40 or more Ds in a sequence of 41 Ds and Rs, is the tail (a one-sided tail) of a probability distribution. The distribution here is a binomial distribution with $n = 41$ and $p = .5$. It counts and assigns probabilities to the total number of Ds in a sequence of length 41. In consequence, F is one of the two tails of this binomial distribution. F is therefore an event denoted by not just a specification but a type of specification whose corresponding event F equates with a specification-induced rejection region R of the sort suitable for eliminating chance. Specifically, R is identified through the modes and tails approach to rejection regions described in Section 3.6.

In any event, the main question left unanswered by the court is not the probability of the rejection region R or its underlying specification, but why a 1 in 50 billion probability should be regarded as small enough to justify a design inference in this case. A 1 in 50 billion improbability will strike most "persons of reason," in the court's manner of speaking, as staggeringly small. But, as stressed in Chapter 4, small probabilities are small only in relation to the relevant probabilistic resources, which give the number of opportunities to witness an event like it.

Nicholas Caputo was the Essex County Clerk in New Jersey responsible for the ballot lines in his county. But how long have county clerks been setting the ballot lines in their counties? And how many counties are there with elections that require determining ballot lines? If billions of county clerks over billions of years have been setting the ballot lines in their counties, then upstanding county clerks should expect on occasion to see 40 out of 41 ballot lines go to their party. In that case, with sufficiently many probabilistic resources, the court would be in its rights to give Caputo the benefit of the doubt, dismissing the 1 in 50 billion improbability that might otherwise suggest he was cheating.

What in fact are the relevant probabilistic resources in the Caputo case? In the United States, there are 3,143 counties or county equivalents. If we imagine an election every month since the Constitution of the United States went into operation in 1789, that means $12 \times 234 = 2,808$ county election cycles to the present (a vast overestimation of election cycles). Thus, in the United States, there would have been at most $3,143 \times 2,808 = 8,825,544$ elections where ballot lines could be selected.

An upper-bound estimate on the probabilistic resources for Caputo getting at least 40 Ds in 41 ballot lines would now treat every one of these 8,825,544 elections as an opportunity to see at least 40 out of 41 Ds. This is in fact a vast overestimation since 40 out of 41 Ds will require 41 elections, not just one. A simple calculation now shows that with all these 8 million plus opportunities factored in, getting 40 out of 41 Ds in 41 elections, even just once, comes to roughly .000177, or less than 1 chance in 5,000.[3] Even with all these probabilistic resources factored in, we should therefore not expect to see ballot line drawings as extremely lopsided as Caputo's.

Did we factor in enough probabilistic resources? As indicated in Chapter 4, for relative probabilistic resources and local probability bounds, these are always set pragmatically to avoid both false positives and false negatives (i.e., mistakenly eliminating chance when it is operating and mistakenly retaining chance when it is not operating). By increasing probabilistic resources enough, any seemingly small probability becomes so large as to retain (or avoid eliminating) chance. Still, we could ask what would happen if we factor in all two-party ballot elections worldwide, thereby increasing our probabilistic resources still further, albeit within a reasonable limit.

The U.S. is the oldest representative democracy, and the U.S. is 4 percent of the world's population. A reasonable estimate is thus that, in the world as a whole, there would have been at most $(3,143 \times 2,808)/.04 = 220,638,600$ elections where ballot lines could be selected. The same calculation as before now shows that with all these 220 million plus opportunities factored in, getting 40 out of 41 Ds in 41 elections, even just once, comes to roughly .00441, or less than 1 chance in 225.

So, are there enough global probabilistic resources to point to chance as the reason for Caputo's ballot lines? Clearly no. Our initial focus with Caputo was on the U.S. and not on the world as a whole. For a U.S.-based court trying to determine whether a given sequence of ballot lines was fairly selected by chance, the probabilistic resources we gave (in terms of number of counties and number of possible county elections) seemed extremely generous. Thus, if the New Jersey Supreme Court were working with our full-fledged design-inferential apparatus, it might well have identified the U.S.-based probabilistic resources that we did, and then applied them to conclude that getting 40 out of 41 Ds in 41 elections, even just once, had probability of less than 1 in 5,000 and that this was enough to secure a design inference. But even on a global scale, the probability of Caputo's ballot selections is less than 1 in 225, making the odds remain suspiciously slim.

Let's now review where we are with the Caputo example. Faced with Nicholas Caputo assigning the Democrats the top ballot line 40 out of 41 times, the New Jersey Supreme Court rejected the chance explanation ("confronted with these odds, few persons of reason will accept the explanation of blind chance"). Only one chance explanation was on the table, namely that the Ds and Rs were probabilistically independent and equally probable. To eliminate chance therefore seemed inescapably to point to design. Left with no other option, the court therefore drew a design inference, concluded that Caputo was cheating, and threw him in jail.

Well, not quite. The court did refuse to attribute Caputo's golden arm to chance. Yet when it came to giving a positive explanation of Caputo's golden arm, the court waffled. The court knew something was amiss. For the Democrats to get the top ballot line in Caputo's county 40 out of 41 times, especially with Caputo solely responsible for the ballot line selections, something had to be fishy.

Nevertheless, the New Jersey Supreme Court was unwilling to explicitly charge Caputo with corruption. Of the six judges, Justice Robert L. Clifford was the most suspicious of Caputo, wanting to *order* Caputo to institute new guidelines for selecting ballot lines. The actual ruling, however, simply *suggested* that Caputo institute new guidelines in the interest of "public confidence in the integrity of the electoral process." The court therefore stopped short charging Caputo with outright dishonesty.

Did Caputo cheat? Certainly, that seems to be the best explanation of Caputo's golden arm. Nonetheless, the court stopped short of convicting or even reproving Caputo. In the court's defense, it had no clear mandate for dealing with highly improbable ballot line selections. Such mandates exist in other legal settings, as with discrimination laws that prevent employers from appealing to chance in trying to justify demographic disparities in hiring that are too extreme. But in the absence of such a mandate, the court needed a verified causal story of how Caputo cheated if the suit against him was to succeed. And since Caputo obscured how he selected the ballot lines, no such causal story was forthcoming. The court therefore went as far as it could.[4]

The court adopted a Fisherian approach to probabilistic reasoning, rejecting the chance hypothesis in response to an event falling inside a small-probability specification-induced rejection region. In a Bayesian approach to probabilistic reasoning, the analysis would proceed differently. It would attempt to weigh the relative merits of a chance hypothesis H (according to which Caputo used a fair random process to select the ballot lines) against a design hypothesis D (according to which Caputo cheated). In the

Bayesian approach, the event E, namely, the observed sequence of ballot lines, would then be used to adjudicate between these two hypotheses.

In particular, the Bayesian probabilist would focus on how the likelihood ratio $P(E|H)/P(E|D)$ changes the evidential support for H and D by means of the following key equation relating prior and posterior probabilities (see Appendix A.4):

$$\frac{P(H|E)}{P(D|E)} = \frac{P(E|H)}{P(E|D)} \times \frac{P(H)}{P(D)}.$$

Neither the prior probability of H nor the prior probability of D will be close to zero. We assume that politicians are normally honest in selecting ballot lines (if only because they don't want to get caught tampering with them). So, $P(H)$ will be large and reasonably close to one. But politicians have also been known to cheat to aid their own party, so $P(D)$ will be non-trivial. It will be smaller than $P(H)$, but it won't be anywhere near zero. Yet given how extreme E is in favoring Caputo's own Democratic party $P(E|D)$ will be much larger than $P(E|H)$, the latter being 2^{-41}.

The likelihood ratio $P(E|H)/P(E|D)$ will therefore significantly favor D over H. This means that compared to the prior probability $P(D)$, the posterior probability $P(D|E)$ will go up a lot. And it also means that compared to the prior probability $P(H)$, the posterior probability $P(H|E)$ will go down a lot. Will these posteriors change enough that their ratio is less than one, thereby causing us to prefer D over H? Bayesians will need to run the numbers, and those numbers will depend on the degree to which E seems suspicious and self-serving. In most people's estimation, it seems much more likely that E was produced by something other than fair random chance.

This is how Bayesians would argue that Caputo cheated. But as we argue throughout this book, Bayesians also implicitly use specification. The event E matches a salient, easily described pattern. That's why it seems suspicious and raises red flags in the first place. Moreover, the event seems clearly self-serving, aiding Caputo's own Democratic party to the detriment of the Republicans.

All these considerations confer a substantial likelihood on the design hypothesis $(P(E|D))$, and thus substantially raises its posterior probability $(P(D|E))$. Most Bayesians will thus conclude that Caputo was cheating, updating their posterior probabilities to seriously favor D over H. Strict Bayesians may come short of outright rejecting H, always seeing comparison of hypotheses as a matter of degree, dependent on how likelihood ratios translate evidential support from prior to posterior probabilities. But practically speaking, Bayesians and Fisherians reach the same conclusion in the Caputo case.

One final point needs to be made before we move ahead. Design inferences as characterized in this book require a convergence of suitable patterns (specifications as determined by short description length) and suitable probabilities (small probabilities as determined by relevant probabilistic resources). As such, a design inference makes a circumstantial argument. It is a form of effect-to-cause reasoning, noting that certain types of effects exhibit certain features (specification and small probability), and from there concluding that a designing intelligence was responsible for those features.

Yet as emphasized earlier, this form of reasoning offers no insight into the particulars of the intelligence involved or its actions. Even in the present example, where a design inference can be validly drawn for Caputo's ballot line selections, the design inference cannot decisively identify Caputo as the cheating intelligence that deliberately circumvented chance. Certainly, that's the most plausible explanation. But perhaps Caputo's assistant sabotaged the capsules that Caputo swished around to determine the ballot line selections. Caputo in that case would have been trying to be honest, but his assistant would have been cheating.

Short of precise causal details of what actually happened, a design inference therefore tells us that an intelligence was involved and acted to contravene chance. But a design inference cannot by itself tell us which intelligence was involved or how it was involved. To answer such questions requires further inquiry and additional evidence. Design inferences should therefore be viewed not as the end but as the start of inquiry. Once a design inference is drawn, we need to dig deeper into the designer's identity, activity, and motives.

5.2 The Generic Chance Elimination Argument (GCEA)

The Generic Chance Elimination Argument (or GCEA) lays out the common form of reasoning that underlies chance-elimination arguments. Chance elimination arguments in general and design inferences in particular are drawn by subjects in light of their background knowledge, context of inquiry, needs, and interests. The GCEA is therefore relativized to subjects (human or, in general, finite rational agents). Drawing on concepts explained in Chapters 3 and 4, the GCEA takes the following form:

#1 A subject S learns that an event E has occurred and notes that E belongs to a reference class of possibilities Ω.

#2 In light of the circumstances under which E occurred, S identifies a chance hypothesis H and a probability measure P—and thus a probability distribution $P(\cdot|H)$ over Ω—that could have been operating to produce E.

#3 S, in light of a language $\mathscr{L}$ and a description length metric $|\cdot|$ defined over $\mathscr{L}$ (and thus in light of a minimum description length metric $D(\cdot)$), sets an upper bound m on the length of descriptions that S will count as short enough to constitute specifications. S also calculates $M = L(m)$, the number of descriptions in $\mathscr{L}$ with minimum description length less than or equal to m.

#4 S discovers a description V in $\mathscr{L}$ and notes that E conforms to V (i.e., V^*, the event identified by V, includes E); at the same time, S confirms that the length $|V|$ is bounded above by m (i.e., $|V| \leq m$). V is thus a specification.

#5 Based on probabilistic resources relevant to S's needs and interests in assessing whether E happened by chance (and, in particular, to avoid mistakenly rejecting H when H actually is responsible for E), S sets an upper bound α (a local or universal probability bound) on the size of probabilities below which they will be considered as small.

#6 S identifies a specification-induced rejection region R that is a subset of Ω, that depends on m, α, V and H, that includes V^*, and that factors in all specifications relevant to assessing E's chance occurrence with respect to H.

#7 S determines that $P(R|H) \leq \alpha$.

#8 S is warranted in inferring that E did not occur according to the chance hypothesis H.

The GCEA encapsulates everything from Chapters 3 and 4 on how chance elimination arguments use specifications and small probabilities to eliminate chance. Its justification was accomplished in those chapters. We could recap that justification here by running through the GCEA point by point in abstract theoretical terms. But, in our view, better insight is gained by providing a detailed example that illustrates the GCEA. Richard Swinburne offers such an example in critiquing the selection-effect interpretation of the anthropic principle.

The anthropic principle states that the universe is structured to allow for the existence of life, and in particular of intelligent observers like ourselves. As such, the anthropic principle is a truism—how could it not be true since we inhabit just such a universe? To the anthropic principle, the selection-effect interpretation adds that the conditions allowing intelligent life to exist require no special explanation because only in a universe satisfying those conditions would we be capable of observing and questioning those conditions in the first place.

The selection effect attempts to make sense of the surprise that a lottery winner experiences in winning a lottery, balancing that surprise against the fact that someone was bound to win the lottery. We seem to be extremely fortunate to have a home in this universe. But we wouldn't exist without such a home. The selection effect thus plays down the surprisal in being the lottery winner, acknowledging that the surprisal may be (emotively) real but then affirming that nothing probabilistically inappropriate is happening to cancel a valid chance explanation. In critiquing this interpretation of the anthropic principle, Swinburne tells the following story about a mad kidnapper:

> Suppose that a madman kidnaps a victim and shuts him in a room with a card-shuffling machine. The machine shuffles ten packs of cards simultaneously and then draws a card from each pack and exhibits simultaneously the ten cards. The kidnapper tells the victim that he will shortly set the machine to work and it will exhibit its first draw, but that, unless the draw consists of an ace of hearts from each pack, the machine will simultaneously set off an explosion that will kill the victim, in consequence of which he will not see which cards the machine drew. The machine is then set to work, and to the amazement and relief of the victim the machine exhibits an ace of hearts drawn from each pack. The victim thinks that this extraordinary fact needs an explanation in terms of the machine having been rigged in some way. But the kidnapper, who now reappears, casts doubt on this suggestion. "You ought not to be surprised," he says, "that the machine draws only aces of hearts. You could not possibly see anything else. For you would not be here to see anything at all, if any other cards had been

drawn." But of course the victim is right and the kidnapper is wrong. There is indeed something extraordinary in need of explanation in ten aces of hearts being drawn. The fact that this peculiar order is a necessary condition of the draw being perceived at all makes what is perceived no less extraordinary and in need of explanation.[5]

The kidnapper explains the kidnap victim's survival by appealing to chance. The kidnap victim thinks this is absurd. Who's right? To settle this question, let us recast this example as a Generic Chance Elimination Argument. We begin by identifying the kidnap victim with a subject S who is clearly interested in eliminating chance if it deserves to be eliminated, and thus is eager to determine whether it does indeed deserve to be eliminated. The underlying probability space Ω consists of all possible sequences of playing cards, each sequence being ten in length. The event E, whose explanatory status is in question, is then the drawing of an ace of hearts from each of the ten decks of playing cards by the card-shuffling machine. To S's relief, S learns that E did indeed happen and thus that S's life has been spared. (Hence #1.) The mad kidnapper, however, wants to convince S that E happened by chance (contrary to #8).

To refute the kidnapper's claim, S proceeds step by step through the Generic Chance Elimination Argument. By inspecting the card-shuffling machine, S determines that so long as there is no tampering, each of the ten decks was shuffled independently (i.e., shuffling one deck does not affect the others), and that within a deck, each card had the same probability of being drawn (we assume a standard deck of playing cards consisting of 52 cards, so that the probability of the ace of hearts being drawn from any one deck is 1/52). The card-shuffling device is thus a chance mechanism that operates according to a chance hypothesis H for which the ten decks are stochastically independent and each card from a given deck is equiprobable. (Hence #2.)

By informing S in advance of precisely which event from the card-shuffling machine will save S's life (namely, the ten aces of hearts), the kidnapper provides S with a prespecification V that matches E.[6] V is presumed to belong to a language $\mathscr{L}$, which we'll take to be English and for which the description length metric simply counts the number of words. We can take V to correspond to "ten aces of hearts," in which case V not only describes E but identifies it exactly, so that $V^* = E$.

Let's assume that $\mathscr{L}$ consists of 100,000 base words and that any finite concatenation of words constitutes a description. Let's say that S is being stingy with what S is willing to count as a specification and thus sets $m = 5$ as the maximum description length complexity for a description to count as a specification. As it is, m is dispensable in this analysis because V is a prespecification. If V were treated as the basis specification for a specification-induced rejection region, we would need to make use of it—and we'll provide some pointers about how we would have made use of it. As it is, $|V| = 4 \leq m$. Also, $M = L(m)$ is roughly $100,000^5$, or 10^{25}. (Hence #3 and #4.)

Even though S intuitively rejects the kidnapper's claim that E happened by chance, S doesn't want to be hasty. After all, highly improbable events happen by chance all the time. What if S is not the only victim ever kidnapped by the madman. Suppose that prior to S being kidnapped, the madman kidnapped multitudes of other hapless victims, placing them all inside rooms with identical card-shuffling machines, and that in each case the card-shuffling machine failed to deliver ten aces of hearts, thereby exploding and killing the victims. S might therefore be an incredibly lucky survivor amidst a stream of carnage.

And let's not forget all the other mad kidnappers out there who didn't imprison their victims next to exploding card-shuffling machines, but still subjected them to probabilistic experiments in which only an incredibly unlikely prespecified event would spare the victim. When all these other victims are factored in, might not S's luck be properly explained as a selection effect, as with what happens in a lottery? After all, even though it's highly unlikely any one individual will win a lottery, that the lottery should have a winner tends to be assured. Lottery winners are invariably surprised by their own good fortune at winning a lottery, but their sense of surprise is hardly a reason to doubt that the lottery was run fairly and that its outcome was therefore due to chance.

Throughout recorded history, the number of humans has not exceeded a trillion, or 10^{12}. S sees this number as vastly over-estimating the number of human beings who might be kidnapped by mad kidnappers and placed inside rooms with randomized killing machines. S therefore sees 10^{12} as a highly generous upper bound on the number of opportunities for the event E to occur, E being the event of drawing an ace of hearts from each ten packs of cards and thus preventing the card-shuffling machine from exploding. In line with our rule that probabilistic resources corre-spond to probability bounds by taking the reciprocal (recall Section 4.6), S therefore sets a local probability bound of $\alpha = 10^{-12}$. (Hence #5.)

Because the description V does not merely specify but in fact prespecifies the event E, and explicitly so, the specification-induced rejection region R corresponding to V is just V^*, which also is exactly equal to E. In other words, $R = V^* = E$ (see Section 3.5 and the beginning of Section 3.6). The probability of R is therefore the probability of independently dealing aces of hearts from all ten decks, which comes to 1 in 52^{10}, or roughly 1 in 1.446×10^{17}. This is the probability $P(R|H)$ and it is less than or equal to α. (Hence #6 and #7.)

Having successfully traversed #1 through #7, S now has in hand a rigorous demonstration of why he need not take seriously the kidnapper's appeal to chance. Rather, S can, with Richard Swinburne, conclude that the kidnapper is mistaken and that chance was not responsible for E. Having satisfied #1 through #7, and having fixed a generous set of probabilistic resources for elimi-nating chance, S is therefore warranted in inferring that E did not occur according to the chance hypothesis H. (Hence #8.)

It might now be asked what would have happened if S had not seen that E was prespecified but rather had only seen that E conformed to the specification V corresponding to the descrip-tion "ten aces of hearts." In that case, S would have seen that E was still specified, but the specification-induced rejection region R that S would have constructed would also have had to factor in

$L(m) = M \approx 10^{25}$ possible descriptions with respect to the basis specification V (recall Sections 3.6 and 3.7). And so, to eliminate chance, the probability $P(R|H)$ would be bounded by $P(V^*|H) \times M$, which would be roughly 10^7. (Yes, you are seeing this number correctly—not 10^{-7} but 10 million!) This is way bigger than the probability bound α. Indeed, it is not even a probability.

It would thus not be possible, given such a rejection region R, for S to conclude that $P(R|H) \leq \alpha$ (as in #7). In consequence, S could not conclude that E did not happen according to the chance hypothesis H, and would therefore not be justified in rejecting chance (as in #8). The lesson here is that the specification that S identifies to try to eliminate chance can drastically affect the probability of the associated specification-induced rejection region R. Thus, chance could be eliminated for one such specification (as with the prespecification we considered) but not for another (as with the specification we just sketched involving $M \approx 10^{25}$). An important part of this lesson is that prespecifications are to be preferred in such chance elimination arguments, leading to rejection regions of smaller probability than rejection regions based on specifications that are not also prespecifications.

Let's step back now to consider what rationally justifies the GCEA. Certainly, in practice, people use the GCEA to eliminate chance, although usually they do so tacitly rather than explicitly, as laid out here. Thus, by using the GCEA, the kidnap victim seemed right in concluding that chance was not responsible for the card-shuffling machine outputting the one highly improbable arrangement of cards compatible with the victim's survival. Likewise, the New Jersey Supreme Court used the GCEA—though just stopping short of a full-blown design inference—to conclude that Nicholas Caputo's ballot line selections were not by chance and that his method of selecting them needed to be upgraded.

But simply because people use a method in practice doesn't mean the method is sound, and that concern applies as well to the GCEA. The GCEA is an expanded version of Ronald Fisher's theory of statistical significance testing applied to specifications in general rather than just to prespecifications, as in his theory (for a recap of Fisher's theory, see Appendix A.5). The key conceptual difficulty in Fisher's theory, and one he never resolved, was to justify a "significance level" α (always a positive real number less than one) such that whenever an event E falls within a rejection region R and $P(R|H) \leq \alpha$, then the chance hypothesis H can be legitimately rejected as the explanation of E. The problem within Fisher's theory has always been that any proposed value for α has seemed arbitrary, lacking what Colin Howson and Peter Urbach call "a rational foundation."[7] Howson and Urbach elaborate:

> Fisher seems to have seen in significance tests some surrogate for the process of refutation, and he frequently went so far as to say that a theory can be "disproved" in a significance test... [Fisher implied] that tests of significance can demonstrate the falsity of a statistical theory [and] that statistical hypotheses may be actually falsified; but the experimental results used in a significance test clearly do not logically contradict the null hypothesis.
>
> Fisher was, of course, aware of this, and when he expressed himself more carefully, his justification for significance tests was rather different. The force of a test of significance, Fisher then claimed, "is logically that of the simple disjunction: *Either* an exceptionally rare chance has occurred, *or* the theory of random distribution [i.e., the null hypothesis] is not true."... Inevitably, therefore, the occurrence of a significant result is either a "rare chance" (an improbable event) or the null hypothesis is false, or both. And Fisher's claim amounts to nothing more than this necessary truth. It certainly does not allow one to infer the truth or falsity of any statistical hypothesis from a particular result...
>
> Expositions of Fisherian significance tests typically vacillate over the nature of the conclusions that such tests entitle one to draw. For example, [the statistician Harald] Cramér contended... that although a rejected theory could in fact be true, when the significance level is sufficiently small, "we *feel* practically justified in disregarding this possibility" (original italics altered). No doubt such

feelings do often arise... but Cramér offered no grounds for thinking that such feelings, when they occur, were generated by the type of reasoning employed in tests of significance, nor was he able to put those feelings onto any systematic or rational basis.[8]

We submit that the probabilistic-specificational apparatus developed in the last two chapters and encapsulated in the GCEA provides such a systematic and rational basis for why we can "feel practically justified" in eliminating chance in a test of statistical significance. As Howson and Urbach rightly note, there is never a logical contradiction in refusing to eliminate chance in a test of statistical significance. It is always a logical possibility that an event E occurred due to the chance hypothesis H, regardless of whether E falls within or outside a rejection region R. But precisely because this is a strictly logical point, one does not have to be a statistician to appreciate it. Indeed, the statistician's task is to tell us when to reject chance in the absence of logical certainty.

Is there, then, a systematic and rational way the statistician can properly explain E in reference to H when the rejection region R does not have zero probability? The Generic Chance Elimination Argument provides such a way. In Fisher's theory, a statistician/ subject S is justified eliminating H as the explanation of E whenever (1) E falls within a rejection region R, and (2) the probability of R given H is sufficiently small, (i.e., $P(R|H) \leq \alpha$ for some α-level). Although this needs to be part of the story in rejecting H, it is not the whole story. The whole story consists in taking Fisher's theory of statistical significance testing and embedding it in the GCEA, which provides a systematic and rational basis for eliminating chance. The GCEA incorporates all the elements in Fisher's theory but also adds to them, notably the concepts of specification and probabilistic resources.

The crucial addition in shoring up Fisher's theory is therefore to justify the significance level (i.e., probability bound) α in terms of probabilistic resources. Fisher and his followers were ready to put forward concrete α-levels, such as .05 or .01 or .001, but these were always arbitrary, meant to fan intuitions about improbability, but without any clear rational grounding. By contrast, from the vantage of the GCEA, any choice of α-level is not arbitrary but obtains a definite meaning in reference to probabilistic resources.

Probabilistic resources are always chosen to avoid mistakenly eliminating chance when it was operating (false positives) and mistakenly retaining chance when it was not operating (false negatives). Because of a lack of sufficient background knowledge, we may simply not see that a pattern has a short description. We may therefore miss that it is a specification. And so we may fail to eliminate chance and thus fail to infer design. False negatives are therefore always a concern. But false positives are always the far bigger concern with chance elimination arguments based on small probabilities and specifications, and thus with design inferences in particular. We need to firmly restrict false positives. Only then can we be confident that when chance is eliminated, chance really was not operating.

Given an event E, a rejection region R, a chance hypothesis H, and a significance level α, Fisher's theory does little more than press the fundamental intuition that H ought not to be viewed as responsible for E so long as E falls within R and $P(R|H)$ is small (meaning $P(R|H) \leq \alpha$). But how small is small enough? The account of probabilistic resources given in Chapter 4 answers that question. Tidy-looking α-levels that appear everywhere in applications of Fisher's theory, such as .05 or .01 or .001, are, from the vantage of probabilistic resources, ridiculous. Probabilistic resources need always to be assigned on a case-by-case basis depending on the relevant number of opportunities for an event to happen by chance. By the reciprocal rule, probabilistic resources (of size K) then induce a corresponding probability bound ($\alpha = 1/K$). That's how α-levels are given a rational foundation.

Cramér's feeling of being practically justified in eliminating a chance hypothesis makes perfect sense in the light of probabilistic resources. The greater the number of probabilistic resources factored in, the smaller the α-level, and thus the more difficult it is for an event E to fall within a rejection region R when $P(R|H) \leq \alpha$. Smaller α-levels, as it were, give chance increasingly less of a chance. As a feeling of practical justification, this sensibility in eliminating chance comes in degrees rather than being all or nothing (as in the case of proof and refutation).

The greater the probabilistic resources, the smaller the corresponding probability bound, and the greater the confidence that the event in question did not happen by chance. The underlying logic here is not the binary logic of truth and falsehood but rather the gradated logic of confirmation and epistemic support, which always comes in degrees. This logic is fallible, since well-confirmed claims can nonetheless be wrong. But for many purposes, binary logic is inadequate and a confirmational logic is the best we can do. The rejection of chance in the conclusion of the GCEA is an instance of such confirmational logic, confirmation in this case being based on probabilistic resources.

The idea that causal explanations come with varying degrees of confirmation will be familiar to Bayesians. Indeed, Bayes' theorem provides a way to understand confirmatory evidence. Suppose, then, we are given that an event R has occurred (R coinciding here with the rejection region in the GCEA). Suppose further that the probability of R under the particular chance hypothesis H is low, namely, $P(R|H) \leq \alpha$ for some small α.

The rejection region R is assembled from events that are improbable under the chance hypothesis H but that are easily described. Any alternative hypothesis, including a design hypothesis, that would more readily produce these easily described events will therefore confer greater probability on R than H. Suppose, then, that H' is such an alternative hypothesis that more readily accounts for how R could have been assembled. In that case $P(R|H)$ should be much smaller than $P(R|H')$.

By Bayes' theorem, the following equation now holds:

$$\frac{P(H|R)}{P(H'|R)} = \frac{P(R|H)}{P(R|H')} \times \frac{P(H)}{P(H')}.$$

On the left of this equation is the ratio of posterior probabilities. On the right is a product: the ratio of likelihoods times the ratio of prior probabilities. Supposing that the prior probability $P(H)$ does not completely overwhelm the prior probability $P(H')$ (why should it?), and given that H' was chosen so that $P(R|H)$ is much smaller than $P(R|H')$, $P(H'|R)$ will be greater than $P(H|R)$. Accordingly, the ratio of posterior probabilities on the left will be less than one, thus constituting evidence in favor of H' as compared to H. This evidential disparity against H and in favor of H' is consistent with our conclusion in the GCEA.

5.3 Key Concepts and Predicates

Our main task in this chapter is to frame the design inference as a valid deductive argument, which we do in the next section. But first, we need to lay out certain key concepts along with certain key predicates based on those key concepts. The GCEA, which underlies the design inference, presupposes these concepts. We shall refer to these concepts collectively as the requisite precondition for chance elimination arguments based on specifications and small probabilities, or simply the *requisite precondition*. Most of these concepts were made explicit in the last section on the GCEA, but a few were left implicit. We now make all of them explicit:

S The subject who will be assessing whether an event happened by chance.

E The event observed by the subject, and whose chance occurrence stands in question.

Ω The reference class of possibilities (also called the possibility or probability space) that gives rise to the event E and other related events.

$\mathscr{B}$ The subsets of Ω that can serve as events and to which probabilities can be assigned. This collection of subsets has been implicit whenever Ω was previously mentioned, though it is only now made explicit.[9]

H The chance hypothesis that sets out the precise form of chance to be eliminated or retained, as the case may be.

P The probability measure defined on Ω, assigning probabilities to each of the elements of $\mathscr{B}$, and doing so with respect to the chance hypothesis H, so that $P(\cdot)$ may also be denoted by $P(\cdot|H)$ or $P_H(\cdot)$.

$\mathscr{H}$ The collection of chance hypotheses that might characterize the chance occurrence of the event E. Thus far, $\mathscr{H}$ has been implicit and has always contained but one element, namely, H. In a full-fledged design inference, which sweeps the field clear of relevant chance hypotheses, $\mathscr{H}$ can contain more than one chance hypothesis.

$\mathscr{L}$ A language that consists of a finite collection of base elements, and whose generic elements consist of finite sequences of these base elements. The point of $\mathscr{L}$ is to describe the events of $\mathscr{B}$.

$|\cdot|$ The description-length metric defined on $\mathscr{L}$ that assigns 1 to the base elements of $\mathscr{L}$ and counts the number of base elements strung together in any sequence in $\mathscr{L}$.

$*$ The asterisk function that maps $\mathscr{L}$ to $\mathscr{B}$ so that if W describes an event in the language $\mathscr{L}$, then W^* is the corresponding event described. Note that the asterisk $*$ may be a partial function in that some elements of $\mathscr{L}$ may not describe any event in $\mathscr{B}$.

$D(\cdot)$ The minimum description length metric that maps from $\mathscr{L}$ to the natural numbers and assigns to a description V in $\mathscr{L}$ the length of the shortest description W in $\mathscr{L}$ such that $V^* = W^*$, i.e., $D(V) = \min\{\,|W| \mid W \in \mathscr{L} \text{ and } V^* = W^*\,\}$. It follows immediately that $D(V) \leq |V|$.

$L(\cdot)$ The descriptive complexity bound that maps natural numbers to natural numbers and assigns to any natural number n the number of descriptions in $\mathscr{L}$ with descriptive length less than or equal to n, i.e., $L(n) =$ cardinality of $\{\, W \in \mathscr{L} \mid |W| \leq n \,\}$.

m A particular complexity bound chosen by S in light of S's interests, needs, background knowledge, and context of inquiry that determines what level of descriptive length will count as a specification.

M M is the least upper bound on the number of descriptions whose length is less than or equal to m. By definition, $M = L(m)$.

K The total number of probabilistic resources. We've tended to leave this number implicit, instead moving to the next concept, on which it is based.

α The probability bound (local or universal) that corresponds to the probabilistic resources K. By the reciprocal rule (see Section 4.6), $\alpha = 1/K$.

V A specification in $\mathscr{L}$ to which E conforms. V therefore satisfies $|V| \leq m$ and $E \subset V^*$.

R The specification-induced rejection region, based on V, by which S will attempt to eliminate H (or, more generally, all hypotheses in $\mathscr{H}$) for failing to adequately explain E by chance. R depends on other concepts on this list, but is crucially dependent on H (and $\mathscr{H}$).

This list of concepts, which together form the requisite precondition, seems more complicated than it actually is. Its concepts should be clear to readers who have understood Chapters 3 and 4. Half of the list is simply what's needed to form the rejection regions of conventional statistical significance testing. The other half is the descriptive-language apparatus needed to define specifications in general (rather than just prespecifications) and use them to form rejection regions capable of eliminating chance. Conventional statistical significance testing does not require this apparatus since it depends on prespecifications, which are easier to deal with. The requisite precondition is needed to handle the combination of specification and small probability in full generality.

With the requisite precondition in hand, we can now define the predicates that will be needed to formulate the design inference as a deductive argument. In these predicates, we avoid mention of S, though S's role in confirming that these predicates are satisfied is always implicit. Indeed, some subject must always be available to confirm these predicates. Thus, for $oc(E)$, the predicate that asserts that the event E has occurred, it could as well be interpreted as affirming that S has confirmed that E has occurred. The following predicates therefore presuppose a subject S, which is the first concept listed above in the requisite precondition. But these predicates also presuppose the other concepts in the requisite precondition. Here are the predicates:

$oc(E)$ The event E has occurred.

$ch(E)$ The event E is the result of chance. In making this assertion, the predicate tacitly assumes that a particular chance hypothesis H characterizes the occurrence of E. $ch(E)$ may therefore be rewritten more fully as $ch(E;H)$.

$des(E)$ The event E is not the result of any chance hypothesis H from the set of relevant chance hypotheses $\mathcal{H}$ that might explain this event. In symbols, $des(E)$ is logically equivalent to $(\forall H \in \mathcal{H}) \sim ch(X;H)$.

$spec(V,E)$ The description V is a specification of the event E. Since specifications are typically formulated in reference to a chance hypothesis H, $spec(V,E)$ may be rewritten more fully as $spec(V,E;H)$.

smp(*V*)   The rejection region *R* formed on the basis
               of *V* has small probability, meaning *R* has
               probability less than or equal to α. The pre-
               dicate *smp*(*V*) depends on the chance hypo-
               thesis *H*, so it may be rewritten more fully as
               smp(*V;H*).

Some commentary on these five predicates is in order. The predicate *oc*(*E*) is straightforward. It simply asserts that *E* has occurred. Indeed, *E* better have occurred if its chance occurrence is to be an object of inquiry. Next is the predicate *ch*(*E*). *ch*(*E*) asserts that the event *E* happened by chance, and specifically with respect to a chance hypothesis *H*, so that *ch*(*E*) can be more fully written as *ch*(*E;H*). As long as *E* has positive probability with respect to *H*, i.e., $P(E|H) > 0$, there is no logical contradiction in asserting that *E* happened by chance according to the chance hypothesis *H*. The challenge is to justify negating the chance hypothesis, which is to say justifying the assertion of ~*ch*(*E;H*). Crucial in this regard will be the upcoming Law of Small Probability.

In the first edition of this book, we included a sixth predicate, namely, *nec*(*E*), which asserts that the event *E* is the result of a necessity, physical or otherwise, rendering it non-contingent and guaranteeing its occurrence. Common usage distinguishes necessary events, which are bound to happen (such as a double-headed coin landing heads) from contingent events, which could happen but are not compelled to happen (such as a fair coin landing heads). Contingent events are the sort that we typically ascribe to chance. Even so, *nec* can be conveniently assimilated to chance by including within the predicate *ch* hypotheses of the sort *H'* in which all probabilities collapse to zero and one. Thus *nec*(*E*) can be identical to *ch*(*E;H'*) where *H'* assigns a probability of one to *E*.

Rather than keep *nec* separate from *ch*, we prefer to assimilate *nec* to *ch*. This approach has the advantage of simplicity and generality. Some evolutionists, for instance, will refer to evolution as proceeding by chance and necessity, and will even insist that everything that exists in the universe is a product of chance and necessity.[10] Assimilating necessity to chance bypasses this distinction without any loss of conceptual insight or explanatory power.

This broader conception of chance avoids imprecision in the term's use. Richard Dawkins provides a case in point. As noted earlier, he has asserted that the "belief that Darwinian evolution is 'random' is not merely false. It is the exact opposite of the truth. Chance is a minor ingredient in the Darwinian recipe, but the most important ingredient is cumulative selection, which is quintessentially nonrandom."[11] But Dawkins is here using a limited conception of chance. Chance, in the form of random mutations, obviously plays a key role in Darwin evolution. Filtered by natural selection, random mutations exhibit chance even if they don't exhibit "pure randomness." Given our broader conception of chance, Dawkins could have avoided the imprecision generated by his usage.

The predicate ch, in our usage, thus includes anything we might mean by chance. It includes necessity. It includes pure chance, as with uniform random sampling. And it includes any combination of chance and necessity, including chance mutations shaped by natural selection. Indeed, ch in this broad conception includes anything characterized by a probability distribution.[12] It therefore includes all stochastic processes. And since any process of nature, insofar as it can be modeled mathematically, can be modeled by a stochastic process, it follows that the predicate ch, as we define it, casts a very wide net indeed.

Next, consider the predicate $des(E)$. It might have been defined causally by asserting that an intelligent agent is responsible for E, or—if we want to bring in the subject S—by asserting that S is warranted in concluding that E is the product of an intelligent agent. As we've defined it, however, $des(E)$ is merely the negation of all relevant chance explanations. This none-of-the-above approach to the design predicate is useful in that it treats design as a purely logical rather than causal category. It thus allows chance and design to be mutually exclusive and exhaustive, making it especially convenient to reason with these predicates via a disjunctive syllogism (negating one disjunct affirms the other).

In the next section we show why this approach to design as a logical category gives us solid grounds to think that it matches up with design as a causal category. What bridges these two categories is, perhaps unsurprisingly, specification, which is not only crucial for eliminating chance but also invariably results from an intelligent act in that it takes an intelligence to specify an event. Thus, while it may seem tendentious to refer to $des(E)$ as a design predicate given that it was defined as a negation of chance, in light of the concept of specification, this usage will ultimately justify itself.

The predicate $spec(V,E)$ is straightforward. It affirms that the description V in $\mathscr{L}$ is indeed a specification and specifies the event E, meaning that $V^* \supset E$. For V to be a specification, it could be a prespecification, and so no descriptive complexity of it need be calculated. Alternatively, it may be a specification whose description length is less than or equal to the descriptive complexity bound m, i.e., $|V| \leq m$. To anticipate the next chapter, it could also be that V, in the way it combines descriptive complexity and probability, has high specified complexity, and in this way constitutes a specification. There are thus several ways V can satisfy the predicate $spec(V,E)$. Whether V is a specification can, as in the definition of specified complexity in the next chapter, depend on an underlying chance hypothesis H. So $spec(V,E)$ can also be written more fully as $spec(V,E;H)$.

The predicate $smp(V)$ is perhaps the trickiest to unpack. It includes tacit reference to the event E because this predicate is asserted in combination with $spec(V,E)$, which guarantees that $V^* \supset E$. $smp(V)$ affirms that the rejection region R determined by V has small probability. Given as probabilistic resources K and the corresponding probability bound $\alpha = 1/K$, this means that $P(R|H)$ will be less than or equal to α. The tricky part here, however, is the precise form that R will take. If V is a prespecification, as in Section 3.5, R is just V^*. If V is a specification associated with a probability density function (Section 3.8) or a perturbation neighborhood (Appendix B.3), the associated rejection region is again straightforward.

Once we move to specifications in general, things get more complicated. For a specification V that satisfies $|V| \leq m$ and for which $L(m) = M$, the corresponding rejection region R, as we defined in Section 3.7, takes the following form: the union of all events W^* from Ω where W belongs to $\mathscr{L}$ such that $D(W) \leq m$ and $P(W^*) \leq \alpha/M$, provided that $P(V^*) \leq \alpha/M$. Determining from this definition exactly what's inside and outside R can be difficult to impossible. Nevertheless, because there are at most M events of probability less than or equal to α/M in this union, the probability of R cannot exceed α. In other words, $P(R|H) \leq \alpha$, which is precisely what we need to assert that $smp(V)$ obtains.

In the next chapter, we consider descriptions V that can be described in s bits and for which the probability of the corresponding event V^* is identified with t bits (t can be fractional; $P(V^*|H) = 2^{-t}$). The specified complexity of V is then defined as $t - s$ bits (presupposed here are certain minimal constraints on the language $\mathscr{L}$). If we then define the rejection region R as the union of all events with descriptions having at least $t - s$ bits of specified complexity, then the probability of R is 1 in 2^{t-s}, or $P(R|H) = 2^{s-t}$. Provided this probability is less than or equal to α, $smp(V)$ will therefore be satisfied for the rejection regions arising out of specified complexity.

We've recalled here the different types of specification-induced rejection regions associated with a basis specification V. We've argued that $smp(V)$ then holds so long as these rejection regions have probability bounded by α. We think we've been thorough in laying out the different types of specification-induced rejection regions. But it could be that there remain other ways of using specifications V to induce rejection regions R. If so, the predicate $smp(\cdot)$ can accommodate them provided that with relevant probabilistic resources K factored in, it follows that $P(R|H) \leq \alpha = 1/K$, making it highly unlikely that any of the available opportunities to match R will in fact match R.

5.4 The Design Inference as a Deductive Argument

Given the predicates described in the last section, we are now in a position to lay out a deductive form of the design inference. To do so, however, we first need a more rigorous formulation of the Law of Small Probability (or LSP). The Law of Small Probability, stated intuitively (recall Section 1.2), asserts that specified events of small probability do not occur by chance. Let's now formalize this intuition. Suppose E denotes an event and V a description whose corresponding event subsumes E (i.e., $V^* \supset E$), and suppose a chance hypothesis H stands in question. In that case, using ordinary symbolic logic, it seems reasonable to formulate the Law of Small Probability as follows:

$$[oc(E) \ \& \ spec(V,E;H) \ \& \ smp(V;H)] \rightarrow \sim\!ch(E;H)$$

In words, if E occurred and if V is a specification of E for which the rejection region associated with V has small probability, then E did not occur by chance. Chance here is understood with respect to the chance hypothesis H.

This statement of the Law of Small Probability is fine as far as it goes, but it is tied specifically to the event E and to the specification V. Since these can vary, it is best to substitute variables for them and then quantify over those variables. In that case, the Law of Small Probability takes the following form, using the universal quantifier $\forall$ ("for all") and the existential quantifier $\exists$ ("there exists"):

$$(\forall X \in \mathcal{B})[oc(X) \ \& \ (\exists W \in \mathcal{L})[spec(W,X;H)$$
$$\& \ smp(W;H)] \rightarrow \sim\!ch(X;H)]$$

In words, for any event X, if X has occurred and if there exists a description W such that W specifies X and induces a rejection region of small probability, then X did not result from chance, chance here being characterized by the chance hypothesis H. This statement of the Law of Small Probability is also fine as far as it goes. As it stands, it encapsulates the GCEA, laying out the conditions whereby a chance hypothesis H might rightly be rejected for an arbitrary event E. But we face one more complication.

What if we need to reject not just a single chance hypothesis H, but rather a relevant set of chance hypotheses, the set denoted by $\mathcal{H}$ (this set appears in the requisite precondition)? In that case, to reject chance will, for an arbitrary event X, require the conclusion $(\forall H \in \mathcal{H})\sim ch(X;H)$. But this, recall, is logically equivalent to $des(X)$. Moreover, to affirm chance will then be the negation of this statement and thus take the form $(\exists H \in \mathcal{H})ch(X;H)$.[13] Thus, with a range of chance hypotheses $\mathcal{H}$ in play, $des(X)$ will also be logically equivalent to $\sim(\exists H \in \mathcal{H})ch(X;H)$.

Because in the GCEA, the role of the chance hypothesis H was unconstrained, we can now also add the universal quantifier $\forall H \in \mathcal{H}$ to the previous form of the LSP. In that case, we now arrive at the following final form of the LSP:

$$(\forall X \in \mathcal{B})(\forall H \in \mathcal{H})[oc(X) \ \& \ (\exists W \in \mathcal{L})[spec(W,X;H)$$
$$\& \ smp(W;H)] \rightarrow \sim ch(X;H)]$$

In words, for every event and every relevant chance hypothesis, if such an event has happened and if there's a description that's a small-probability specification with respect to such a chance hypothesis, then that event did not happen in accord with that chance hypothesis. This reads clunky. Indeed, the symbolic logic here does a better job of articulating the Law of Small Probability. But to reiterate, this law is simply an encapsulation of the GCEA.

Given this formulation of the Law of Small Probability, the following is now a valid deductive argument:

Premise 1: $oc(E)$

Premise 2: $(\forall H \in \mathcal{H})(\exists W \in \mathcal{L})[spec(W,E;H) \ \& \ smp(W;H)]$

Premise 3: $(\forall X \in \mathscr{B})(\forall H \in \mathscr{H})[oc(X)$ &
               $(\exists W \in \mathscr{L})[spec(W,X;H)$
               & $smp(W;H)] \rightarrow \sim ch(X;H)]$

Premise 4: $(\exists H \in \mathscr{H})ch(E;H) \vee des(E)$

Conclusion: $des(E)$

To see that this is indeed a valid deductive argument, note that premises 1 and 2 supply the antecedent for premise 3 (once the quantifier $\forall X \in \mathscr{B}$ is peeled away). Premise 3 is the Law of Small Probability. From these three premises, $(\forall H \in \mathscr{H})\sim ch(X;H)$ follows, which is logically equivalent to $\sim(\exists H \in \mathscr{H})ch(X;H)$. Premises 1 to 3 therefore give us the negation of the first disjunct of premise 4. Premise 4 is a dichotomy rule, which, given the way we defined the predicates *ch* and *des*, ensures that the disjuncts in this premise are mutually exclusive and exhaustive. An application of disjunctive syllogism then leads to the conclusion, namely, $des(E)$.

Given that this deductive argument ends in the conclusion $des(E)$, it can legitimately qualify as a deductive form of the design inference.[14] That said, our final form of the design inference modifies one of the premises. In practice, we tend to use a stronger version of premise 2 in which a specification is chosen independently of the relevant chance hypotheses. Often, we are lucky to identify even one specification for a given event. So, in practice, we identify a single specification that will hold across all relevant chance hypotheses up for elimination.

This means changing premise 2. We do this by reversing the dependence of its quantifiers. Right now premise 2 reads $(\forall H \in \mathscr{H})(\exists W \in \mathscr{L})[spec(W,E;H)$ & $smp(E;H)]$. We want to reverse the quantifiers so that it reads $(\exists W \in \mathscr{L})(\forall H \in \mathscr{H})[spec(W,E;H)$ & $smp(E;H)]$. This latter expression is stronger than the former in the sense that the latter entails the former. We can therefore substitute it into the previous deductive argument without sacrificing its validity. Making this change means looking for only one specification to invalidate all the relevant chance hypotheses (i.e., all H in $\mathscr{H}$), which is what typically happens with design inferences. This, then, is our final version of the design inference formulated as a deductive argument:

Premise 1: $oc(E)$

Premise 2: $(\exists W \in \mathscr{L})(\forall H \in \mathscr{H})[spec(W,E;H)$ & $smp(W;H)]$

Premise 3: $(\forall X \in \mathscr{B})(\forall H \in \mathscr{H})[oc(X)$ & $(\exists W \in \mathscr{L})[spec(W,X;H)$ & $smp(W;H)] \rightarrow \sim ch(X;H)]$

Premise 4: $(\exists H \in \mathscr{H})ch(E;H) \lor des(E)$

Conclusion: $des(E)$

Briefly, in words, the argument starts by noting that an event E has happened. Next, it notes that there is some description that is a specification of the event, that this specification induces a small probability rejection region, and that it induces such a rejection region irrespective of which chance hypothesis is considered from a relevant set of chance hypotheses. These facts, combined with the Law of Small Probability in premise 3, ensure that E did not happen by chance. Premise 4 then asserts dichotomy, which, via disjunctive syllogism, leaves as the only option $des(E)$, the conclusion of the argument.

Let's now step back and consider what gives the design inference its persuasive force when it is unpacked as a deductive argument. As an argument within the conventional predicate logic, it clearly is valid. Indeed, it is straightforward to see that the conclusion follows from the premises via truth-preserving inference rules. Thus, if the premises are true, the conclusion must be true as well. But in what sense are the premises true? Here some clarification is called for, at least with one of the premises, namely, premise 3 (the Law of Small Probability). Premises 1, 2, and 4 are, by contrast, unproblematic, at least with respect to their truth status within the argument.

Premise 1 is straightforward: its truth is confirmed by observation. In other words, a subject needs to observe that the event E has occurred.

Premise 2 is where all the heavy lifting in a design inference takes place. Indeed, identifying a specification that makes premise 2 true is where a subject in drawing a design inference will expend the most effort. This premise has an existential quantifier in front: $(\exists W \in \mathcal{L})(\forall H \in \mathcal{H})[spec(W,E;H)$ & $smp(W;H)]$. But in practice, it is confirmed by finding a particular specification V that satisfies the expression without the existential quantifier: $(\forall H \in \mathcal{H})[spec(V,E;H)$ & $smp(V;H)]$.

The job of premise 2 is to establish specification and small probability, which factor critically into the Law of Small Probability. Perhaps we don't have enough information to confirm smallness of probability with respect to the range of chance hypotheses in $\mathcal{H}$. Perhaps the event E can be specified but we simply have not seen the salient pattern with the short description that reveals it to be specified. There are all sorts of ways we can fail to confirm premise 2 (even if with more knowledge we might be able to confirm it). But if we can confirm this premise, its role in the argument is unproblematic.

Premise 4, the dichotomy premise, is true because it is a tautology. Indeed, we defined $des(E)$ as $(\forall H \in \mathcal{H}) \sim ch(E;H)$. Premise 4 is therefore true as a matter of pure logic.

That leaves premise 3, the Law of Small Probability. Is it true? It is true, but not as a metaphysical claim about ultimate reality. Rather, it is true as a *regulative principle* that we need to navigate wisely through life's challenges. The Law of Small Probability is prudentially true rather than metaphysically true. We need it to exercise sound practical reason in everyday experience. To say that the Law of Small Probability is a regulative principle is therefore to say that we need it for sound judgment and reasonable action in the world.

We use such regulative principles all the time. Suppose someone is sick and faces two treatment options. Which option should be taken? Clearly, the one with the better record of success in overcoming the sickness. Behind such advice is a regulative principle, namely, to choose the effective treatment, which is to say the treatment with the greater probability of success. Now it can happen that the less effective treatment yields a good outcome (the patient got lucky). And it can happen that the more effective treatment yields a bad outcome (the patient got unlucky). But the prudential course is, other things being equal, to go with the treatment option having the better record of success. All our work in motivating, articulating, and explaining the Law of Small Probability has been to justify it as a sound regulative principle for guiding our use of probabilities.

How does the Bayesian fare with the deductive form of the design inference? The Bayesian faces no contradiction in accepting its logic and conclusion. If all the premises, taken jointly, are very likely to be true, it follows that the conclusion must also very likely be true (for the details about why this is so, see Section 5.7). Nevertheless, the Bayesian might prefer to recast this deductive argument in Bayesian terms. In that case, the Bayesian could identify a collection of chance hypotheses $\mathcal{H}$ as well as a generic design hypothesis D. Together, the probability that at least one of these hypotheses holds (D as well as those in $\mathcal{H}$) would be 1, or very close to 1. Thus, we assume that the Bayesian is confident that if some chance hypothesis is the correct explanation, it has been identified as one of the chance hypotheses in $\mathcal{H}$. There may be other conceivable chance hypotheses, but their prior probability would then be extremely low.

The observed event E will then be such that its probability under any single chance hypothesis H is very low, which is to say $P(E|H)$ is very close to 0, or $P(E|H) \approx 0$. Because E is specified, it exhibits a salient design-conducive pattern, and so its probability is relatively high under the design hypothesis: $P(E|D) \gg 0$. The update rule for apportioning Bayesian evidence (see Appendix A.4) now equates the ratio of posterior probabilities (the term on the left of the following equation) with the product of the ratio of likelihoods (first term on the right of the following equation) times the ratio of prior probabilities (second term on the right of the following equation):

$$\frac{P(D|E)}{P(H|E)} = \frac{P(E|D)}{P(E|H)} \times \frac{P(D)}{P(H)}.$$

Unless there are extremely compelling independent reasons for thinking that $P(D)$ must be vastly less than any of the prior probabilities $P(H)$ (for H in $\mathcal{H}$), the ratio of priors $P(D)/P(H)$ will not be determinative of what this equation signifies. What will be determinative is the ratio of likelihoods. But this ratio, as we've noted vastly favors D over any H in $\mathcal{H}$, making $P(E|D) \gg P(E|H)$. In consequence, we should think that for any H in $\mathcal{H}$, the ratio of posterior probabilities $P(D|E)/P(H|E)$ will be bigger than 1—indeed, much bigger than 1. Since in a posterior probability, the event being conditioned on serves as evidence for the hypothesis in question, this means that the probability of D in light of the evidence E is much bigger than the probability of any H in $\mathcal{H}$ in light of the same evidence. We should therefore conclude that D is much better supported than any of the hypotheses H in $\mathcal{H}$, and that the design hypothesis is therefore very likely to be true.

Up to this point in our account of the design inference as a deductive argument, an important question has been left unresolved. We defined the design predicate as the negation of chance, that is, $des(E) = (\forall H \in \mathcal{H})\sim ch(E;H)$. But what does this have to do with design in the conventional sense as an intelligent cause? Our next task is therefore to make clear why this logical definition of design, as a negation of chance, connects meaningfully with a causal definition of design in the sense of intelligent agency.

5.5 From Design to Agency

In the deductive form of the design inference of the last section, all the hard work is in establishing premise 2. To establish that premise, specification and small probability need to be jointly confirmed. Premise 2 operates as a criterion for determining whether chance was operating. Consequently, we call it the specification/small-probability criterion, or SP2 criterion for short.[15] The SP2 criterion is crucial for inferring design. Yet, from the way we defined the predicate *des(E)*, concluding design in the deductive form of the design inference is logically equivalent only to eliminating chance. It may therefore be asked why such an inference yields not merely design as the negation of chance (purely as a matter of logic), but rather a full-throated conception of design attributable to an intelligent agent. The challenge, in other words, is to connect design as a logical category to design as a causal (or agentive) category.

To see why the specification/small-probability (or SP2) criterion is well suited for recognizing intelligent agency, we need to understand what it is about intelligent agents that reveals their activity. The principal characteristic of intelligent agency is *directed contingency*, or what we typically call *choice*. Whenever an intelligent agent acts, it chooses from a range of competing possibilities. The very etymology of the word *intelligent* makes this clear. The word derives from two Latin words, the preposition *inter*, meaning between, and the verb *lego*, meaning to choose or select.[16] Thus, according to its etymology, intelligence consists in choosing between. For an intelligent agent to act is therefore to choose from a range of competing possibilities. This narrowing of possibilities is the defining feature of information (Appendix A.9), which is why the principal thing that intelligences do is create information.[17]

This choosing among options, which is the defining characteristic of intelligence, holds not just for humans, but also for animals as well as for possible alien intelligences. A rat navigating a maze must choose whether to go right or left at various points in the maze. In trying to detect an extra-terrestrial intelligence, SETI researchers assume such an intelligence could choose from a range of possible radio transmissions, and then attempt to match the observed transmissions with patterns regarded as sure indicators of intelligence. Whenever a human being utters meaningful speech, a choice is made from a range of possible sound combinations that might have been uttered. Intelligent agency always entails discrimination, choosing certain things and ruling out others.

Given this characterization of intelligent agency, the crucial question is how to recognize it. Intelligent agents act by making a choice. How, then, do we recognize that an intelligent agent has made a choice? A bottle of ink spills accidentally onto a sheet of paper; someone takes a fountain pen and writes a message on a sheet of paper. In both instances ink is applied to paper. In both instances one among an almost infinite set of possibilities is realized. In both instances a contingency is actualized and others are ruled out. Yet in one instance we ascribe agency, in the other chance.

What is the relevant difference? Not only do we need to observe that a contingency was actualized, but we ourselves need also to be able to specify that contingency. The contingency must conform to an independently given pattern, and we must be able independently to formulate that pattern. A random ink blot is unspecified; a message written with ink on paper is specified. True, the exact message may not be specified, but it is specified as a meaningful message in a known language.

The philosopher Ludwig Wittgenstein emphasized this aspect of meaningful messages when he wrote, "We tend to take the speech of a Chinese for inarticulate gurgling. Someone who understands Chinese will recognize *language* in what he hears."[18] That is, someone who does not understand Chinese, in hearing a Chinese utterance, recognizes that a particular utterance from a range of possible utterances was actualized but is unable to specify the utterance as coherent speech. It will therefore sound like so much random gibberish—unspecified nonsense. In contrast, the person who understands Chinese not only recognizes that a single utterance from a range of all possible utterances has been actualized but is also able to specify the utterance as coherent Chinese speech.

The actualization of one among several competing possibilities, the exclusion of the rest, *and* the specification of the actualized possibility encapsulates how we recognize intelligent agents. Actualization-Exclusion-Specification—this triad—provides a general scheme for recognizing intelligence, be it animal, human, or extra-terrestrial. Actualization establishes that the possibility in question is the one that actually occurred. Exclusion establishes that there was genuine contingency (i.e., that there were other live possibilities, and that these were ruled out). Specification establishes that the actualized possibility conforms to a pattern given independently of its actualization.

A pattern's independence from the actualized event ensures that the pattern was not simply read off of the event. With chance events, we can always read off patterns to which those events conform. For instance, a long random sequence of coin tosses can be represented as a pattern of zeros and ones that match it, zeros for tails, ones for heads. But such a pattern, by simply being read off of the event, tells us nothing about its chance occurrence. Only if the pattern can in some sense be derived from our prior background knowledge without reference to (and thus independently of) the actualized event can we use the pattern to rule out chance for the event. That's why the language of "detachability" was applied to specifications in the first edition of this book. Specifications were "detachable patterns" in the sense that they could be identified or articulated apart from the events they described.

Specifications fulfill this independence of patterns from events that detachability was meant to capture. We define specifications as patterns that have a short description length. It's this shortness of description length that shows specifications to be independent patterns. When the description length is short, we can run through all the short descriptions to attempt to match them with the event that was actualized. In other words, with a specification, we are, in principle, able to construct a pattern that matches the event in question, yet without reference to it. That's because there tend to be only a few short descriptions, so searching through them is manageable. By contrast, the patterns with long description lengths are so numerous that it is unmanageable to run through enough of them to match the actualized event. That, then, is the intuition behind specifications and why they work in the Actualization-Exclusion-Specification scheme to signal intelligence.

Where does choice, the defining characteristic of intelligent agency, figure into this scheme? The problem is that we never witness choice directly. Instead, we witness actualizations of contingency that might be the result of choice (i.e., directed contingency), but that also might be the result of chance (i.e., blind contingency). Now there is only one way to tell the difference—specification.

Specification is the only means available to us for distinguishing choice from chance, directed contingency from blind contingency. Actualization and exclusion together guarantee we are dealing with contingency. Specification guarantees we are dealing with a directed contingency. The Actualization-Exclusion-Specification scheme is therefore precisely what we need to identify choice and therewith intelligent agency.

Psychologists who study animal learning and behavior have known of the Actualization-Exclusion-Specification triad all along, albeit implicitly. [19] To learn a task an animal must acquire the ability to actualize behaviors suitable for the task as well as the ability to exclude behaviors unsuitable for the task. Moreover, for psychologists to recognize that an animal has learned a task, they must not only observe the animal performing the appropriate behavior, but also specify this behavior. Thus, to recognize whether a rat has successfully learned how to traverse a maze, a psychologist must first specify the sequence of right and left turns that conducts the rat out of the maze.

True, a rat randomly wandering a maze also discriminates a sequence of right and left turns. But by randomly wandering the maze, the rat gives no indication that it can discriminate the appropriate sequence of right and left turns for exiting the maze. Consequently, the psychologist studying the rat will have no reason to think the rat has learned how to traverse the maze. Only if the rat executes the sequence of right and left turns specified by the psychologist will the psychologist recognize that the rat has mastered the maze. Now it is precisely the learned behaviors we regard as intelligent in animals. Hence it is no surprise that the same scheme for recognizing animal learning recurs for recognizing intelligent agents generally—namely, actualization, exclusion, and specification.

This general scheme for recognizing intelligent agents is but a thinly disguised form of the specification/small-probability (SP2) criterion: one event happened (what was actualized), other events were possible (but they were excluded), and the event that happened was specified (it conformed to an independently given pattern). So the match between how this general scheme recognizes intelligent agency and how the SP2 criterion infers design is exact.

One loose end remains—the role of small probabilities. Although small probabilities figure prominently in the SP2 criterion, their role in the Actualization-Exclusion-Specification scheme for recognizing intelligent agency is not immediately apparent. In this scheme, one among several competing possibilities is actualized, the rest are excluded, and the possibility which was actualized is specified. But where in this scheme are the small probabilities?

They are there implicitly. To see this, consider again a rat traversing a maze, but this time take a very simple maze in which two right turns conduct the rat out of the maze. How will a psychologist studying the rat determine whether it has learned to exit the maze? Just putting the rat in the maze will not be enough. Because the maze is so simple, the rat could by chance just happen to take two right turns and thereby exit it. The psychologist will therefore be uncertain whether the rat actually learned to exit the maze or simply got lucky.

But contrast this with a complicated maze in which a rat must take just the right sequence of left and right turns to exit the maze. Suppose the rat must make one hundred appropriate discriminations, and that any mistake will prevent the rat from exiting the maze. A psychologist who sees the rat take no erroneous turns and in short order exit the maze will be convinced that the rat has indeed learned how to exit the maze, and that this was not dumb luck. With the simple maze there is a substantial probability that the rat will exit the maze by chance; with the complicated maze, this is exceedingly improbable.

It's now clear why the SP2 criterion is so well suited for recognizing intelligent agency: its use in inferring design coincides with how we recognize intelligent agency generally. In general, to recognize intelligent agency we must establish that one from a range of competing possibilities was actualized, determine which possibilities were ruled out, and then specify the possibility that was actualized. Moreover, the competing possibilities that were ruled out must be live possibilities, sufficiently numerous so that specifying the possibility that was actualized rules out chance. In terms of probability, this just means the possibility that was specified has small probability.

All the elements in the general scheme for recognizing intelligent agency (i.e., Actualization-Exclusion-Specification) find their counterpart in the SP2 criterion. And so, the SP2 criterion formalizes what we have been doing right along when we recognize intelligent agents, and indeed pinpoints how we recognize intelligent agency. As premise 2 of the deductive form of the design inference, the SP2 criterion therefore ensures that when the deductive form of the design inference yields a design conclusion, the inference does not merely eliminate chance but also convincingly implicates the activity of an intelligent agent.

The Metaphysics of Design and Information

The focus of this section has been logical and empirical, focusing on the mechanics of inferring design based on certain logical forms of reasoning and certain types of observational evidence. But this section also has metaphysical implications. As noted, ultimately what the design inference identifies is information. But not any old type of information. The design inference identifies information that is both specified and improbable, satisfying the SP2 criterion. This form of information (also called specified complexity or complex specified information) suggests a mode of causation separate from ordinary physical causation.

In this vein, the late physicist and theologian John Polkinghorne distinguished between "physical causality" and "top-down causality." Top-down causality could, according to Polkinghorne, influence the physical world via "active information," which he never fleshed out but which seems naturally to match up with complex specified information. In any case, for Polkinghorne, a transcendent intelligent cause could be expected to act entirely through "information input," influencing the physical world but without disrupting its nexus of physical causality. Top-down causality would thus be the work of active information.[20]

Physicist Wolfgang Smith takes a similar line, referring to physical causality as "horizontal causation" and top-down causality as "vertical causation." Smith contrasts what he calls "reducible wholeness," in which a whole is composed of parts and is nothing more than the sum of its parts, with what he calls "irreducible wholeness," in which the whole is greater than the sum of its parts. He argues that horizontal causation can give rise to reducible wholeness, but that for irreducible wholeness, horizontal causation is inadequate and vertical causation is required.

A careful reader of *The Design Inference* and other works by Dembski, Smith interprets the design inference metaphysically, as showing that horizontal causation cannot produce complex specified information (which he refers to by that very term). Moreover, he takes this form of information to decisively implicate vertical causation in the production of irreducible wholeness.[21] Irreducible wholeness, it's worth noting, provides a metaphysical counterpart to Michael Behe's notion of irreducible complexity.[22]

Theologian John Haught likewise sees in information something distinct from ordinary physical causality:

> In its encounter with the informational component of nature, modern science has come up abruptly against something quite distinct from the mechanical or material causes with which it has been most at home. Information, in the broad sense in which I am using the notion here, quietly orders things while itself remaining irreducible to the massive and energetic constituents that preoccupy conventional science. Its effects are clearly verifiable, but information itself slips through the wide meshes of science's mechanistic nets. Information, like mathematics, has a certain aspect of timelessness to it. It does not originate out of the historical past, nor does it undergo transformation in the same way that matter and energy do. It abides patiently in the realm of "possibility," waiting to be actualized in time. It is almost as though its various configurations have *always* dwelt

"somewhere," anticipating appropriate opportunities to become incarnate.[23]

Of these thinkers, only Smith could be regarded as embracing intelligent design. Polkinghorne was always ambivalent, shying away from it in biology but sympathetic to it in cosmological fine tuning. And Haught has always been outright critical of it. Haught's opposition to intelligent design makes his statement here about information all the more remarkable, since there's nothing here for a design theorist to disagree with. All three of these thinkers have a theological bent. Their metaphysical views, therefore, lie outside the philosophical mainstream, which is still largely committed to naturalism. Compare, for instance, how the philosophical mainstream understands the metaphysics of causation apart from God, as in the work of John Mackie.[24]

Yet to the degree that the design inference is not only valid but also widely applicable, the three thinkers cited here anticipate how the design inference can be metaphysically significant. For a sustained treatment of the metaphysics of information from a design-theoretic perspective, see my [WmD's] book *Being as Communion*.[25]

5.6 The Explanatory Filter

The Explanatory Filter is a flowchart that provides a user-friendly version of the design inference. It is what people best remember about the first edition of this book. Because in the present edition we have assimilated necessity into chance, the Filter as presented here is simpler than the earlier version. The earlier version had three decision nodes. This newer version has two. Logically, however, this and the earlier Filter are equivalent.

The Filter summarizes how the design inference arrives at a conclusion of design. It should be viewed as a shorthand. The design inference proper is captured in the specification/small-probability criterion of Section 5.5 and more comprehensively still in the full deductive argument of Section 5.4. So, if there is a fault in the design inference, the fault needs to be located in its logical underpinnings, as laid out in those two sections, and not in its distillation, as laid out in this section. Without further ado, here is the Filter:

The Explanatory Filter

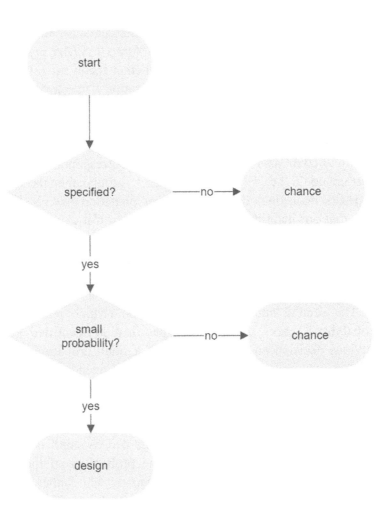

The Filter's operation is straightforward and, by now, ought even to be self-explanatory. As always with design inferences, to get the ball rolling we need an event E whose explanation stands in question. The event E is therefore fed into the Filter at the start node. Doing so affirms premise 1 in the deductive form of the design inference, namely, $oc(E)$, that the event E has occurred. E is then fed to the first decision node of the Filter.

The Filter's two decision nodes correspond to the specification/ small-probability (or SP2) criterion. Specifically, these two nodes correspond to confirming premise 2 of the deductive form of the design inference, namely, $(\exists W \in \mathcal{D})(\forall H \in \mathcal{H})[spec(W,E;H) \& smp(W;H)]$. The first decision node, which asks whether the event E is speci- fied, requires finding a particular specification V that, across all the chance hypotheses under consideration (i.e., $\forall H \in \mathcal{H}$), satisfies $spec(V,E;H)$.

Accordingly, V must have sufficiently low descriptive com- plexity to count as a specification and its corresponding event V^* must include E (i.e., $V^* \supset E$). If no such V is found, then the answer at the first decision node is no, and the explanation given is chance. Note that our inability to find such a V, even if it exists, will take us from the first decision node to a terminal node marked "chance." The danger of false negatives is therefore endemic to design inferences—we may miss a valid design conclusion simply for not knowing enough or not being perceptive enough.

The second and final decision node (which reads "small probability?") asks whether the specification identified at the previous decision node also renders E a small probability event. What counts as a small probability here depends on a set of probabilistic resources that are always made explicit in the full formalism of the design inference but are left unstated and implicit in the Filter.

Note that the probability to be regarded as small is not the probability of E and need not be the probability of the event associated with the specification V, namely V^*. Rather, for the probability to be small, it must be the probability of a rejection region induced by V. What's more, there can be multiple such rejection regions if more than one chance hypothesis must be eliminated, because rejection regions can vary with chance hypotheses. Hence, if even one relevant chance hypothesis fails to induce a rejection region having small probability, then the answer is no at the second decision node, and the Filter will conclude chance rather than design. Only if the answer is yes at both decision nodes does the Filter land at the terminal node labeled design.

The Explanatory Filter stands or falls with the full technical version of the design inference developed earlier in this chapter. Yet because the Explanatory Filter is what people tend to think of first if they think about the design inference, it is worth reviewing two main criticisms against the Filter: (1) we can never adequately grasp the full range of chance processes that could account for a natural phenomenon, and so the Filter is inherently unreliable; (2) the Filter makes chance and design mutually exclusive when instead these categories of explanation can overlap.

John Wilkins and Wesley Elsberry made the first criticism in an article titled "The Advantages of Theft Over Toil."[26] The theft here is supposed to be on the part of intelligent design proponents. It's an ironic title because, in practice, design inferences require a lot more work, and therefore toil, than inferences that merely draw a conclusion of chance. Wilkins and Elsberry attempt to invalidate the Explanatory Filter by claiming that we might always miss some chance process, and so we can never be justified in sweeping the field clear of such explanations to leave design as the only viable alternative. They thus charge the Explanatory Filter with committing an argument from ignorance, drawing a design conclusion from what we don't know rather than from what we know.

To see that their criticism fails, consider how they attempted to rebut design for an example from the first edition of this book. The example is about a combination lock that allows 10 billion combinations, only one of which can open the lock. With ordinary combination locks, any combination is potentially as good as any other for opening the lock. It therefore makes sense to assign each combination the same probability, or 1 in 10 billion. With this assumption and given probabilistic resources at no more than a billion opportunities to open the lock,[27] the Explanatory Filter would conclude design if someone walked up to the lock, spun a combination, and the lock opened (which is the event *E*). And indeed, who seriously would think of attributing such an event to chance?

Nonetheless, Wilkins and Elsberry question a design inference in even so simple and stylized an example as this by raising the specter of a poorly constructed lock for which the probability of opening it by chance is much larger than 1 in 10 billion. Granted, this might happen. But it might also happen that the lock requires dialing the right combination so precisely that opening it by chance has far smaller probability than 1 in 10 billion. Further investigation of the lock could therefore not only reverse but also reinforce a design inference.

The prospect of further knowledge upsetting a design inference poses a risk for the Explanatory Filter. But it is a risk that faces all scientific inquiry. Science is an inherently fallible enterprise. If we make faulty assumptions, we may draw faulty conclusions. Notwithstanding, the mere possibility of getting things wrong should never stop us from doing the best we can with what we do know—and reasoning accordingly from there. For Wilkins and Elsberry, no amount of investigation into a phenomenon is enough to rule out natural chance-driven processes as its proper explanation (and this holds even with our liberal understanding of chance, which also includes necessity). Yet if design in nature is real, such a recommendation ensures it will ever go undetected.

The risk of further knowledge upsetting a design inference is a feature and not a bug of scientific inquiry in general. This risk therefore has nothing to do with the Filter's reliability. The Filter's reliability is all about its accuracy in detecting design once we have set out the concepts and premises (Section 5.3) needed to make it work. Notably, this means we need a good grasp of the underlying chance hypotheses (i.e., the set $\mathcal{H}$). And from there it follows that if we don't know enough, we may not be able to apply the Filter. But contrary to Wilkins and Elsberry, knowing enough to apply the Filter does not mean being omniscient. We can know enough to apply the Filter and infer design without knowing and confirming every eventuality that might cause us to question the inference.

The second main criticism to the Explanatory Filter is that it artificially separates explanations into chance and design when they could overlap (or necessity, chance, and design when, in the first edition, we still separated necessity from chance). Philosopher of biology Michael Ruse makes this criticism. Citing Ronald Fisher, Ruse argues that all three modes of explanation—necessity, chance, and design—might need to be "run together." Ruse therefore writes:

> [Fisher] believed that mutations come individually by chance, but that collectively they are governed by laws (and undoubtedly are governed by the laws of physics and chemistry in their production) and thus can provide the grist for selection (law) which produces order out of disorder (chance). He cast the whole picture within the confines of his "fundamental theory of natural selection," which essentially says that evolution progresses upwards, thus countering the degenerative processes of the Second Law of Thermodynamics. And then, for good measure, he argued that everything was planned by his Anglican God![28]

Thus, we see chance, in the form of mutations, mixed together with necessity (what Ruse calls law, such as the laws of physics and chemistry), and then still further blended with the planning of the Anglican God, and thus with design.

But is Ruse addressing an actual weakness of the Explanatory Filter? The Filter always requires some event to be fed into it. But what event is Ruse actually referring to here that will cause the Filter to falter, not assigning an event just to design or just to chance (for Ruse, chance and necessity) but assigning it to both? His example from genetics leaves that question unanswered. If the event is that one strand of DNA is matched up with a complementary strand of DNA, then that seems to happen by pure chemistry, and so the proper explanation seems to be necessity (or as we frame it in this edition, chance where the probabilities are all zero and one).

If the event is a random change in a DNA base pair, then that event can be referred to chance (in the sense of an event with probability neither zero nor one). Indeed, no one denies that mutations happen. But what about the event that arranged the structure of a strand of DNA that enables it to code for a brand new protein that does really cool things inside the cell? Perhaps it is just the result of a sequence of mutations that could be explained by chance. But perhaps instead it is the result of design, a possibility that the Filter is able to countenance and whose a priori prohibition seems nothing more than a naturalistic prejudice.

Using the Filter requires clarity regarding what event is fed into it. You can't just stick a rusted old car into the Filter. A rust spot on the car's roof may rightly be referred to chance. Its sagging shocks will be due at least in part to gravity, and thus to necessity (which we are treating as a special case of chance). But the shape of its body or the structure of its chassis will clearly be the result of design. Without such clarity of what we're feeding into the Filter, we hamper our use of the Filter in accurately distinguishing between chance and design.

Consider an embossed sign reading "Eat at Joe's" that blows over during a storm and falls face down into the snow. The sign's fall is due to chance. The impression of the sign in the snow, however, is a matter of necessity (the sign would impress on the snow whatever was embossed onto it). Yet the lettering in the snow traces back to the construction of the sign, which is the result of design. The Filter is able to sort through all these options, but only if it is fed properly. GIGO—or garbage in, garbage out—applies to science generally and to the Filter specifically.

Attempts to discredit the Explanatory Filter always evince special pleading. People in general are happy to apply it, whether tacitly or explicitly, to the situations of life. Indeed, we cannot make sense of the world apart from the design inference, and the Explanatory Filter constitutes a convenient rational reconstruction of how we draw design inferences. If there's a problem, it's not with the Filter as such, but with what to do when it leads us to undesirable design conclusions. Overwhelmingly, those who resist the Explanatory Filter do so because it could have non-naturalistic implications that they would rather avoid.

5.7 Probabilistic *Modus Tollens*

In the logic of conditionals, two main inference rules exist, *modus ponens* and *modus tollens*. In *modus ponens*, given the conditional *if A, then B* and given also *A*, the conclusion *B* follows. In *modus tollens*, given the conditional *if A, then B* and given also *~B* (the negation of *B*), the conclusion *~A* (the negation of *A*) follows. For instance, given the conditional *if it's raining outside, it will be wet outside*, and given the claim *it's raining outside*, we may, from *modus ponens*, conclude *it's wet outside*. However, given that it's wet outside, we cannot conclude that it's raining outside because it could be wet outside without it raining outside (e.g., if sprinklers are running). The fallacy of inferring that it's raining outside from it being wet outside is known as *affirming the consequent*. *Modus tollens*, by contrast, rightly infers from it not being wet outside that it's not raining outside.

Philosopher of biology Elliott Sober has raised the concern that the design inference aspires to be a probabilistic form of *modus tollens*, and that in this respect it fails and shows itself to be invalid.[29] But the argument Sober makes here fails, and it is instructive to see how it fails. To set the stage, we need to distinguish between two concepts of logic, namely, strict and partial entailment. The design inference was formulated in Section 5.4 as a valid deductive argument. And yet, in most applications, the premises of the design inference will only be probable rather than certain or true. This fact about design inferences, however, does not undercut their widespread applicability.

In a valid deductive argument, as with the design inference, not only does the truth of the premises guarantee the truth of the conclusion; rather, it's also true that high probability of the premises (taken jointly) guarantees high probability of the conclusion. Thus, if C is a conclusion validly drawn from the premises $Q_1, Q_2, ..., Q_n$, not only is C true if $Q_1, Q_2, ..., Q_n$ are all true, but the probability of C is at least as large as the joint probability of $Q_1, Q_2, ..., Q_n$, which is to say $P(C) \geq P(Q_1 \;\&\; Q_2 \;\&\; \cdots \;\&\; Q_n)$.[30] Valid deductive arguments thus render a conclusion probable if the premises themselves are probable. Logicians denote this probabilistic relation between premises and conclusion as *partial entailment*. Entailment (or *strict entailment*) means that the truth of the premises guarantees the truth of the conclusion. Partial entailment means that the probability of the premises taken jointly guarantees at least as much probability for the conclusion.[31]

To see that strict entailment automatically yields partial entailment, it's enough to note that if several premises jointly entail a conclusion, then the ways those premises can together be true cannot exceed the ways the conclusion can be true. To return to the example of rain and wetness, the premise "it's raining outside" entails the conclusion "it's wet outside." Now every way for it to be wet outside includes every way for it to be raining outside, but might also include some other ways (such as a sprinkler being on). It follows that the probability of it being wet outside is at least as great as the probability of it raining outside. This simple example generalizes to all valid deductive arguments. Entailment therefore confers not only truth, but also probability. From this it follows that the design inference is robust, preserving probability, and thus is entirely suited to scientific inquiry and the uncertainties that so often attend scientific claims.

Given this connection between the logic of entailment and the logic of partial entailment, it follows that the standard logical inferences of *modus ponens* and *modus tollens* hold probabilistically. Thus, if we let $\rightarrow$ denote what logicians call the material conditional, which is to say that $E \rightarrow F$ is logically equivalent to $\sim E \lor F$, the following two inferences hold as a matter of strict entailment and thus also as a matter of partial entailment (the symbol $\therefore$ denotes "therefore" and thus signifies the conclusion from the two preceding premises):

$$\underline{Modus\ Ponens}$$
$$E \rightarrow F$$
$$E$$
$$\therefore F$$

Modus Tollens
$$E \rightarrow F$$
$$\sim F$$
$$\therefore \sim E$$

Of these two inferences, *modus tollens* seems particularly relevant to design inferences. That's because *modus tollens*, when construed probabilistically, depends on finding an event of small probability, call it *F*, that depends on some prior condition, call it *E*. This in turn implies that *~F* has large probability, and thus calls into question the condition *E* from which *F* is supposed to logically follow, thereby leading to the conclusion *~E*. Because the premises of both *modus ponens* and *modus tollens* strictly entail their conclusions, the premises also partially entail the conclusions, and so probabilistic versions of both *modus ponens* and *modus tollens* hold true.

And yet, Elliott Sober argues that "no probabilistic version of *modus tollens* is to be had" and that "there is no probabilistic analog of *modus tollens*."[32] In making these claims, Sober is not denying that from high-probability premises, *modus tollens* confers high probability on its conclusion. Rather, he is denying that any reliable inference can be drawn from this fact. According to him, *modus tollens*, as a strict logical entailment, requires that the conclusion be accepted given the truth of the premises. But he denies that its conclusion should be accepted given only the high probability of its premises.

Sober sees his refutation of probabilistic *modus tollens* as a refutation of the design-inferential logic developed in this book. But in fact, Sober's refutation of probabilistic *modus tollens* fails and, with it, his critique of the design inference. To see this, consider that Sober does not recast *modus tollens* as a partial entailment. In particular, the conditional that appears as the first line in Sober's version of *modus tollens* is not a material conditional (as is needed to make partial entailment a "probabilistic analog," as he puts it, of strict entailment). Instead, it is a Bayesian conditional that depends on conditional probabilities. We'll denote this conditional by $\Rightarrow$ (in place of the material conditional, which we denoted by $\rightarrow$).

Thus, in Sober's approach to *modus tollens*, he forms a conditional that depends on a hypothesis *H*, and then he forms a conditional probability based on this hypothesis. His version of *modus tollens* therefore looks as follows:

Modus Tollens

$H \Rightarrow E$

$\sim E$

$\therefore \sim H$

Moreover, the relevant probabilities here are $P(E|H)$ for the first premise, and $P(\sim E)$ for the second premise, and $P(\sim H)$ for the conclusion.

Now it should be noted that even in this Bayesian reformulation of *modus tollens*, high probability of the premises confers high probability on the conclusion. To see this, suppose $P(E|H)$ and $P(\sim E)$ are both high probabilities—in other words, both close to 1. Because $P(\sim E) = P(\sim E|H)P(H) + P(\sim E|\sim H)P(\sim H)$, and because $P(\sim E|H)$ must be close to 0 given that $P(E|H)$ is close to 1, it follows that the only way $P(\sim E)$ can be close to 1 is for $P(\sim E|\sim H)P(\sim H)$ to be close to 1, and that means that $P(\sim H)$ must be close to 1, implying that $P(H)$ is close to 0 and thus that *H* has small probability.

Nonetheless, Sober argues that *H* should not for that reason be rejected. The problem is that once $\sim E$ is observed, and thus seen to have large probability (rendering the probability of *E* small), the probability *P*, as applied to *H* and $\sim H$ in the conclusion, must now be updated in light of the non-occurrence of *E*, that non-occurrence being treated as evidence that affects the probability of *H* and $\sim H$. The prior probabilities $P(H)$ and $P(\sim H)$ thus, for Sober, no longer apply to the conclusion. Accordingly, the probability of *H* could now go way up because, properly conceived, as far as Sober is concerned, it is temporally indexed by new evidence. But this temporal indexing of probabilities, as a way of reassigning and updating them, is entirely handwaving—Sober offers no solid grounds or procedure for doing so.

To understand what is really going on here, let's consider Sober's main counterexample against probabilistic *modus tollens* and thus against rejecting *H* in such probabilistic arguments. Note that Sober's intention in not rejecting *H* is to defeat the design inference as it might apply to evolutionary biology: *H*, which would correspond to a chance hypothesis involving natural selection and mutation, would therefore not need to give way to a design inference that yields a conclusion of design. In reading this counterexample, think of *H* as the chance hypothesis for spinning a fair roulette wheel:

> Consider a roulette wheel in which we distinguish only double-zero and not-double-zero as possible outcomes. A perfectly satisfactory theory of this device might say that the probability of double-zero is 1/38 and the probability of not-double-zero is 37/38 on each spin. Suppose we spin the wheel 3,800 times and obtain a sequence of outcomes in which there are 100 double zeros. The probability of this exact sequence of outcomes is $(1/38)^{100} \times (37/38)^{3,700}$, which is a tiny number. The fact that the theory assigns this outcome a very low probability hardly suffices to reject the theory.[33]

The problem with this counterexample is that Sober has focused on the wrong event to assess the tenability of the chance hypothesis (i.e., the hypothesis assigning probability of 1/38 to double-zero and 37/38 to not-double zero). Sober asks us to consider spinning a roulette wheel 3,800 times and obtaining a sequence of outcomes in which there are exactly 100 double zeros. But spinning a roulette wheel 3,800 times and getting 100 zeros is not the same event as getting an "exact sequence of outcomes" having 100 zeros. The latter event is a particular instance of the former, but the former includes many more such exact sequences. Sober in effect gestures at doing one thing, but then ends up doing something very different: he first identifies the event of getting 100 zeros, but then calculates the probability not of this event but of getting one particular exact sequence that's included in this event.

The event whose probability Sober should have calculated is the one he identified at the first when he wrote, "Suppose we spin the wheel 3,800 times and obtain a sequence of outcomes in which there are 100 double zeros." This is the event of getting 100 double zeros in any order. Calculating the probability of this event requires multiplying the probability he calculated by all combinations of 100 double zeros among 3,800 spins of the roulette wheel, a number denoted by $\binom{3,800}{100} = \frac{3,800!}{100! \times 3,700!}$, which is equal to roughly 2.739×10^{199}. The probability Sober calculated was $(1/38)^{100} \times (37/38)^{3,700}$, which is equal to roughly 1.475×10^{-201}. Multiplying these numbers therefore yields the probability of seeing 100 double zeros in 3,800 spins and comes to .040, or 4 percent, which is 1 in 25.[34]

That's not a huge probability, but it is sizable. It's a smaller probability than getting 4 heads in a row, but a larger probability than getting 5 heads in a row. And who has not seen 5 heads in a row at some point in their coin tossing career? The probability Sober calculated, however, is 1.475×10^{-201}, whose probability falls between tossing 667 and 668 heads in a row. And where in the history of coin tossing has anyone seen that many heads in a row? A probability of 1 in 25 is sufficiently large that it's hard to imagine any circumstance where such a modest level of improbability would cause us to decisively reject a chance hypothesis H. Granted, some scientific research rejects null hypotheses at an improbability level of .05, which is 1 in 20 and so is bigger. But that's more a matter of casting doubt on H rather than actually eliminating it, to say nothing of therewith inferring design.

A hundred double zeros is the most likely number of double zeros to see in 3,800 spins of this roulette wheel. It is, as probabilists would say, the expected value. If this is our event E and if we substitute it into *modus tollens*, it's no longer the case that the probability of E is tiny (very close to 0) or that the probability of $\sim E$ is huge (very close to 1). Given these revised probabilities (i.e., .04 in place of Sober's 1.475×10^{-201}), the second premise in Sober's probabilistic *modus tollens* argument fails, and so his entire probabilistic *modus tollens* argument fails. Rejecting H, or concluding $\sim H$, now simply no longer follows. It's not that probabilistic *modus tollens* fails as an argument, as Sober suggests, but that the conditions for this argument to succeed were simply not met.

It's helpful to compare Sober's example to a similar one in which rejecting H is warranted. Suppose instead of 100 double zeros, no double zeros at all appeared in 3,800 spins of the roulette wheel. The probability in this case would come to $(37/38)^{3,800}$, or roughly 10^{-44}. There are no combinations to be factored into this probability because there are no double zeros here at all. And yet, even with double zeros having only a $1/38$ probability of appearing on any spin of the roulette wheel, we expect at least some double zeros in 3,800 spins. This departure from expectation is, to say the least, suspicious. The probability of 10^{-44} in this case is damning, suggesting that H, in characterizing a fairly spun roulette wheel, was not in fact operating. In running these two examples through probabilistic *modus tollens*, we see that the first doesn't lead us to reject H whereas the second does.

Sober offers still another example to refute probabilistic *modus tollens*. This time he considers urns containing white balls, with a different hypothesis attaching to each urn depending on probabilities of selecting different colored balls from it:

> Suppose I send my valet to bring me one of my urns. I want to test the hypothesis that the urn he returns with contains 2% white balls. I draw a ball and find that it is white. Is this evidence against the hypothesis? It may not be. Suppose I have only two urns—one of them is as described, while the other contains 0.0001% white balls. In this instance, drawing a white ball is evidence in favor of the hypothesis, not evidence against.[35]

Yes, in this case, observing a white ball would favor the 2% urn over the 0.0001% urn. But what if 100 random samples with replacement were taken from one of the urns, and in each sample only white balls were observed? The probabilities of 100 white balls for either urn would be vanishingly small. With the 2% urn, the probability for 100 random samples, with the ball returned each time and the urn shaken, would be 1.268×10^{-170}. With the 0.0001% urn, it would be 10^{-600}. In consequence, we should reject that either urn was used.

Sober, by contrast, would have to say that these 100 random samples decisively confirmed the 2% urn. Sober is committed to a form of Bayesian probabilistic reasoning in which all hypotheses have to be laid out in advance and then one must be preferred.[36] But confronted with probabilities on the order of 10^{-170} and 10^{-600}, the right answer here is not to prefer even one of these hypotheses—the right answer is "none of the above." Indeed, a more robust Bayesianism needs to be open to novel hypotheses that were not all laid out in advance. Thus, in the present example, we should think the valet substituted some third urn whose probabilistic behavior is very different from the urns Sober thought that his valet might bring him. That is, Sober's valet brought him an urn brimming with white balls.

As a matter of course, Bayesians need to assign significant prior probability to hypotheses that account for events matching specified patterns. Why? The very fact that these patterns are specified and thus easily described suggests that an event corresponding to it could reasonably be expected to happen. Thus, in Sober's example, a Bayesian ought to give some prior probability to the hypothesis of an urn containing only white balls. Even if that prior probability were somewhat low, after the hundredth white ball, its posterior probability would need to be quite high.

In sum, Sober's attempt to refute probabilistic *modus tollens* fails on its own terms. It therefore has no force against the logic of the design inference as laid out in this chapter.

CHAPTER 6

SPECIFIED COMPLEXITY

6.1 A Brief History of the Term and Idea

SCIENTISTS AS EMINENT AS FRANCIS CRICK, PAUL DAVIES, LESLIE Orgel, and Richard Dawkins have all endorsed specified complexity, sometimes using the very term, at other times using the terms *complexity* and *specification* (or *specificity*) in the same breath. All of them have stressed the centrality of this concept for biology and, in particular, for understanding biological origins. The actual term *specified complexity* goes back to 1973, when biologist Leslie Orgel used it in connection with origin-of-life research: "Living organisms are distinguished by their specified complexity. Crystals such as granite fail to qualify as living because they lack complexity; mixtures of random polymers fail to qualify because they lack specificity."[1]

In his book on the origin of life, *The Fifth Miracle*, physicist and science writer Paul Davies suggests that any laws capable of explaining the origin of life must be radically different from scientific laws known to date. The problem, as he sees it, with currently known scientific laws, like the laws of chemistry and physics, is that they cannot explain the key feature of life that needs to be explained, namely, the simultaneous occurrence of specification and complexity. As he puts it, "Living organisms are mysterious not for their complexity *per se*, but for their tightly specified complexity."[2]

Francis Crick, the co-discoverer with James Watson of the helical structure of DNA, didn't actually use the term specified complexity, but he talked around it and grasped the underlying idea. Crick as far back as 1958 wrote, "By information I mean the specification of the amino acid sequence of the protein."[3] To set the stage for this claim, Crick cited, with approval, A. L. Dounce: "The problem is to find a reasonably simple mechanism that could account for specific sequences without demanding the presence of an ever-increasing number of new specific enzymes for the synthesis of each new protein molecule."[4] Such a simple mechanism would thus explain the ever-increasing complexity of new enzymes and proteins without requiring small probability. In any case, specified complexity is evident even in this early work of Crick.

Last, consider biologist Richard Dawkins, who likewise grasps the underlying idea even if he doesn't use the exact term specified complexity. Thus, in his defense of Darwinism, *The Blind Watchmaker*, he writes:

> We were looking for a precise way to express what we mean when we refer to something as complicated. We were trying to put a finger on what it is that humans and moles and earthworms and airliners and watches have in common with each other, but not with blancmange, or Mont Blanc, or the moon. The answer we have arrived at is that complicated things have some quality, specifiable in advance, that is highly unlikely to have been acquired by random chance alone.... [M]y characterization of a complex object—statistically improbable in a direction that is specified not with hindsight—may seem idiosyncratic... [But] whatever we choose to call the quality of being statistically-improbable-in-a-direction-specified-without-hindsight, it is an important quality that needs a special effort of explanation.[5]

Dawkins here is defining complexity as improbability in the sense of pure random chance, and thus apart from natural selection. Our treatment of specified complexity, as we will see, takes a more general approach to complexity, focusing on improbability as it arises from any relevant probability distribution, even (or especially) one in which natural selection operates.

Technical nuances aside, each of these scientists admits that there is something crucially important for biology in the conjunction of specification and complexity. Indeed, they all imply that specified complexity as a concept (regardless of whether they use the exact term) is crucial for understanding biological innovation and origins. Yet none of them developed specified complexity with the theoretical precision needed to make it an effective analytic tool for science. Rather, they have been content to treat it as a suggestive, pre-theoretic concept. In consequence, none of them sees in specified complexity a reason to infer design. Rather, by keeping specified complexity ill-defined and under-developed, they effectively prevented specified complexity from challenging naturalistic understandings of biological innovation and origins.

The Reason for Specified Complexity's Demotion

Given the initial enthusiasm among mainstream scientists for specified complexity, why didn't they further expand or refine the concept? And what led them ultimately to withdraw their support for it? The most plausible explanation is that the emerging intelligent design movement started using specified complexity to denote the type of information emblematic of intelligent activity. Thus, by the early 2000s, only intelligent design proponents were using the term favorably.

Charles Thaxton, Walter Bradley, and Roger Olsen used the term in the late 1970s and early 1980s to characterize the type of information that was stymieing origin-of-life research. Though themselves proponents of intelligent design, they were circumspect during this period in what they wrote about specified complexity. In their 1984 book titled *The Mystery of Life's Origin: Reassessing Current Theories*, they addressed a mainstream scientific audience. The book elucidated the current naturalistic origin-of-life scenarios and the problems facing these scenarios.

Aside from the book's epilogue, *The Mystery of Life's Origin* was right in line with other books back then that were skeptical of existing naturalistic origin-of-life scenarios. A case in point was NYU biologist Robert Shapiro's 1986 book *Origins: A Skeptic's Guide to the Creation of Life on Earth*. This book dismissed all existing naturalistic origin-of-life scenarios to date as abject failures. Where *Mystery* struck new ground was in raising the possibility of intelligent design for the origin of life. It did this modestly in the book's epilogue, suggesting design as an option but not pressing for its adoption.

Ten years later, however, Bradley and Thaxton were more explicit about connecting specified complexity to intelligent design.[6] In 1994 they argued that the one constituted evidence for the other: "The only evidence we have in the present is that it takes intelligence to produce the second kind of order [i.e., specified complexity]."[7] But this remark appeared in an anthology titled *The Creation Hypothesis* and was published, unlike *Mystery*, with a religious press. *The Mystery of Life's Origin* had appeared with Philosophical Library, a well-regarded secular press that had published works by more than twenty Nobel laureates.

In the late 1990s, I [WmD] started to write about specified complexity, directly connecting it to intelligent design, at times also referring to it as *complex specified information*.[8] Yet the watershed event that scuttled use of the term among the scientific mainstream occurred in 2002. In that year, I published a book titled *No Free Lunch*. It was subtitled *Why Specified Complexity Cannot Be Purchased without Intelligence*. Clearly, this subtitle threw down the gauntlet.

Widely reviewed, including in *Nature* (by Brian Charlesworth), that book stilled any enthusiasm for the concept among mainstream naturalistic biologists. Up until this book's publication, specified complexity had been an intriguing notion acceptable to many in the scientific mainstream. Thereafter, specified complexity came to be regarded as a suspect notion, acceptable only to those outside the scientific mainstream.

Representative of the subsequent backlash against specified complexity is its Wikipedia entry. It's hard to imagine how the Wikipedia entry on specified complexity could be more dismissive of the concept. The entry begins, at the time of this writing in April 2023, as follows: "Specified complexity is a creationist argument introduced by William Dembski, used by advocates to promote the pseudoscience of intelligent design." As is obvious from our review of the term's earlier use among leading biologists, that claim is untrue. The account of specified complexity in that entry is not only biased but also about twenty years out of date. Provided unbiased editors at Wikipedia are able to gain control of this entry, they could use the material in this chapter to rehabilitate it.

By contrast, the aim of this book is to take specified complexity as a pre-theoretic concept and put it on solid footing as a theoretic concept for science in general and for biology in particular. Thus, we want to bring analytic precision to what is vague and fuzzy in statements by Orgel and others about specified complexity, empowering specified complexity to become a precise tool for scientific inquiry. We've already accomplished much of this task through our work on specification, small probability, and their coupling in the logic of the design inference (Chapters 3 through 5). The formal treatment of specified complexity in this chapter expands on specification and small probability as developed in earlier chapters.

The term *specified complexity* denotes an information-theoretic entity. The word *specified* matches up precisely with our understanding of specification as developed in Chapter 3. The word *complexity*, on the other hand, needs to be understood probabilistically, though recast in information-theoretic terms as bits. The more bits, the more complex, the more improbable. Thus in particular, the probability of an event E, namely $P(E)$, corresponds to the number of bits associated with E, namely $I(E) = -\log_2 P(E)$ (see Appendix A.11 for details on this connection between probability and information as measured in bits).

In a sense, then, everything we have done so far to unpack design inferences in terms of specification and small probability applies equally well to specified complexity. Just as the specification/small probability criterion, or SP2 criterion (see Section 5.5), is a marker of design, so specified complexity now becomes a marker of design. To exhibit specified complexity is to exhibit both a specification and require a long sequence of bits, where such a long sequence is a negative logarithm of a probability, and thus corresponds to a small probability. We can thus speak equally well of specified complexity as a criterion for design. Specified complexity may therefore be understood as encapsulating the design inference.

That said, specified complexity, as developed so far in this book, is an amalgam rather than a unification of specification and complexity. Specified complexity, to this point in our treatment, does not constitute a unified information measure. Nor for that matter does it make precise what we mean by description length in defining the concept of specification. Specifications, as we have so far defined them, are short descriptions of events. In motivating this definition intuitively, we focused on English language examples involving short description lengths. But such an approach to description length lacks generality and falls short of a rigorous mathematical analysis. We therefore need a precise mathematical formalism for specified complexity, and for description length in particular. One benefit of such a formalism is that we can then use specified complexity to shed light on less formal design inferences.

We are therefore faced with two understandings of specified complexity—generic specified complexity and algorithmic specified complexity. *Generic specified complexity* is simply the combination of specification and small probability, or equivalently specification and complexity, as has played out in our account of the design inference so far. Generic specified complexity is technically sound as far as it goes. It follows the formalism laid out in Chapter 5 on the logic of the design inference. Yet, as treated there, specified complexity is an amalgam of specification and complexity (or small probability). Moreover, it leaves description length in the definition of specification informally defined, and thus lacking full rigor.

In this chapter, by contrast, we develop specified complexity as a unified information measure that brings together specification and complexity (or small probability) in a seamless whole. In some of our published work, we have referred to specified complexity in this sense as *algorithmic specified complexity*.[9] Algorithmic specified complexity is a special case of generic specified complexity. In algorithmic specified complexity, description length is fully formalized and, just as with complexity, characterized in terms of bits. Moreover, as a unified information measure, algorithmic specified complexity is the arithmetic difference of two quantities in bits: complexity in the sense of improbability (in bits) minus minimum description length (in bits).

Intuitively, what this means is that complexity, in the sense of improbability, gets controlled by the length of the minimum description needed to describe the event. The longer this description, the more artificial or factitious it will be, and the more it will detract from the improbability of the event, rendering chance elimination increasingly implausible. Indeed, any event, however improbable, is describable by some description if the description is made elaborate enough. It's the improbable events captured with short descriptions that catch our attention and cause us to question whether they happened by chance. Algorithmic specified complexity captures this intuition deftly and makes it rigorous.[10]

As a unified information measure, algorithmic specified complexity has much to commend it. Algorithmic specified complexity is measured entirely in bits. An event E that measures n bits of specified complexity is going to have probability less than 2^{-n} because n will equal m minus k, where m is the complexity in bits and k is the description length in bits, both of which are positive. It follows that m must exceed n, and so the probability of E, namely 2^{-m}, must be less than 2^{-n}.

Also, if we consider the union of all events F with specified complexity n or greater, the probability of that union will be precisely 2^{-n} (given certain reasonable assumptions about the underlying language and about the minimum description length metric applying to that language). This union, an event in its own right, will then constitute a rejection region R for eliminating chance as the explanation of E. Given a probability bound α (local or universal), the inequality $P(R) = 2^{-n} \leq \alpha$ will then decide whether the probability is small enough to warrant the elimination of chance.

Moving forward, we will use the term specified complexity to denote both generic specified complexity and algorithmic specified complexity. From context, it will be clear which we intend, the giveaway being whether the language on which we are calculating description length has the formal properties required by algorithmic specified complexity. Generic specified complexity will then just be a shorthand for the combination of specification and small probability at the heart of the design inference as laid on in previous chapters. We apply generic specified complexity when we lack all the full technical apparatus of algorithmic specified complexity, though even then we usually are able to do so confidently and cogently. For the remainder of this chapter, however, we will develop that full technical apparatus, and so the focus here will be on algorithmic specified complexity.

6.2 Description Length

In the design inference as laid out in the previous chapters, the concept of probability is well understood and rigorously developed. Nonetheless, the other key concept of the design inference, specification, calls for further development. As previously discussed, we can model simplicity of patterns in terms of length of descriptions. Simple patterns have short descriptions whereas complicated and contrived patterns require long descriptions.

For our purposes, descriptions describe events rather than individual outcomes. A description can thus fit many different outcomes. For example, the description "three of a kind," a card hand in poker, fits many different combinations of cards. Mathematically, if Ω is the set of all possible outcomes, an event is defined as a subset of Ω. Note that outcomes, which are individual elements of Ω, can be formed into events by forming singleton sets of them. Thus, for $x \in \Omega$, $\{x\}$ is the event corresponding to that outcome (see Appendix A.1).

We will assume that our descriptive language allows only a countably infinite number of distinct sequences of symbols. At the same time, in probability theory, uncountably infinite probability spaces frequently arise. These spaces then also give rise to an uncountably infinite number of events. It follows that for such spaces, it is impossible that a description exists for every possible outcome or event (there are too few descriptions for too many outcomes and events). We therefore make the simplifying assumption that the probability spaces we consider in this chapter are countably infinite. It is a fact about the convergence of probability measures that for the probability spaces that arise in practice, any probability measure can be approximated to arbitrary precision with probability measures that concentrate all probability on just finitely many outcomes (i.e., convex linear combinations of finitely many point masses).[11] Thus, countably infinite probability/search spaces are sufficient for most practical cases of interest.

A description only exists in the context of a particular language. Without a language, a description is meaningless. Furthermore, the same sequence of symbols may have different meanings in different languages. For example, in American English the phrase "public school" refers to free government-run schools that provide education to all citizens, whereas in British English the same phrase refers to fee-charging schools that mainly appeal to the rich.

In common parlance, British and American English are considered the same language even if they do not assign the same meaning to every word or phrase. Even so, for our purposes we will consider any difference in the meaning of any combination of symbols to constitute a different language. This means not only that British and American English are different languages, but also that no two people speak the exact same language. No two people will interpret in the exact same way all the sequences of symbols they use in ordinary communication. Consequently, every person uses a subtly different language from anyone else.

Furthermore, people using the same languages in different contexts will intend different meanings. The word "straight" has one meaning when describing a hand of cards. It has another meaning when discussing the state of a road. It has yet another meaning when describing a person's character. And there are still other meanings. The context of a communication will dictate what linguistic meanings and conventions are used.

Mathematically, we will characterize a language $\mathscr{L}$ as a mapping from strings of symbols to events in Ω. In other words, for any particular string of symbols considered valid in $\mathscr{L}$, there is one and only one event in Ω described by that string of symbols, or description. If W is a description, then, in line with our asterisk notation of the previous chapters, we denote the event that the description describes as W^*. The length of the description is then denoted by $|W|$.

Evaluating description length in this way is central not just to algorithmic specified complexity as developed in this chapter but also to the field of algorithmic information theory.[12] This field belongs to information theory and theoretical computer science. It evaluates the complexity of data by the shortest possible computer program that, when run, outputs the data. This shortest program length is called the Kolmogorov complexity of the data. The Kolmogorov complexity of a binary string x is therefore defined as follows:

$$K(x) = \min_{W \ where \ W^{\#}=x} |W|$$

In this equation, we take W to be a program and $W^{\#}$ to be the output produced when the program is run. The minimization is then over all programs W such that $W^{\#} = x$.

To handle effectively the languages ($\mathscr{L}$) used to describe events, we introduce three simplifying assumptions about those languages. Following the lead of algorithmic information theory, we require that the language in question must be binary, prefix-free, and Turing complete.[13] Binary languages use only two symbols, 0 and 1. This is not a serious restriction on languages because any other system of symbols can be encoded as binary. All that is required is to assign a unique binary sequence to each possible symbol. Such a binary convention ensures that all languages considered are effectively using the same unit, namely bits, which makes it straightforward to compare lengths between them.

A prefix-free language is one in which no valid description is a prefix for another valid description. If a given binary string is a valid description, no other valid descriptions can be formed by adding onto the end of the string. In other words, the language must be self-delimiting: the logic of the language determines when we have reached the end of the description. No external factor, such as the gratuitous addition of more symbols, should be necessary in determining whether we have reached the end of a description.

It is straightforward to convert a non-prefix-free language into a prefix-free language either by embedding the length of the description in the description itself or by adding a stop signal to indicate the end of the description. Both of these methods will make the description slightly longer. Indeed, descriptions in prefix-free languages will be slightly longer than descriptions in non-prefix-free languages. But by limiting ourselves to prefix-free languages, we ensure consistency and comparability between different languages. This is similar to the consistency obtained by requiring the languages to be binary.[14]

A Turing complete language is one that is able to perform any algorithm, calculation, or computation that could be performed on a Turing machine. A Turing machine is a simple, but completely general, model of a computer. Thus, anything that can be computed on any computer can be computed on a Turing machine. Turing completeness ensures that it must be possible for the language $\mathscr{L}$ to describe any pattern that a computer could be programmed to recognize. The purpose of this assumption is to ensure that the language is sufficiently powerful to handle all the patterns we might face. We want to avoid languages that lack the ability to describe a full range of complicated patterns.[15]

Two approaches exist to characterize a Turing machine that recognizes a particular pattern. First, the Turing machine may take, as input, a binary encoded description of an outcome in the probability space Ω and either reject or accept that outcome depending on whether it fits the pattern. Alternatively, a Turing machine may, when run, output all sequences of binary encoded descriptions that fit the pattern. Either approach is equivalent, and the difference need not concern us.

The requirement that a language be Turing complete is not onerous. Very simple languages can be Turing complete. It is, in fact, difficult to avoid a language becoming Turing complete. Non-Turing complete languages tend to be trivial, unable to engage in computations that we cannot do without. The requirement of Turing completeness essentially comes down to insisting that the descriptive language be non-trivial.

Inspired by Kolmogorov complexity, we now define the minimum description length D of a given event X as the shortest program or description in a particular language that describes that event:

$$D(X) = \min_{W \text{ where } W^* = X} |W|.$$

This is very similar to the definition of Kolmogorov complexity except that it is defined for events in Ω rather than for binary strings. The minimum, here, is over all descriptions W in the language $\mathscr{L}$ for which the associated event, W^*, equals X. If no description W in $\mathscr{L}$ equals X, then $D(X)$ will be infinite—in other words, $D(X) = \infty$.

Even though we require that the language $\mathscr{L}$ be a binary, prefix-free, and Turing complete, this requirement leaves considerable variation in the possible languages that could be used. As discussed, no two people use the same language, and even the same person will use different languages in different contexts. These different languages might require descriptions of different lengths to identify the same pattern. Consequently, an agent evaluating a design inference using one language might come to a different conclusion from another agent using a different language.

It might seem disconcerting that the logic of the design inference could lead two agents to draw different conclusions about an event, one concluding chance, the other design. Such potential for disagreement is, however, found in any non-deductive argument, which always allows for judgments and thus for divergent conclusions. In the case of a deductive argument, the conclusion follows unavoidably from the premises. But even though the logic of the design inference can be formulated deductively, its premises are the result of non-deductive arguments, which determine the underlying probability distribution, descriptive language, etc. So, a design inference is never a purely deductive argument. It always has non-deductive elements and thus in the end constitutes a nondeductive argument.

Divergences of judgment in design inferences tend to be language dependent, with different languages leading to different evaluations of descriptive complexity for particular patterns. Language dependence is built into design inferences so that in the same circumstances one language may allow a design inference to succeed whereas another may cause it to fail. Indeed, without adequate linguistic resources, design inferences cannot be drawn.

Different schools of thought exist for making sense of the judgments inherent in non-deductive inference. The more objective schools propose that there is an objectively correct standard for these judgment calls and that human judgment calls are correct insofar as they approximate the true standard. On the other hand, the more subjective schools reject the existence of such a standard, instead affirming the acceptability of different agents drawing different conclusions.

We won't attempt to resolve that longstanding debate. Nonetheless, we will show that, in practice, different languages used by different agents will assign commensurate description lengths for the same event. That is, even though different agents will use different languages, they will not assign radically different description lengths to the same events. The description length of an event in the language used by one agent will be approximately the same as the description length used by other agents in their respective languages.

An intuitive confirmation of this claim comes from considering the relative lengths of works translated into different languages. A sentence translated from English into Spanish does not become a book. A treatise in Mandarin does not become a single word when translated into Latin. There certainly are differences in the lengths of the same work translated into different languages. Nevertheless, a sentence translated into another language remains the same general length: a sentence. A book translated into another language is still a book.

Given an objective view of the judgment calls underlying non-deductive inference, the natural approach to the design inference would be to propose a single objective language to be used as the standard for all design inferences. However, if the description lengths of practical languages are commensurate, this objective standard should be well approximated by measuring and comparing description lengths in the languages used by individual agents.

Given a subjective view, the natural approach would be to allow agents to use their own language and accept that conclusions will differ from agent to agent. However, if the description lengths of these languages are commensurate, an event with a short description in one language will be short across most languages. The length of a description in any one agent's language is a good approximation of the length across any other agent's language.

This means that whatever school of thought is found compelling, the commensurability of description lengths indicates that they will be roughly the same regardless of the choice of language. Indeed, languages that we use in reasoning about the world seem to share a common expressive power and to preserve concision (short descriptions) under translation. As such, if a design inference is valid according to one agent's language, it will tend to be valid for other agents' languages as well. And if it isn't valid in a different language, often what's needed are new coinages that enable what previously required long descriptions to be stated more succinctly. Languages are dynamic and can develop the linguistic resources needed to draw design inferences that otherwise would elude them.

When a design inference leads to a conclusion of design in one language and a conclusion of chance in another, the difference may be viewed as a shortcoming in the language that concludes chance. Other things being equal, design inferences conclude design more readily when shorter rather than longer descriptions describe the event whose design is in question. When a design inference fails to conclude design in one language but succeeds in another, the former language is unable to describe the event whose design is in question with the same concision as the latter language. For instance, if a language necessitates the complete and unabridged listing of every coin toss, then a thousand heads in a row will require a lengthy description, rendering a design inference unable to conclude design. Of course, our ordinary language allows for a much shorter description of a thousand heads, thereby underwriting a successful design inference.

Differences in languages may therefore alternately validate or invalidate a putative design inference. Moreover, there can be borderline cases where the inference is marginal and small difference in an agent's assessment of the description length of a pattern will make or break a design inference. However, for many interesting cases, the inference will not be marginal and a small difference in description length will not make or break the inference to design.

Algorithmic information theory faces a similar problem in that the shortest program with a given output will have different lengths depending on the language. To circumvent this language relativity, algorithmic information theory appeals to an invariance theorem. This theorem limits how much difference in length can exist between the shortest program in one language compared to another. In plain English, this theorem says that the difference between the two is negligible.[16] Because Kolmogorov complexity and description lengths are parallel concepts, the same invariance applies to description lengths with respect to the types of languages we are considering (i.e., binary, prefix-free, Turing complete).

For a description in one language, we can describe the same thing in another language by explaining in that other language how to understand the first language. It's then a simple matter to apply that explanation to the description in the first language to form a description in the second language. Descriptions can thus be moved freely from language to language by incorporating within each other explanations of other languages (which, in essence, are rules for translating between languages).

For example, if we had a description given in English, we could produce a description in French by first explaining in French how to understand English and then by giving the original description in English. The same idea works across the board to produce descriptions in Spanish, Mandarin, or any other language. Granted, this is a roundabout way to move a description from one language to another. But it indicates a limit for how much longer a description could be in one language versus another. The maximum difference will then be the length of the description of one language inside another.

Even so, it is rare that the shortest way to write a new description would be by describing the language of the original description. In practice, languages are usually similar enough that a more direct and efficient translation is possible. Recall that we required the languages in question to be Turing complete. The Church-Turing thesis states that all Turing complete languages can describe the same algorithms. All different Turing complete languages incorporate algorithmically equivalent constructs, thereby allowing equivalent descriptions to be constructed (what can be described in one language can be described in the other).

Another thing to bear in mind is that languages arise, in practice, from a common context of inquiry. They are not arbitrarily constructed. Instead, they are developed precisely in order to describe those sorts of patterns that agents expect to encounter. Consequently, agents, especially insofar as they share a common context of inquiry, will develop similar languages in which the description lengths will also be similar.

These considerations suggest that description lengths will be commensurate between different choices of language. The length of a description in some language will approximate the length in any other language. The degree of approximation will depend on the particularities of the languages. However, in most practical situations we expect the approximation to be good enough for all agents to reach the same conclusions in drawing or failing to draw design inferences.

6.3 Practical Approximation of Description Length

Theory is all well and good, but how do we, practically speaking, estimate the lengths of descriptions? One approach is to determine the elements required in a description, estimating how many bits each element would require to encode, and then add these bits for each element in the description together. This approach gives us an easy and effective way to estimate the number of bits required to form a description.

For example, a description could be formed from the set of approximately 200,000 words that effectively make up the English language.[17] A straightforward encoding of a word could be constructed by assigning a binary string to each of these 200,000 words. Such an encoding would require $\log_2 200,000 \approx 18$ bits. We can from there construct two-word descriptions, which will then have $200,000^2 = 40$ billion possibilities. Encoding these two-word descriptions would then require $\log_2 200,000^2 = 2 \log_2 200000 \approx 36$ bits. Multi-word descriptions are then handled similarly.

Nevertheless, such a straightforward encoding does not take into account that many of those 200,000 words are used rarely and that others are used frequently. An efficient encoding of the English language would take this frequency information into account to keep down the average number of bits needed for descriptions.

A useful connection between prefix-free languages and probability distributions pertains to this discussion of description length. A binary prefix-free encoding can be converted into a probability distribution by assigning a probability of $2^{-|W|}$ to the any description W (recall that $|W|$ is the length of W). Because the code is prefix-free, the extended Kraft inequality applies:[18]

$$\sum_{W} 2^{-|W|} \leq 1$$

A probability distribution must sum to one when summed over all possible outcomes.[19] A code converted into a probability distribution will, however, sum to at most one. If the code is complete, so that every possible sequence is either a valid code or a prefix to a valid code, then it will sum to one. On the other hand, if there are certain binary strings that are invalid in the sense of not mapping to valid descriptions, the probability distribution will apply incompletely, and so some of the probability will not be assigned to certain binary strings.

We just reviewed converting a prefix-free code into a probability distribution. It is also possible to reverse the process, converting a probability distribution into an approximately equivalent prefix-free code. In this scheme, the length of the code in bits matches, to as high a degree as possible, the probability of the associated outcome (r bits corresponding in the best case to a probability of 2^{-r}). Matching up codes and probabilities in this way can be done using a technique called Shannon coding.[20] Accordingly, we assign binary codes to each possible outcome so that the length of the code for an outcome X is approximately $-\log_2 P(X)$.[21]

In consequence, we can convert back and forth between prefix-free encodings and probability distributions. Thus, when looking at words from the English language, we can estimate the probabilities according to the frequency of words in English. Given data provided by Google in a trillion-word corpus,[22] the word "patriotic" appears approximately 4 million times in a trillion words, and "speech" appears 31 million times. We can thus calculate the length of the joint encoding ("patriotic speech") of these words:

$$-\log_2 \left[\left(\frac{4,000,000}{10^{12}} \right) \times \left(\frac{31,000,000}{10^{12}} \right) \right] \approx 33 \text{ bits.}$$

Why should we expect this probability distribution to provide a sensible approximation of the description length? In the case at hand, this approach tells us how many bits it would take to encode the phrase "patriotic speech" with respect to a particular probability distribution. Yet a complete standalone description would also have to specify the probability distribution that is used.

In practice, however, the appropriate probability distribution to be used is obvious and can be described succinctly (e.g., uniform distribution or binomial distribution). In such cases, specifying the distribution will make little contribution to the description length. Its effect will thus be negligible, and as a matter of course it can therefore be ignored from the calculation.

It is important to reiterate that assessing description lengths by summing the number of elements required to make up a description is an exercise in approximation. Indeed, as noted, all descriptions lengths are approximations, either in an ideal description language or in the subjective description language of other agents. Even so, in practice we find that the approximations are close enough to each other and to any presumed true value that the conclusions we draw tend to be consistent.

6.4 Specification and Complexity

Before inferring to design to explain an event, we must eliminate the alternative to design, namely chance. A number of distinct chance hypotheses might be relevant to explaining the event. For example, a page of text might be explained by ink randomly spilled, by monkeys banging on a keyboard, by randomly pasting together small chunks of text, or by many other random means. Each such explanation corresponds to a distinct chance hypothesis. Our approach to the design inference, then, is to use specified complexity to reject incorrect chance hypotheses, sweeping the field clear of them.

How then do we decide whether a given chance hypothesis H accounts for an event E? The crucial probability for answering this question is the probability of the event E given that it was produced in accord with the chance hypothesis H, which we write as $P(E|H)$. Note that it does not, strictly speaking, make sense to talk about the probability of E as such. Rather, we must always speak of the probability under some assumed hypothesis that might account for the event. Thus, when we see a probability like $P(E)$, an underlying chance hypothesis has simply been left unexpressed.

In Section 6.2, we not only characterized specification in terms of minimum description length but also gave a precise definition of this concept. There we defined $D(E)$ as the number of symbols required for the shortest possible description of the event E in the underlying language $\mathscr{L}$. Accordingly, the complexity (or improbability) of the event E is measured as $I(E|H)$ ($= \log_2 P(E|H)$) and its specification is measured as $D(E)$.

We need the language being used, $\mathscr{L}$, to be independent of the event being considered, E. To understand why, imagine that you drop your cell phone and, as a result, a complex network of cracks forms across the screen. Many different possible networks of cracks could form on your cell phone's screen, and any particular cracks you see will be highly improbable. Nevertheless, they now have a short description for you, namely, "my phone's cracks." But this concise description is only possible because your language now depends on the event in question. Instead, we require the language used for evaluating description length to be independent of the event under consideration.[23]

Ensuring strict probabilistic independence between the language $\mathscr{L}$ and the event E, however, may be difficult. Past events affect us in various ways, and that can include shaping the language we use to evaluate specified complexity. However, we have argued for the commensurability of description lengths between different languages. Insofar as the descriptions lengths assigned by $\mathscr{L}$ are commensurate to lengths in other languages that are independent of E, conclusions drawn using $\mathscr{L}$ will still follow.

The combined action of specification and complexity is helpfully illustrated in Figure 6.1. This figure recaps our work on the logic of the design inference in Chapter 5. Thus, in judging whether an event happened by chance, we bring together these two criteria, specification and complexity. Typically, we have some intuitive sense of a probability or complexity limit. Events more probable than this limit are comfortably attributed to chance. Only if the event is less probable than this limit will we have reason to reject the chance hypothesis. The vertical axis of Figure 6.1 characterizes this probability limit, cashed out in terms of an information measure $I(X|H)$.

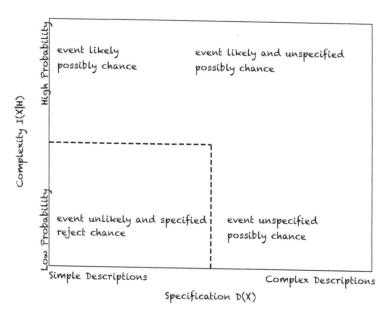

Figure 6.1

But we need more than complexity in the sense of improbability to reject chance. We also need specification in the sense of a low pattern complexity. Typically, we have some intuitive sense also of a pattern complexity limit. The horizontal axis of this Figure 6.1 characterizes this pattern complexity limit, cashed out in terms of a minimum description length measure $D(X)$.

We tend to pay little attention to patterns with a complexity greater than that limit. Such patterns require long-winded descriptions. They will thus seem artificial, and we will tend to dismiss them. But when an event is specified, which is to say it has a short description falling below that limit, and when it is also highly improbable (below the probability limit), then we feel justified in rejecting the chance hypothesis.

Mathematically, we may define a particular complexity limit $\bar{C}$ characterizing how improbable an event must be. And we must define a specification limit $\bar{S}$ characterizing how specified an event must be. We reject the chance hypothesis for events where $I(E|H) \geq \bar{C}$ and $D(E) \leq \bar{S}$. The following single event, comprising the union of all these events, then makes up the relevant rejection region:

$$R_{\bar{C},\bar{S}} = \bigcup \{E \mid I(E|H) \geq \bar{C} \ \& \ D(E) \leq \bar{S}\}.$$

This rejection region is none other than a specification-induced rejection region, described in Sections 3.6 and 3.7.

This approach to chance elimination, however, does not adequately capture all the cases for which we might reject a chance hypothesis. As it turns out, the following figure captures what's missing in the previous approach:

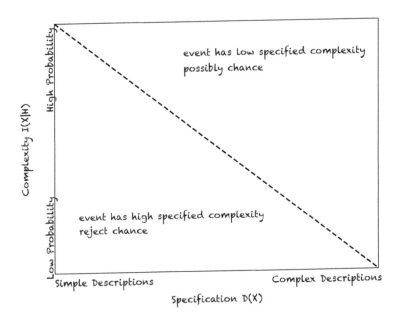

Figure 6.2

Figure 6.2 depicts the fundamental tension in chance-eliminative reasoning between lower probability on the one hand and longer description length on the other. Specifically, in eliminating a chance hypothesis to explain an event, we need to lower the probability of an event to the degree that we raise the length of its description. Chance elimination can live with longer descriptions provided that the probability is low enough.

For instance, in most circumstances, we would reject chance when getting heads in twenty consecutive coin flips. The probability is about 1 in a million, which for most purposes will seem quite small. Moreover, the description "all heads" for this pattern of coin flips is very short. So here we have a fairly small probability combined with a really short description length.

Nonetheless, we'd be more hesitant to reject chance when getting twenty consecutive coin flips that correspond to the bits of binary numbers written in ascending lexicographical order: 01000110110000010100. This is the Champernowne sequence, which was examined in Section 3.4. The probability of obtaining the first twenty bits in the Champernowne sequence is low—about 1 in a million, or the same as 20 heads in a row. But the pattern in the Champernowne sequence seems complicated. It requires a description not nearly as short as "all heads" but, instead, a longer description like "arrange binary digits in ascending lexicographical order." This complication in the description length makes us less inclined to reject chance for the first twenty bits of the Champernowne sequence.

That said, we would comfortably reject chance as an explanation for the first 100 bits of the Champernowne sequence:

$$0100011011000001010011100$$
$$1011101110000000100100011$$
$$0100010101100111100010011$$
$$0101011110011011110111100$$

Why would we almost certainly reject chance for these 100 bits but be far less sure about rejecting chance for its first twenty bits? The answer is that the longer description ("arrange binary digits in ascending lexicographical order") is offset by the greater improbability with 100 bits (one in 10^{30} in this case, rather than one in 10^6 for the twenty-bit sequence).

We can model this trade-off between probability and description length by combining specification and complexity into a single unified metric. This metric, which we refer to as *algorithmic specified complexity*, or simply as *specified complexity* when clear from context, is defined as

$$SC(E|H) = I(E|H) - D(E).$$

Note that from this definition $SC(E|H)$ gets bigger to the degree that $I(E|H) = -\log_2 P(E|H)$ gets bigger, which correlates inversely with the degree to which $P(E|H)$ gets smaller. Note also from this definition that $SC(E|H)$ gets bigger to the degree that $D(E)$ gets smaller, by thus subtracting fewer bits away from $I(E|H)$.

Events with low probability and short description lengths therefore have high specified complexity. On the other hand, events that are probable will have low specified complexity. Moreover, events that have only long descriptions will tend to have low specified complexity unless their improbability is extreme enough to counteract the long description length.

All these relationships between probability and description length in the specified-complexity metric are depicted in Figure 6.2, and underscored especially in its top-left to bottom-right diagonal. Whereas specification and complexity are represented on the horizontal and vertical axes as separate metrics, specified complexity is represented as a unitary metric on the diagonal.

Figure 6.2 captures an important feature of specified complexity that is absent from Figure 6.1. In Figure 6.1, the combination of specification and small probability is represented as a rectangle. This rectangle delivers a limited form of specified complexity in making specification and small probability constrain an event separately. But it's the diagonal in Figure 6.2 that captures the full understanding of specified complexity, balancing trade-offs between probability and description length so that a longer description length can always be offset with sufficiently small probability.

Combining improbability (in the form of $I(E|H)$) and minimum description length (in the form of $D(E)$) into a single unitary metric induces a new type of rejection region. Thus, for a specified complexity bound σ, its induced rejection region is

$$R_\sigma = \bigcup \{E \mid SC(E|H) \geq \sigma\}.$$

This is the union of all events E whose specified complexity with respect to the chance hypothesis H is at least σ.

As we'll see in the next section, the probability of R_σ is bounded above by $2^{-\sigma}$. It follows that a large specified complexity induces a rejection region of small probability ($2^{-\sigma}$ is small to the degree that σ is large). This is as it should be if specified complexity is going to eliminate chance based on a specification-induced rejection region (in line with Chapters 3 and 4).

The beauty of this specified complexity metric SC is its conceptual unity, simplicity, and elegance. Instead of working separately with two metrics and verifying that together they induce a valid rejection region, we now need to work with only a single metric that directly induces a rejection region (i.e., R_σ as opposed to $R_{\bar{C},\bar{S}}$).

As we will see in the examples of specified complexity at the end of this chapter, this feature of the SC metric allows it to effectively decide between chance and design in situations where otherwise so deciding might stand in doubt. Note that there is nothing wrong with using the rejection region $R_{\bar{C},\bar{S}}$ to assess specification and small probability, and therewith to reject chance. But the rejection region R_σ arises from a deeper conception of specified complexity.

6.5 Frequentist Interpretation

Specified complexity is relevant to the testing of chance hypotheses. It applies both to frequentist hypothesis testing, as in this section, and to Bayesian hypothesis testing, as in the next section.

In frequentist hypothesis testing, an experimenter identifies a null hypothesis H to be tested. The experimenter calculates the probability of an observed event E according to that null hypothesis—in other words, $P(E|H)$. This probability is called a p-value. If this p-value is sufficiently small (typically less than some probability bound α), the experimenter rejects the null hypothesis and so rejects chance.

With specified complexity, we need to evaluate two quantities, minimum description length (in relation to specification) and smallness of probability (treated as an information-theoretic complexity measure). We reject the chance hypothesis if the event falls into the rejection region defined by the complexity bound $\bar{C}$ and the specification bound $\bar{S}$ (recall the last section). In order to compute a p-value for a given choice of $\bar{S}$ and $\bar{C}$, we need to determine how probable it is that the outcome would fall within the region $R_{\bar{S},\bar{C}}$, which is defined by those bounds.

Given $\bar{S}$ as a bound on the length of descriptions in a prefix-free binary encoding, there are at most $2^{\bar{S}}$ possible descriptions satisfying that bound. Furthermore, the events corresponding to the descriptions included in the region have the probability bounded by $2^{-\bar{C}}$. These events are not, in general, mutually exclusive. Nevertheless, a bound on the total probability can be given by multiplying the number of possible events $2^{\bar{S}}$ and the maximum probability of each event, $2^{-\bar{C}}$. Thus,

$$P\left(R_{\bar{S},\bar{C}}\right) \leq 2^{\bar{S}}2^{-\bar{C}} = 2^{-\bar{C}+\bar{S}}.$$

This calculation gives the p-value for an event. The p-value is small when the event is both highly complex (in the sense of having a small probability) and highly specified (in the sense of having a short description). Within a frequentist statistical testing framework, the p-value being small justifies rejecting the null or chance hypothesis. This characterization of chance elimination is logically equivalent to the one described in Sections 3.6 and 3.7, the only difference being that we use an information-theoretic idiom here, a more straight-up probabilistic idiom back in those earlier sections.

One pitfall that frequentist statistical testing needs to guard against is *p-hacking*. *P*-hacking refers to performing numerous statistical tests, looking for one that gives a desired result. Once one observes the desired result, all the negative results are then simply ignored. One pretends they never happened, putting them away in a "file drawer." This is why the flipside of *p*-hacking is referred to as the *file drawer effect*. *P*-hacking looks for a given positive outcome by hiding negative outcomes in the proverbial file drawer, hoping they will be forgotten there. With a large enough file drawer, any event, however improbable, can be made to seem as though it occurred by chance.

Every individual statistical test has some chance of a false positive. Consequently, running many tests, as in *p*-hacking, increases the chance of a false positive. The specified complexity framework largely avoids *p*-hacking, and thus false positives, because it strictly limits the number of statistical tests. Choose a complexity bound $\bar{C}$ and a specification bound $\bar{S}$, and all the multiple tests, in the form of probabilistic resources, will automatically be factored in and given their probabilistic due. The only way to engage in *p*-hacking then is by varying the choice of complexity and specification bounds. This is why, in Section 3.7, we insisted on choosing these bounds in advance of any statistical analysis.

But there's an even better way to avoid *p*-hacking, and that's to use the specified complexity metric *SC*, which combines specification and complexity into a single metric. This approach yields a single rather than a joint test statistic, paralleling conventional statistical test metrics. In this case, we reject a chance hypothesis if the specified complexity exceeds some limit σ. The relevant *p*-value, then, is the probability of the union of all events with a specified complexity above that limit. In other words, we need to calculate, as in the last section, the probability of the rejection region R_σ.

To that end, we must consider all events making up the rejection region R_σ. These are the events E that satisfy $SC(E|H) = I(E|H) - D(E) \geq \sigma$. With algebraic manipulation, we can see these are the events bounded by $P(E|H) \leq 2^{-D(E)-\sigma}$. The total probability of all these events is then bounded by the following summation:

$$P(R_\sigma) \le \sum_{E \in R_\sigma} P(E|H) \le \sum_{E \in R_\sigma} 2^{-D(E)-\sigma} = 2^{-\sigma} \sum_{E \in R_\sigma} 2^{-D(E)} .$$

As we saw in Section 6.3, by the Kraft inequality, prefix-free codes have the property that summing over all descriptions yields at most one. It follows that

$$\sum_{E \in R_\sigma} 2^{-D(E)} \le 1.$$

Consequently, the above summation is bounded above by $2^{-\sigma}$—in other words:

$$P(R_\sigma) \le \sum_{E \in R_\sigma} P(E|H) \le 2^{-\sigma} .$$

The probability of obtaining σ bits of specified complexity is thus $P(R_\sigma) \le 2^{-\sigma}$. This means that the p-value for σ bits of specified complexity is an elegant $2^{-\sigma}$. Moreover, this p-value is immune to p-hacking. Thus, for large σ we are justified in rejecting the chance hypothesis.

6.6 Bayesian Interpretation

An alternative approach to testing the chance hypothesis H uses Bayes' theorem:

$$P(H|E) = \frac{P(E|H)}{P(E)} P(H)$$

In a Bayesian approach, $P(H)$ signifies confidence in the truth of a particular hypothesis H before taking into account an event E. $P(H|E)$ then signifies the revised confidence in H after taking into account that event E, treated as data or evidence. The factor $P(E|H)/P(E)$ then denotes how the event should affect our confidence in the chance hypothesis H.

The underlying intuition here is that when a particular event E is highly probable under a particular hypothesis H (which is to say $P(E|H)$ is large), but when the event E is generally improbable across relevant chance hypotheses (which is to say $P(E)$ is small), then it is unlikely that we would observe E unless the hypothesis is true. The ratio $P(E|H)/P(E)$ will thus be large because the occurrence of the event constitutes evidence that the hypothesis is true. Alternatively, if the event is improbable under a particular hypothesis (which is to say $P(E|H)$ is small), but the same event is probable in general (which is to say $P(E)$ is large), then the ratio $P(E|H)/P(E)$ will be small, constituting evidence against the hypothesis.

We are interested in the second case, where the probability of the event given the chance hypothesis, namely $P(E|H)$, is small. This makes the numerator of $P(E|H)/P(E)$ small. It's tempting to conclude that E therefore provides evidence against the hypothesis because we then expect $P(E|H)/P(E)$ to be small as well. But we cannot conclude that $P(E|H)/P(E)$ is small simply because $P(E|H)$ is small. If $P(E)$ is close to $P(E|H)$, the factor $P(E|H)/P(E)$ will instead end up being close to one. This happens in the case that the observed event is improbable under all relevant hypotheses. And so, its improbability under the hypothesis H will not constitute evidence against that hypothesis.

Given an explicitly identified chance hypotheses H, it's typically straightforward to evaluate $P(E|H)$. Moreover, we are assuming that $P(E|H)$ is small. But how can we evaluate $P(E)$? We could use the law of total probability (see Appendix A.2), evaluating the total probability of the event under all possible hypotheses (with the caveat that they are sufficiently fine-grained to avoid overlap and thus are mutually exclusive):

$$P(E) = \sum_{X} P(E|X)P(X)$$

Calculating this sum over all possible hypotheses, however, is impractical. For instance, consider an alternative hypothesis A that guarantees the observed event E: $P(E|A) = 1$. Suppose the prior probability of A is larger than the probability of the event under the chance hypothesis: $P(E|H) < P(A)$. This makes sense since we are considering the case where $P(E|H)$ is small, and thus positing that this alternative hypothesis has a probability at least as large. In that case,

$$P(E) = \sum_{X} P(E|X)P(X) \geq P(E|A)P(A) = P(A).$$

$P(A)$, the prior probability of this alternative hypothesis, then forms a lower bound for $P(E)$.

But this approach hinges on the prior probability of the hypothesis A, namely, $P(A)$. As it is, Bayesians are not agreed on how to evaluate priors. A plausible method for evaluating priors could be Occam's razor. This parsimony principle says we should prefer simpler theories over more complex ones (provided they have the same explanatory power). Moreover, a plausible approach to quantifying this principle would then employ description length of events.

These lengths will depend on the choice of language. Let's assume we use the same language as that used to evaluate the specification of events. Thus, we can define a prior distribution over possible alternate hypotheses A satisfying

$$P(A) \geq 2^{-D(E)}.$$

As a consequence, the general probability of E (in other words, $P(E)$, the probability of E without this event being conditioned on any particular hypothesis) can be bounded using description length. Thus, we can expect a shorter description $D(E)$ to imply a bigger probability $P(E)$:

$$P(E) \geq P(A) \geq 2^{-D(E)}.$$

In that case,

$$\frac{P(E|H)}{P(E)} \leq \frac{P(E|H)}{2^{-D(E)}}.$$

If we now take the negative logarithm, we obtain:

$$
\begin{aligned}
-\log_2 \frac{P(E|H)}{2^{-D(E)}} &= -\log_2 P(E|H) + \log_2 2^{-D(E)} \\
&= I(E|H) - D(E) \\
&= SC(E|H),
\end{aligned}
$$

implying that

$$-\log_2 \frac{P(E|H)}{P(E)} \geq SC(E|H).$$

Thus, for a parsimoniously described event, a high quantity of specified complexity constitutes Bayesian evidence for rejecting the chance hypothesis. In that case, the ratio $P(E|H)/P(E)$ that's the multiplier in Bayes' theorem will be quite small. The intuition here is that if an event has a short description, we expect that some parsimonious explanation exists for that event. As such, we should not accept that the event was instead produced by chance.

An alternative Bayesian approach to Occam's razor exists. It compares two different hypotheses by applying Bayes' theorem to the ratio of their posterior probabilities:

$$\frac{P(H|E)}{P(A|E)} = \frac{\dfrac{P(E|H)}{P(E)} \times P(H)}{\dfrac{P(E|A)}{P(E)} \times P(A)} = \frac{P(E|H)}{P(E|A)} \times \frac{P(H)}{P(A)}.$$

This approach has the advantage of removing the troublesome general probability of E, that is, $P(E)$. Thus, we do not need to be concerned about the total probability of the event, but only the probability of the event for the two hypotheses under consideration. The disadvantage of this formulation is that it can only tell us the merits of hypotheses H and A relative to each other, thus perhaps missing some important hypothesis. The advantage is that by not being swamped with other hypotheses, we can adjudicate clearly between the two.

The quantity $P(E|H)/P(E|A)$ is the Bayes factor, and also known as the likelihood ratio (see Appendix A.4). It quantifies how the event denoted by E, thought of as data or evidence, ought to shift our confidence between the hypotheses H and A. It is often useful to consider the log Bayes factor:

$$
\begin{aligned}
-\log_2 \frac{P(E|H)}{P(E|A)} &= -\log_2 P(E|H) - \log_2 P(E|A) \\
&= I(E|H) - I(E|A) \\
&= SC(E|H) - SC(E|A).
\end{aligned}
$$

We can compare a particular chance hypothesis, call it H, with that of a generic agent hypothesis, call it A. We assume, as seems reasonable, that a generic agent is more likely to take actions that will result in an event with a short description. After all, the agent will be operating with a language and thus be more likely to attend to shorter descriptions and try to instantiate them.

We therefore assume that in bringing about the event E, the agent chooses a short description and then causes the event answering to that description, essentially converting our language into a model for plausible agent activity. In particular, we assume, in line with Shannon's coding theorem (see Section 6.3), that the agent chooses a description of E with probability $2^{-D(E)}$, so that $P(E|A) = 2^{-D(E)}$.

If we evaluate the log Bayes factor for the two hypotheses H and A, we obtain:

$$
\begin{aligned}
I(E|H) - I(E|A) &= I(E|H) - \log_2 2^{-D(E)} \\
&= I(E|H) - D(E) \\
&= SC(E|H).
\end{aligned}
$$

Here again we espy the specified complexity metric. The underlying intuition is that for events with short descriptions, an agent is much more likely to cause those events than would happen under random chance. As such, a high quantity of specified complexity provides Bayesian evidence for a generic agent hypothesis A over against a chance hypothesis H.

We thus have two different ways to interpret specified complexity within a Bayesian framework. We can take the description length to give an estimate of probability based on either an agent or a parsimony assumption. In either case, specified complexity takes on quantitative values that indicate evidence against a chance hypothesis.

It's important to remember that within a Bayesian framework, different languages give rise to different priors. It follows that once a minimum description length metric is associated with a Bayesian framework via a given language, the applicability of specified complexity as a guide to Bayesian reasoning becomes unavoidable. The only way to block this logic of specified complexity is then to object to a particular prior.

A Bayesian might therefore insist that a proffered prior is incorrect and that a different one is required. Nonetheless, any prior is going to depend on some choice of language involving description lengths. As such, any Bayesian has to accept the basic logic of specified complexity. Their only recourse against this logic is to then object to the description language used for drawing probabilistic inferences. And most such languages are effectively equivalent.

Thus, it makes little difference if Bayesians settle on a language consistent with their idea of the correct prior probability. Insofar as that prior exemplifies parsimony, as it seems a plausible prior would, it should still be consistent with priors derived from description lengths. As such, we expect that the description lengths in a Bayesian's chosen language will remain comparable to those used by non-Bayesians. Specified complexity's role in eliminating chance and inferring design thus applies to Bayesians and non-Bayesians alike.

6.7 Contextual Factors

Specified complexity can accommodate various contextual factors, including background knowledge, scope independence, repeated trials, and transformation. We consider these next.

Background Knowledge

Previously, we argued that different possible languages used to describe events will be commensurate in that they will assign description lengths that are approximately the same. However, this conclusion can break down when some languages incorporate background knowledge that is absent from others. Such background knowledge, distinct from the language under study, may be related to the event being considered and revise its minimum description length.

To illustrate the problem here, consider Mount Rushmore, a large carving of four U.S. presidents into the side of a mountain. We are able to give the shape of Mount Rushmore a short description because of our knowledge of the U.S. presidents, or least of human faces. Consequently, an American whose language contains references to presidents or people is able to assign a short description to the monument. However, an alien unfamiliar with either presidents or people would be unable to give it a short description.

For the purposes of evaluating the specified complexity of Mount Rushmore, is it valid for humans to use their knowledge of presidents and people to assign Mount Rushmore a short description? After all, an alien would not be able to do the same. As such, the description lengths assigned by the alien versus the American would not be commensurate. However, we can show that the same conclusion would be reached by the aliens when they take into account the relevant background knowledge.

To that end, we can define a minimum conditional description length metric, which generalizes our ordinary minimum description length metric $D(E)$. Thus, $D(E|B)$ will denote the length of a shortest possible description of E given the background knowledge B. Accordingly, the minimum conditional description length $D(E|B)$ may be shorter than, and cannot be much longer than, the ordinary minimum description length $D(E)$ because $D(E|B)$ will be able to make use not just of all the resources of the language in question, which we usually denote as $\mathscr{L}$, but also of the background knowledge B.

In the example we are considering, the conditional description length of Mount Rushmore given knowledge of people or presidents will be short. On the other hand, the unconditional description length would be longer without that background knowledge. In essence, B expands our language $\mathscr{L}$ to a new language $\mathscr{L}_B$. This expanded language incorporates the expressive resources of B and thereby allows E to be more briefly described.

It is convenient to treat events and background knowledge within a more general framework, treating them both under the umbrella of *propositions*. In philosophy, a proposition is an abstract entity that has a meaning (sense) and makes a truth claim (reference). Typically, propositions are represented linguistically via statements that express the propositions. An event can then correspond to a proposition asserting the event's occurrence. And background knowledge can then correspond to a proposition asserting the background knowledge.

Given this more general framework, we can then assign probabilities to propositions (corresponding to the probability of their truth). Moreover, because statements are linguistic entities used to express propositions, we can use our minimum description length measure to determine the shortest statement that characterizes a proposition. Statements are to propositions as descriptions are to events, and so we can use our minimum description length metric (both conditional and unconditional) for statements of propositions just as we used it for descriptions of events.

Accordingly, it makes sense to think of a complete description of an event conjoined with background knowledge (i.e., E & B) as decomposing into a description of the background along with a further conditional description of the event given the background knowledge. Note that because all entities here are propositions, we can apply logical connectives to them, such as conjunction and disjunction. Furthermore, it makes good intuitive sense for the shortest length of this description of the conjunction to be approximately the same length as the shortest description of the background knowledge plus the shortest conditional description of the event given the background knowledge:

$$D(E \mathbin{\&} B) \approx D(B) + D(E|B).$$

In other words, the descriptive cost of E and B jointly is the descriptive cost of B plus the descriptive cost of E given that we can use B to describe E.

By the Shannon coding theorem, the encoding of a description of an outcome follows a probability distribution. We can extend such a probability distribution over descriptions of outcomes (and from there to nonoverlapping events) to a probability distribution over statements of propositions. The encoding of a statement of a proposition will thus likewise follow a probability distribution. We can therefore apply the previous result to background knowledge, characterizing it in terms of the probability distribution over the range of possible items of background knowledge. As such, we can approximate the minimum description length of $E \mathbin{\&} B$ as follows:

$$D(E \mathbin{\&} B) \approx I(B|H) + D(E|B).$$

We can then plug this approximation into the definition of the specified complexity metric:

$$\begin{aligned} SC(E \mathbin{\&} B|H) &= I(E \mathbin{\&} B|H) - D(E \mathbin{\&} B) \\ &\approx I(E \mathbin{\&} B|H) - I(B|H) - D(E|B). \end{aligned}$$

It can happen that the event and background knowledge will not be probabilistically independent, as when the background knowledge is simply read off of an event. But in case the event and background knowledge are probabilistically independent, as when humans have knowledge of U.S. presidents or human faces independently of Mount Rushmore, the following equation will then hold:

$$SC(E \,\&\, B|H) \;\approx\; I(B|H) + I(E|H) - I(B|H) - D(E|B)$$
$$= \; I(E|H) - D(E|B).$$

On the very right of this equation is almost the specified complexity metric $SC(E|H)$, but with $D(E|B)$ in place of $D(E)$. This means that as long as the background knowledge is independent of the event under consideration, using the background knowledge to describe the event does not decrease its specified complexity. Furthermore, it is not required that the event and background knowledge be independent under all possible hypotheses, but only under the hypothesis being considered.

Thus, incommensurability in the description lengths assigned by different languages as a consequence of independent background knowledge will still produce approximately the same, or even more, specified complexity. Except for artificial circumstances in which background knowledge is directly read off of events, such independence is typical. Thus, the variability of different languages in assigning different description lengths to events based on the presence of absence of different background knowledge will typically leave specified complexity largely unaffected and the conclusions we draw unchanged.

Scope Independence

In some cases, the same event may be described with varying degrees of specificity. The same animal might be described as a collie, dog, canine, mammal, or tetrapod. Intuition suggests that increasing the specificity of the description should not decrease the specified complexity (and may well increase it). Let's now confirm this intuition.

Consider a general event E. E may be divided into a number of non-overlapping more specific events $E_1, E_2, E_3, ..., E_{n-1}, E_n$. There will be some probability distribution $P(E_k|H,E)$ for $1 \leq k \leq n$ that defines the probability of each more specific event. If, for instance, E denotes mammal, then E_1 could denote dog and E_2 could denote cat and so forth.

Furthermore, using Shannon coding, we can encode the specific event that took place. This encoding gives us a description length of $D(E_k|E) \approx -\log_2 P(E_k|H,E) = I(E_k|H,E)$. As such, an equivalence exists between the conditional probability for the more specific event and the conditional description length for that event.

The (probabilistic or information-theoretic) complexity of E_k, i.e., $I(E_k|H)$, can thus be expressed as $I(E|H) + I(E_k|E,H)$. An upper bound on the description length of E_k, i.e., $D(E_k)$, can then be approximated as $D(E) + D(E_k|E)$ (see the previous subsection for conditional description length). Given this approximation, we can then evaluate the specified complexity of E_k as follows:

$$
\begin{aligned}
SC(E_k|H) &= I(E_k|H) - D(E_k) \\
&\geq I(E|H) + I(E_k|E,H) - D(E) - D(E_k|E,H) \\
&= I(E|H) + I(E_k|E,H) - D(E) - I(E_k|E,H) \\
&= I(E|H) - D(E) \\
&= SC(E|H).
\end{aligned}
$$

An approximate lower bound on the specified complexity of the more specific event E_k is therefore the specified complexity of the more general event E. Increasing the specificity of the pattern therefore maintains or even intensifies the specified complexity of the pattern. For the specified complexity actually to increase requires that the additional detail that narrows down the pattern and thereby increases the specificity maintains a short description. In this case, higher specificity will provide additional grounds for rejecting the chance hypothesis.

Repeated Trials

Consider the case where we have n independently repeated trials of which a single trial produces an event E of high specified complexity. How does n factor into the event's specified complexity? Let's start by calculating the specified complexity if we take only one trial into account. Let p be the probability of the rare event E. In that case,

$$SC(E|H) = I(E|H) - D(E) = -\log_2 p - D(E).$$

Next, let E' denote the event in which the event E of probability p occurred exactly once in n repeated trials. Then, the probability of one occurrence of the rare event among n trials is, by the binomial theorem, $np(1-p)^{(n-1)}$. If we plug this number into the definition of specified complexity, we get,

$$
\begin{aligned}
SC(E'|H) &= I(E'|H) - D(E') \\
&= -\log_2 np(1-p)^{n-1} - D(E') \\
&= -\log_2 n - \log_2 p - (n-1)\log_2(1-p) \\
&\quad -D(E').
\end{aligned}
$$

Nonetheless, if the event E is rare, then $1 - p \approx 1$ and $-\log_2(1-p) \approx 0$. Additionally, the minimum description length of one event among n trials and of a single event by itself is approximately the same, so $D(E) \approx D(E')$. It follows that

$$
\begin{aligned}
SC(E'|H) &\approx -\log_2 n - \log_2 p - D(E) \\
&= SC(E|H) - \log_2 n.
\end{aligned}
$$

Thus, there is a decrease of $\log_2 n$ bits of specified complexity to account for n repeated trials. This makes good intuitive sense. Giving a rare event more independent opportunities to occur will increase its probability and will thus, provided minimum description lengths are roughly the same (as they are in this case), diminish the corresponding specified complexity.

Given n independently repeated trials to produce a rare event E of probability p, we might, instead of asking whether one of the trials produces E, ask whether E occurs on a particular trial. Denote this event by E'. Its description therefore needs to be made more specific by identifying the particular trial that produced the rare event, which will also require a longer description.

Consequently, this approach requires specifying which of the n trials was successful at producing the rare event, and this will take an additional $\log_2 n$ bits. Plugging this into the formula for specified complexity we obtain:

$$
\begin{aligned}
SC(E'|H) &= I(E'|H) - D(E') \\
&\approx -\log_2 p - D(E) - \log_2 n \\
&= SC(E|H) - \log_2 n.
\end{aligned}
$$

Thus, whether we adjust the description or the probability, we obtain the same decrease in specified complexity.

Transformation

Consider the case where we apply a transformation to an event. That is, we have an event E and function f, and we consider the event $f(E)$. What can we say about the specified complexity of that event? First, we can consider the case of including the function as part of the hypothesis. Then we could compute the specified complexity as follows:

$$
SC\big(f(E)|H_f\big) = I\big(f(E)|H_f\big) - D\big(f(E)\big).
$$

In this case, we adjust both the probability and description lengths to take into account the transformation. The same conclusions about specified complexity that hold in general apply to this case. However, we can consider a slightly different situation in which we maintain the same hypothesis but consider the specified complexity of the transformed event:

$$
SC(f(E)|H) = I(f(E)|H) - D\big(f(E)\big).
$$

It can easily happen that the function f decreases the specified complexity. For example, the function could map all possible events to a single point. In that case, the probability of obtaining that point would be 100 percent, and thus it would not exhibit a positive amount of specified complexity.

It can also happen that the function f increases the specified complexity. For example, consider the case where the events are sequences of 100 coin flips. The transformation maps from the space of these sequences of coin flips to a binary space with two choices: "special" and "not-special." Only one random sequence is considered special whereas all the other random sequences are considered non-special. In that case, the event before transformation was random and had low specified complexity in the original space. Yet, under the transformation, it mapped to the rare and specified "special" item, upon which the transformed event exhibited a large amount of specified complexity in the target space.

In this case, however, the increase in specified complexity derives from peculiarities of the transformation that was applied. The particular binary string that is considered special had to be built into the transformation. The sequence was presumably not special in its original context but only gained specialness due to its being the sequence that the transformation selected. Accordingly, describing such a transformation will need to describe the particular binary string and thus will incur a potentially high description length cost (since the vast majority of binary strings are not succinctly described). For more on how transformation can embed description length complexities within themselves, see Appendix B.2.

Is it possible for a transformation to substantially increase the specified complexity of an event when the transformation is simple? No. To see this, consider the event $f(E)$. We can construct an event E' consisting of the union of all events that, when transformed by f, are included in $f(E)$. By definition, $P(E'|H) = P(f(E)|H)$ and therefore $I(E'|H) = I(f(E)|H)$. Furthermore, because the transformation is assumed to be simple, we can describe the event E' using the transformed event $f(E)$. Thus, $D(E') \leq D(f(E))$. By applying the definition of specified complexity, we then discover the following inequality:

$$
\begin{aligned}
SC(E'|H) &= I(E'|H) - D(E') \\
&\geq I(f(E)|H) - D\big(f(E)\big) \\
&= SC(f(E)|H).
\end{aligned}
$$

The event E' therefore exhibits at least as much specified complexity as $f(E)$. Because E is a subset of E', it follows that $P(E|H) \leq P(E'|H)$ and therefore $I(E|H) \geq I(E'|H)$. We might then expect $D(E) \leq D(E')$ because E' is described with reference to E. But in fact, E' may capture some simple feature of E, with the result that $D(E) > D(E')$. Yet, in case E is more simply described than E', so that $D(E) \leq D(E')$, it follows that

$$
\begin{aligned}
SC(E|H) &= I(E|H) - D(E) \\
&\geq I(E'|H) - D(E') \\
&= SC(E'|H).
\end{aligned}
$$

The original event will in that case exhibit at least as much specified complexity as any events associated with it under transformation—provided the transformation is simple and $D(E) \leq D(E')$.

The key takeaway from this subsection is that a transformation can significantly increase specified complexity only insofar as the transformation itself is complex, containing and contributing specified complexity because it was built into the transformation. This fact about specified complexity under transformation has been dubbed conservation of specified complexity or conservation of complex specified information. For further discussion of this topic, see Appendix B.2.

6.8 Examples

Previously in this chapter, we laid out a mathematical framework for specified complexity and proved several results associated with it. We argued that description lengths among different agents are commensurable and yield similar conclusions. We covered both a frequentist and a Bayesian interpretation of specified complexity, either of which can provide evidence for the rejection of a chance hypothesis. Furthermore, we saw that our mathematical framework is robust in handling background knowledge, variations in specificity, and repeated trials.

We turn now to examples in which we use the framework laid out earlier in this chapter to calculate actual values of specified complexity. We will consider some examples in which the correct answer is obvious, and other examples in which the conclusion may be more controversial. We will approximate the probability and description lengths in these examples to determine whether the evidence is enough to reject the relevant chance hypotheses.

The length of a description depends on the language used for that description. Thus, different agents using different languages will end up with different lengths for that description. Even so, as argued above, we expect the lengths to be approximately commensurate. For the purpose of having a simple and practical approach, we will consider descriptions written in English. There are approximately 200,000 words in common usage in English. It takes approximately twenty bits (17.6 bits to be more exact) to select one of these 200,000 words. As such, we adopt a crude approximation of description length as twenty bits per English word.

Texts

As a first example, we consider finding on the wall of a monument the text of the Gettysburg Address, a famous speech beginning "Four score and seven years ago..." In fact, that speech, delivered by Abraham Lincoln, is engraved in stone at the Lincoln Memorial in Washington, DC. This was a short speech that contained fewer than three hundred words. It consists of approximately fifteen hundred characters: letters, spaces, and punctuation marks. If someone found this text engraved into a monument, no one would doubt that it was the deliberate work of an intelligent agent.

But how much specified complexity underlies that conclusion? Let's attempt to approximate the specified complexity in the

Gettysburg Address by calculating $SC(E|H) = I(E|H) - D(E)$. The first term on the right of this equation corresponds to a probability. This probability will in turn depend on a proposed chance hypothesis H for how this text came to be placed on the monument. For now, let's take as this hypothesis that the text was composed by randomly punching keys on a keyboard so that each letter was equally likely to be any letter, a space, a comma, or a period.[24] This sort of scenario is often described in terms of a monkey typing on a keyboard. Of course, we immediately dismiss this as a silly explanation for the engraved text, but for the purpose of elucidating specified complexity, let's carry on.

The proposed mechanism assumes a uniform probability over 29 characters. A uniform probability makes every outcome equally likely. The probability of each individual character being correctly chosen is thus 1 chance in 29. Given the approximately 1,500 characters in the address, that total probability $(1/29)^{1,500}$. Computing the complexity, we obtain: $I(E|H) = -\log_2(1/29)^{1,500} \approx$ 7,286 bits.

Our next task is to approximate the description length $D(E)$. Lincoln's address can be described using two words, namely, "Gettysburg Address." Thus, following our simple approximation rule of twenty bits per English word, we conclude that $D(E) \approx 40$ bits. Putting both terms together: $I(E|H) - D(E) \approx$ $7,286 - 40 = 7,246$ bits. From the earlier work in this chapter, we know that the probability of obtaining 7,246 bits is less than or equal to $2^{-7,246} \approx 10^{-2,181}$. The probability is thus extremely low that the text on the monument could be explained as a sequence of randomly selected characters.

What if, however, the situation is altered so that the text found on the monument does not correspond to a famous speech precisely identified with a simple description? What if instead it is a unique text, written in English, but only found on that particular monument? In that case, we can no longer use a simple description, like "Gettysburg Address," to identify that exact text. But we could use a simple description at least to narrow down the text, such as the description "English text." The word "text" here denotes, as is typical, coherent meaningful written communication.

But to estimate specified complexity in such a case, we will first need to estimate how probable it is that randomly selected characters would produce English text. Ordinarily, we would simply think it obvious that randomly chosen characters would be extremely unlikely to produce English text. However, since our goal here is to perform a calculation of specified complexity in reasonably full detail, we cannot simply conclude what seems obvious but instead will need to perform some calculations.

Determining the exact probability that randomly chosen characters will happen to form meaningful English texts is a practical impossibility. Nonetheless, we can calculate the probability that randomly written characters contain only valid English words. Any valid English text will be made up of valid English words. That said, many possible assemblages of valid English words do not form a meaningful English text, and thus the probability of obtaining valid English text is lower than the probability of merely obtaining a sequence of valid English words. Thus, by using the lower standard of valid English words as a proxy for valid English text, we again greatly understate the improbabilities actually involved. This habit of understatement, as noted in Chapters 4 and 5, safeguards against drawing a design inference where the actual complexity or improbability is insufficient to warrant it.

The probability of obtaining a possible English word (whether valid or invalid letter combinations) of particular length k can be expressed as the product of the probability of typing k letters in a row without any punctation or space, followed by some punctation or space: $(26/29)^k \times (3/29)$. Given our assumption that there are 29 characters of which 26 are letters, this is the fraction of k-length letter combinations that might be English words. The rest can be ruled out before we even check the resulting characters. (So, for instance, if we are focused on 4-letter English words, this calculation discounts such possible results as *fl,b,* since it isn't an unbroken string of four English letters.) We can estimate the fraction of letter combinations that are indeed valid English words by enumerating the valid English words such as used in a spell-checker.[25] The following table shows the resulting enumeration:

Word Length	Number of Valid Words	Probability of Any Word
1	26	9.2746730083%
2	139	1.7097871991%
3	1294	0.5488624747%
4	4994	0.0730432020%
5	9972	0.0050293926%
6	17462	0.0003036891%
7	23713	0.0000142208%
8	29842	0.0000006171%
9	32286	0.0000000230%
10	30825	0.0000000008%

The probability of obtaining a valid word rapidly declines as we consider increasing word lengths. In total, by summing the right column, we find an approximately 11.6 percent probability of randomly typing a valid English word. This is the probability of typing at random, stopping when the first non-letter is typed, and then determining whether the letters typed constitute a valid English word. These probabilities just get the ball rolling, however, looking at combinations of letters to form words. Next, we need consider combinations of words.

Indeed, we are not interested in the probability of typing a single valid word, but the probability of typing a meaningful English text. Let's consider a text that is 300 words long, roughly comparable to the length of the Gettysburg Address, which is exactly 272 words in length. And again, we will set the bar artificially low by demanding only that all the letter combinations form valid English words rather than their forming a meaningful English text. The probability of typing 300 words in a row such that they are all valid English words is $(11.6\%)^{300} \approx 2.1 \times 10^{-281}$. Let's denote this text by E.

Calculating the specified complexity $SC(E|H) = I(E|H) - D(E)$, we obtain $-\log 2.1 \times 10^{-281} - 40 \approx 892$ bits. This is far fewer bits than in the Gettysburg Address. That's because it is much more likely that random typing will produce a sequence of words all of which happen to be valid English words rather than the exact text of the Gettysburg address. However, it is still very unlikely to produce all valid English words, and we are being eminently reasonable to conclude that such a text was not the product of randomly banging on a typewriter.

These examples may seem silly. But closely related examples exist that are less silly, such as the Voynich manuscript. This is a handwritten manuscript from the fifteenth century consisting of 240 pages filled with characters written in a unique script. We are unable to identify the language or translate the text—if indeed it constitutes meaningful text in a proper language.[26] We do know that the Voynich manuscript does not consist of randomly written symbols. This conclusion follows because patterns in the manuscript's written characters are too improbable to appear there by chance.

If the Voynich manuscript consisted of purely random markings, we would not expect to see the same small number of characters used and reused throughout the manuscript. Instead, most of the characters belong to what appears to be an alphabet of twenty to twenty-five different symbols used throughout the manuscript. Obviously, random lines and marks would not produce such a pattern consisting of just a few characters used over and over.

But the manuscript might consist of gibberish constructed by randomly stringing characters together, akin to the hypothetical monkey banging away randomly at a keyboard. The author of the Voynich manuscript might have constructed its alphabet of symbols and simply written its symbols at random to fill up the manuscript. However, we know that this is not the case because the text has many statistical properties that would not be explained by such a random sampling of symbols.

A simple property to check for is whether some aggregates of characters, or words, are much more common than others. In

English, words such as "the" or "a" are very common, making up a large percentage of all words that appear in English texts. We find something similar in the Voynich manuscript. The most common word, which looks like *daiin*, appears approximately 550 times, making up approximately 1.5 percent of the manuscript's words. Together there are five words, each with five letters, that make up 5 percent of the manuscript's word total. It's a common feature of natural languages that certain words are very common while others are rare. This fact about natural languages is a consequence of Zipf's law, named after the American linguist George Kingsley Zipf.[27]

What is the specified complexity associated with this pattern of common words? Let's begin with the probability. What is the probability of randomly choosing five letters and having them be one of these five words? Assuming an alphabet of twenty symbols, the probability is $5/20^5$. However, not all words are five letters, and thus the actual probability will be somewhat less. Exactly how much less will depend on how many words have five letters. Nevertheless, we can use this probability, remembering that it is an upper bound, and that actual probability will be lower.

Five percent of the text is approximately 1,600 words, and the probability of selecting one of these five words 1,600 times is: $(5/20^5)^{1,600}$. However, those words are not required to be all in a row but can be found anywhere in the manuscript. Given approximately 30,000 words in the manuscript as a whole, the different ways these common words could be found come to "30,000 choose 1,600" (which is how mathematicians refer to the number of ways of choosing 1,600 possibilities among 30,000 total possibilities).

The probability that five percent of 30,000 words would fall among these common words is thus

$$\binom{30,000}{1,600} \times \left(\frac{5}{20^5}\right)^{1,600}.$$

But note, this formula gives us the probability of obtaining five specific words with a given frequency. To compute the probability for any five words of five symbols, we need to multiply this probability by the number of possible words of that length, which comes to $(20^5)^5$. The probability of the pattern is thus

$$(20^5)^5 \times \binom{30,000}{1,600} \times \left(\frac{5}{20^5}\right)^{1,600}.$$

The quantities in this formula are very large. If we do the calculation, we obtain $I(E|H) \approx 21,749$ bits.

The pattern that we investigated was "five words of five characters make up five percent of the text." At twelve words this gives us an estimated pattern length of $20 \times 12 = 240$ bits. Thus, the estimated specified complexity is: $I(E|H) - D(E) \approx 21,749 - 240 = 21,509$ bits. This shows that the pattern of common words exhibits a high degree of specified complexity. We conclude that the Voynich manuscript did not originate by randomly choosing letters.

This conclusion, however, does not rule out other, more sophisticated, randomization techniques to produce the Voynich manuscript. One possibility is that the author constructed the manuscript by copying and modifying earlier parts of it.[28] The tendency of some words to appear very frequently might just be a consequence of such repeated copying. We don't know whether the manuscript presents a real but undeciphered language or whether its arrangement of symbols is the product of some sophisticated stochastic algorithm. What is clear is that simple random processes, like the one evaluated here, cannot explain the manuscript. It is either a real meaningful text or the product of someone going to considerable lengths to produce an arrangement of symbols that looks as though it could be meaningful.

Trees

Every fall in Polk County, Oregon a giant smiley face appears in a forest. It is a forest of mostly Douglas firs, which keep their needles year around. However, some larch trees are mixed in. These have needles that turn yellow and drop during the fall. In one particular spot, the larch trees have been arranged so that they form a smiley face. The round face is mostly larch trees that have turned yellow, except for the eyes and mouth, which are Douglas firs and thus remain green.[29]

After the fall of the Berlin wall, a similar phenomenon was rediscovered in Northern Germany. Instead of a smiley face, the figure shown was a swastika. It appears to have been constructed by Hitler youth during the reign of Nazi Germany. For obvious reasons, this was deemed unacceptable, and many of the larch trees were cut down to destroy the figure. But this was not the only case. Several additional cases of forest swastikas have been found in Germany and beyond. In most cases, we have no idea who was responsible for their creation.[30]

There is little doubt that somebody deliberately created the smiley faces and swastikas. We draw this conclusion for two reasons. First, the pattern is very simply described: "smiley face" or "swastika." Second, the probability of those particular images being produced by random chance is extremely small. Indeed, we can say with assurance that the probability of a smiley face or swastika is very small because the fraction of tree configurations that resembles swastikas or smiley faces is very small.

The forest swastika took up 0.36 hectares. Based on an estimate of 1,000 trees per hectare, this suggests that the swastika is made up of approximately 360 trees. For now, we will assume that each tree must be either a fir or a larch, as dictated by the two distinguishing colors in the image. If each kind of tree is equally probable, then the probability of any one tree being correctly positioned to depict the image is .5. The probability of 360 trees being correctly positioned is $.5^{360}$.

The pattern denoted by "swastika" requires one word. Given a dictionary of 200,000 words (Section 6.3), its description length is therefore 18 bits (more precisely, 17.6 bits). But let's be generous, not overly privileging our choice of language, and so estimate its description length at 20 bits. It is therefore a specification. Computing its specified complexity, we obtain that $I(E|H) - D(E) = -\log(.5)^{360} - 20 = 360 - 20 = 340$ bits. We can do a similar calculation for the smiley face. It takes up approximately

0.65 hectares, and thus comprises approximately 650 trees. The pattern is two words long, thus requiring approximately $2 \times 20 = 40$ bits to describe. The specified complexity is thus $I(E|H) - D(E) = 650 - 40 = 610$ bits.

These calculations require some adjustment. Variations in either the face or the swastika could still fit the salient pattern in these images. A few trees could be out of place without the pattern becoming unrecognizable. The pattern could also be rotated, but rotations would keep the same proportion of recognizable to unrecognizable patterns, so we can ignore them. To try to be generous in the amount of variation we permit, let's assume there are 1,000 possible distinct patterns that would fit each description and that up to 10 percent of the trees could be of the incorrect kind without ruining the pattern.

We now need to count how many alternative tree patterns might have matched the pattern in question. We can compute the number of possible patterns of n trees with k incorrect trees as $\binom{n}{k}$. For up to 10 percent of the trees being incorrect, we would therefore compute the following sum: $\sum_{k=0}^{.1n} \binom{n}{k}$. Taking into account the 1,000 posited alternative patterns, we obtain:

$$
\begin{aligned}
I(E|H) &= -\log_2[(.5)^n \times 1,000 \times \textstyle\sum_{k=0}^{.1n}\binom{n}{k}] \\
&= n - \log_2 1,000 - \log_2 \textstyle\sum_{k=0}^{.1n}\binom{n}{k}
\end{aligned}
$$

For 360 trees, this comes to 185 bits, and for 650 trees it comes to 339 bits. This gives specified complexities of respectively 165 bits and 299 bits.

Faces

In 1976, the Voyager 1 spacecraft transmitted a photo of Mars back to Earth. This photograph contained a surprising feature: a rock formation with an uncanny resemblance to a humanoid face including eyes, a nose, and a mouth. Some interpreted this as a monument created by some alien race that once populated Mars.

Nevertheless, most accept that it is a natural formation that just happened to resemble a face.[31]

A variety of similar phenomena exist in which seemingly natural formations resemble human faces or figures. In New Hampshire, there is a rock formation called the Old Man of the Mountain. It resembles a jagged profile of a human face. In Alberta, there is a geomorphic feature called the Badlands Guardian. When viewed from above, it resembles a human head wearing an indigenous headdress. A region on the moon was identified in 2013 by facial recognition software. It is referred to as the Face on Moon South Pole. It resembles an alien face.[32]

On the other hand, cases abound of rock formations that have been deliberately carved to resemble human faces. Mount Rushmore is one example. A number of very large carvings of the Buddha exist in Asia. In Romania, there is a rock relief depicting the face of the last king of Dacia, Decebalus. Constructed after the fall of communism in Romania, it is the largest rock relief in Europe, measuring 180 feet in height.[33]

Often the history of these carvings is well known. For instance, we know that the artist Gutzon Borglum is responsible for the faces of Mount Rushmore, and we even have pictures of him and his crew working on it. However, even if we did not know the origin of Mount Rushmore, a Buddha statue, or the face of a king, we would conclude that they were not natural rock formations but were deliberately carved. Yet the Mars Face, Badlands Guardian, and Old Man of the Mountain are superficially similar to these humanly made carvings. We think that such formations came about naturally rather than from deliberate carving. What explains the differing conclusions?

The answer becomes obvious when these natural formations are carefully examined. Then they appear to be, at best, vague approximations of what they are said to resemble, with the human imagination bridging the gap. This is significant because it is not unlikely for a random geological formation to bear a passing resemblance to a human head. On other hand, it is extremely unlikely that a natural geological formation would resemble the human head as closely as it does with Mount Rushmore or the Romanian rock relief of Decebalus. The resemblance to human features in the carvings is far more detailed than for natural structures. Erosion may well create the Old Man of the Mountain. But Mount Rushmore seems well beyond its remit.

Can we, however, quantify that improbability of forming by natural means a detailed human resemblance in rocks? A precise evaluation of the probability would require a probabilistic model for the formation of different natural structures and a determination of what proportion of those structures would closely resemble a human head or face. Nonetheless, we do not require such a precise evaluation to get a general idea of what the probabilities must be. We can use a rough approximation or back-of-the-envelope calculation.

One such approach is to count the number of identifiable features in the image that contributes to the impression that it looks like a face. For example, the Mars Face contains two eyes, a nose, and a mouth. There are thus four identifiable features of the image that lead to the impression that it looks like a face. If we assume that each of these features has a probability of approximately one in a thousand and that the occurrence of any feature is probabilistically independent of any other (as seems reasonable), the probability of all three features being present together is approximately one in a trillion.

But how many identifiable features are present on Mount Rushmore? Each eye has an identifiable iris, pupil, and sclera. There are two eyelids, one above and the other below the eye. Above each eye is an eyebrow with an eyelash. Thus, we have seven identifiable components to each eye. Each face then has two eyes. In addition, each face has a nose with two nostrils, making an additional three components. Each face also has a chin, a mouth, and hair. This makes up twenty features per face. There are four faces on Mount Rushmore, making eighty total identifiable features. If we assume, again, that each feature has a probability of one in a thousand and is probabilistically independent of the others, the probability of all of these features being present together is one in 10^{240}.

To compute the specified complexity associated with these patterns, we need to consider their descriptive length. Both patterns may be approximately described as "face" or "faces," and thus both patterns can be approximated as requiring 20 bits to describe (which, as we've seen, is more bits than we need to manage a 200,000-word dictionary). This means that the specified complexity of the Mars Face is at least $-\log_2 10^{-12} - 20 \approx 20$ bits. Note that our approach to finding two eyes, a nose, and a mouth coming together to form a face says nothing about how these features are arranged next to each other to form a coherent-looking face. Factoring in the rarity of such coherent arrangements of facial features might intensify the improbability and thus raise the specified complexity. But in fairness, neither are we giving the same benefit to Mount Rushmore for properly arranging its various specified features—we're simply looking at the collocation of the facial features of Mount Rushmore, and not gauging whether they are coherently arranged.

Now 20 bits of specified complexity for the Mars Face may seem like a lot, but it certainly is nowhere near the level of improbability needed for a universal probability bound, which is around 500 bits of specified complexity. Moreover, at the scale of the earth, to say nothing of the scale of the solar system, and given all the places humans have looked, and given that they are primed to see faces in just about anything, a specified complexity of 20, 30, or even 40 bits doesn't seem all that extreme for ruling out chance.

But contrast this specified complexity calculation with that of Mount Rushmore. As with the Mars Face, and so ignoring how coherently the facial features are arranged next to each other and instead focusing simply on their presence, we find that Mount Rushmore has a specified complexity of at least $-\log_2 10^{-240} - 20 = 777$ bits. This corresponds to a very small probability and reflects that we do not observe carvings like Mount Rushmore arising naturally.

`Oumuamua

In October 2017, an astronomer named Robert Weryk discovered an unusual interstellar object passing through the solar system. It is small as interstellar objects go, estimated to be 100–1,000 meters long (Halley's Comet, by contrast, has a mean diameter of 11,000 meters). Instead of being spherical, as is typical of such objects, it is elongated. It lacks the coma (gaseous cloud) typically associated with comets. It exhibits a tumbling rather than a spinning motion. And it accelerates in a way not readily explained by gravitational pulls. These oddities have led some astronomers, notably Avi Loeb of Harvard University, to argue that it is of artificial origin. Nevertheless, most scientists dismiss this possibility, arguing that despite its oddities this object is of natural origin. The object was dubbed `Oumuamua, the Hawaiian word for scout.

In 2018, Loeb wrote an article for *Scientific American* titled "6 Strange Facts about the Interstellar Visitor `Oumuamua."[34] In it, he gave six reasons to think that `Oumuamua might be of artificial extra-terrestrial origin. These reasons can be summarized as follows:

1. If interstellar objects are as rare in the solar system as we thought, happening to observe this one is extremely unlikely.

2. The object is very nearly at rest relative to the motion of nearby stars.

3. In order to be at rest, the object must have received a push or kick nearly exactly opposite to the motion of the solar system it came from.

4. The object is very elongated.

5. The object appears to be very small in size but very shiny for its size.

6. The object does not move strictly according to the action of gravity.

These, then, are supposed to constitute six highly improbable properties of ʻOumuamua. Loeb's case is, in essence, a specified complexity argument: it seems highly improbable that the object would exhibit all these properties, and these properties together seem to specify it. Let's now flesh out this argument.

For the sake of discussion, let's assume that each of these six properties has a one in a trillion chance of being true and that each property is probabilistically independent of the others. This would mean that the probability of all six properties being simultaneously true of an interstellar object is one in $(10^{12})^6 = 10^{72}$, which is one followed by seventy-two zeros, yielding a minuscule probability.

Since we are performing a specified complexity calculation, it is not enough to estimate the probability; we also need to estimate the description length. Each of the six properties laid out by Loeb seems a bit complicated to describe. To use specified complexity to argue for the artificial origin (design) of ʻOumuamua, Loeb would want to maximize the specified complexity calculated. But that would mean minimizing the description lengths for each of the six properties. We might therefore estimate that it would take at least two words to describe each property (this seems generously low). A complete description of the observed pattern would thus require at least twelve words. Given our standard approximation of twenty bits for each word, we can estimate the specified complexity of ʻOumuamua as follows:

$$I(E|H) - D(E) = -\log_2(10^{12})^6 - 12 \times 20 \approx 0 \text{ bits.}$$

Despite the immense improbability of the observed properties, this improbability is wiped away by the long description required to jointly identify these properties. If the estimates used of the probability and description length are roughly correct, there is no basis for drawing a design inference from ʻOumuamua.

The fundamental issue here is that independent properties tend to increase the improbability while simultaneously increasing the required description length. These two effects tend to cancel each other. Consequently, attempting to draw design inferences from lists of independent properties is often ineffective. An effective design inference requires a single pattern with a short description that encompasses a number of improbable properties. But `Oumuamua appears to lack any such short description.

Lottery Fraud

Lotteries are games in which people purchase a small chance at winning a large prize. They trace back to Caesar Augustus, who sold lottery tickets in order to fund repairs of the city of the Rome. As with all forms of gambling, they have a controversial history, being both widely practiced and widely condemned. One of the challenges with lotteries is ensuring that they are run fairly. If the people running the lottery also play the lottery, they face a strong temptation to cheat by ensuring that their tickets are the winning tickets.

In 1980, the announcer for the Pennsylvania Lottery's Daily Number, Nick Perry, replaced the ping pong balls used to determine the winning lottery numbers with weighted ones to ensure that only the numbers four or six would be chosen. The chosen numbers turned out to be 6, 6, 6. Perry and his co-conspirators, however, made the mistake of purchasing a large number of tickets with combinations of six and four. The specified complexity of such an unusually large number of tickets purchased for those digits tipped off the authorities that something unusual had happened. They investigated, and eventually tried and convicted Perry.[35]

Another case of cheating played out in 2006 in the province of Ontario. In this case, however, a widely dispersed group rather than a single individual with co-conspirators was responsible for the fraud. The tip-off to the fraud was that employees of stores that sold lottery tickets won major prizes at an unusually high rate.

Between 1999 and 2006, these employees won approximately 200 of the 5,713 major prizes. Approximately 13 million people lived in Ontario, of which approximately 60,000 worked for stores that sell lottery tickets. This means that about half a percent of the population worked for such stores, and thus would be expected to win about half a percent of the prizes. Instead, they won 3.5 percent of the prizes.

As a mitigating factor against the charge of fraud, employees of stores that sell lottery tickets would, plausibly, be more likely to purchase lottery tickets than the general population. Based on surveys, it was estimated that the employees were twice as likely to play the lottery as the general population. Consequently, they might be expected to win 1 percent of the prizes, not simply half a percent.

But that still leaves us with a question: is the difference between 1 percent expected winnings and 3.5 percent actual winnings enough to justify the inference that the games were not fair and that the employees were stealing prizes? The underlying probability distribution here is the binomial. There were 5,713 prizes and 200 went to insiders who should, collectively, have had a 1 percent chance of winning. The remainder, 5,513, went to other people. The probability is thus

$$\binom{5,713}{200} \times (.01)^{200} \times (.99)^{5,513} \approx 2^{-163}.$$

The underlying pattern here can be described as "two hundred lottery sellers," which takes four words, and which we thus approximate as requiring $4 \times 20 = 80$ bits to describe. Putting these numbers together, we therefore calculate the following specified complexity:

$$I(E|H) - D(E) = -\log_2 2^{-163} - 4 \times 20 \approx 83 \text{ bits.}$$

This is a large number of bits, corresponding to an improbability of flipping more than 80 heads in a row with a fair coin (a probability of one in a trillion trillion). It shows that there was indeed a problem with the lotteries. Specifically, it shows that the lotteries were not being run fairly and that insiders were winning far too disproportionately. As it is, a design inference was in this case fully vindicated: further investigations confirmed a widespread ruse in which employees would tell winning customers that their tickets had not won, upon which they took the discarded tickets for themselves.[36]

Bible Codes

A long tradition exists of attempting to find hidden meanings and codes in religious texts. Texts that have received special scrutiny include the Bible as a whole (both Old and New Testaments), the Torah (the first five books of the Bible, traditionally ascribed to Moses), and the Quran (taken by Muslims as direct divine communication to the prophet Muhammad). Examples of hidden meanings and codes in such texts include finding the names of famous Rabbis in the Torah and finding unusual statistical patterns in the Quran.

One particularly simple and interesting example is finding the numbers π and e embedded into the texts of Genesis 1:1 and John 1:1 respectively. In this pair of verses, John 1:1 echoes Genesis 1:1. In the ESV translation, Genesis 1:1 reads:

In the beginning, God created the heavens and the earth.

In the same translation, John 1:1 reads:

In the beginning was the Word, and the Word was with God, and the Word was God.

There is a long-standing practice of assigning numerical values to letters in ancient languages like Greek and Hebrew. This is called *gematria*. For example, the first word of Genesis 1:1 (and thus of the whole Bible) is בראשית (bə·rê·šîṯ), which is typically translated as "in the beginning". Following the standard assignment of numbers to these six letters, we obtain 400, 10, 300, 1, 200, and 2. If we add these numbers together, we get 913 and if we multiply them together, we get 480 million.

For each verse of the Bible, we can perform a complex calculation: multiply together all of the values for all of the letters in the verse; next, multiply together the sum of the values of each letter in each word in the verse; next, divide the first value by the second value; next, multiply this value by the number of letters in the verse; and finally, divide this last value by the number of words in the verse.

If this calculation is done for Genesis 1:1, the outcome is $3.141554508 \times 10^{17}$. This number bears an uncanny resemblance to the number π, which is familiar to most people and is equal, in its first five decimal digits, to 3.14159. If the same calculation is done for John 1:1, the outcome is $2.718312812 \times 10^{40}$. This number bears an uncanny resemblance to the number e, a vitally important number for mathematics, though less familiar than pi to those not trained in mathematics. In its first five decimal digits, e is equal to 2.71828.

Are these uncanny resemblances simply a coincidence, or is there something deeper going on here? In both cases, the approximation has four correct digits. The probability of randomly picking four correct digits is one in ten thousand. Correctly doing this for two different verses has a probability of 1 in 100 million. That's a fairly low probability.

But what about the relevant description length here? The description would have to incorporate quite a lot of information. It has to mention π, e, gematria, the number of words, the number of letters, the product of the sum of the letters in each word, the product of all of the letters, and an accuracy of one in ten thousand. If we assume that each of these concepts requires at least one word to describe, this means that any description would be at least seven words long. Practically speaking, it would need to be much longer. Using a seven-word description, however, we can estimate the specified complexity to be

$$I(E|H) - D(E) = -\log_2 \frac{1}{100{,}000{,}000} - 7 \times 20 \approx -120 \text{ bits.}$$

The specified complexity in this instance is a large negative number. This means that there is no specified complexity here to speak of. What improbability exists is wiped out in this equation on account of the excessive length of the description. The approximations of π and e in Genesis 1:1 and John 1:1, close though they may seem to the true values of these numbers, are thus readily explained as a mere coincidence for which chance is a perfectly acceptable explanation.[37]

Conclusions

In this section, we have explored a number of applications of specified complexity. We have seen some cases where the reasonable answer was to reject the chance hypothesis and others where it was not. Some cases were straightforward and therefore did not need to be analyzed in detail. Other cases were less obvious, with the mathematical analysis bringing clarity to whether chance should be legitimately accepted or rejected.

We stressed two complementary themes. One theme was the importance of careful and correct estimations of probabilities. If a probability is incorrectly calculated to be very small, we will tend to incorrectly reject a chance hypothesis when it should be retained. Utmost care must therefore be taken in estimating probabilities and ensuring that they are close to the true underlying probabilities.

A second complementary theme is the importance of taking into account description lengths for patterns that match events. Several cases we considered seemed at first blush to yield improbabilities that would defeat chance—until we took into account description length. We found that the descriptions in some situations were quite long, and even though the events being considered were indeed improbable, their improbability was insufficient to offset the lengthiness of the descriptions.

To sum up, smallness of probability and lengthiness of description reside on opposite sides of a seesaw where one typically outweighs the other. Rarely do they balance each other, and when they do, it's an unstable equilibrium. Because specified complexity is a difference between these two numbers of bits (bits associated with small probability minus bits associated with description length), the elimination of chance and any resulting inference to design require that the bits associated with small probability decisively outweigh the bits associated with description length.

Ultimately, what specified complexity does is ensure that small-probability chance-elimination arguments are not subverted because we simply read off the descriptions of events, allowing them to become arbitrarily long, and thereby empowering unspecified events to eliminate chance (which is always a no-no). Events of small probability that are also specified occupy the sweet spot. They exhibit specified complexity in the sense defined in this chapter. They are the events for which chance is legitimately eliminated and for which design is legitimately inferred.

CHAPTER 7

EVOLUTIONARY BIOLOGY

7.1 Insulating Evolution against Small Probabilities

IN THE YEARS SINCE THE FIRST EDITION OF THIS BOOK WAS PUBLISHED, its ideas have permeated evolutionary biology. Increasingly, system after biological system has given way to the design inference. Through its method of design detection, the design inference has convincingly shown that many, if not most, aspects of biology exhibit the tell-tale marks of specification and small probability and so are the result of intelligent activity. Specified complexity has thus come to be seen as posing an insuperable barrier to naturalistic forms of evolution. Intelligent design is therefore well on its way to winning the day and becoming the new orthodoxy for evolutionary biology.

And now back to the real world...

In fact, nothing like this has happened in biology. In other areas of inquiry, the design inference has proven itself useful, and indeed indispensable, for uncovering evidence of intelligent activity (recall Chapter 2). Even so, mainstream evolutionary biologists have denied that design inferential reasoning has any validity when applied to the evolution of biological systems. Whatever its merits elsewhere, the design inference is thus supposed to tell us nothing about the origin and subsequent history of life on Earth. In effect, the biological mainstream has insulated evolutionary biology from the design inference. It is instructive to see how this happened and why such moves to artificially protect evolutionary biology in the end don't work.

To be clear, the design inference does not oppose biological evolution per se. It challenges a naturalistic form of biological evolution, which is both mechanistic and materialistic. Biochemist Michael Behe, for instance, is among the best-known proponents of intelligent design. He also holds to common descent, the view that all organisms have evolved from a universal common ancestor.[1] This is evolution at its most extensive—what philosopher of biology Michael Ruse calls "monad to man" evolution.[2] Where Behe departs from mainstream evolutionary biology is in denying that the evolutionary process is driven by purely material and unguided mechanisms such as natural selection acting on random variations. Instead, he sees clear marks of intelligent design in biological systems, marks for which the design inference is capable of convincingly drawing a conclusion of design.

Unlike Behe, most evolutionary biologists admit into biology no convincing marks of intelligence, such as might be identified through a design inference. Instead, they only see the outworking of natural forces apart from any real intelligent guidance or intervention. Yet how did mainstream evolutionary biologists come to view the design inference as irrelevant to biological origins? Did they painstakingly show that for every biological system that exists, no design inference could legitimately be drawn? No.

The mainstream biological community is convinced that the design inference comes up empty regardless of where it is applied in evolutionary biology. Moreover, it reaches this conclusion by invoking the Darwinian mechanism of natural selection and random variation. To rule out the design inference from biology, evolutionary biologists contend that Darwinian evolution is able to prevent biological systems from becoming too improbable. In effect, this joint mechanism of natural selection acting on random variations is seen to act as a *probability amplifier* to turn otherwise small probabilities into large probabilities incapable of underwriting a design inference. Thus, the small probabilities needed to make a workable design inference in biology are said simply not to arise.

The design inference is about identifying specified complexity or specified improbability—namely, events that are highly complex or improbable (complexity here being an information-theoretic recasting of probability) but also specified (specification here consisting of patterns with short description lengths). Hardly any critic of applying the design inference to biology denies that biological systems are specified.[3] Where they overwhelmingly push back against the design inference is in denying that these systems are highly improbable.

Thus, although mainstream biologists are ready to see biological systems as specified, they are unwilling to see them as improbable.[4] These systems might appear to be improbable. But, we are assured, their probability is not small once we understand that material mechanisms such as natural selection are available to wash away any seeming improbability. This refusal to countenance real objective small probability in these systems underlies how mainstream evolutionary biologists keep the design inference at bay.

In effect, evolutionary biology says that specified complexity in biology is not real but only apparent. Biological systems thus only seem to exhibit specified complexity. Alternatively, these systems are said to exhibit specified complexity but only with respect to probability distributions that are irrelevant to the ones actually in play. With respect to these irrelevant probability distributions, the calculated probabilities may be very small indeed. But once the actual underlying probabilities are accurately assessed, they are never small enough to justify a successful design inference.

The details of how evolutionary biology resists the challenge from small probabilities are instructive. To see what's at stake, let's return to a passage cited earlier, from Richard Dawkins' *The Blind Watchmaker*: "This belief, that Darwinian evolution is 'random,' is not merely false. It is the exact opposite of the truth. Chance is a minor ingredient in the Darwinian recipe, but the most important ingredient is cumulative [or natural] selection, which is quintes-sentially nonrandom."[5]

In context, Dawkins here contrasts cumulative selection (which for him abstracts the essence of natural selection) with what he calls single-step selection. Cumulative selection can break a seemingly vast improbability into a sequence of manageable steps, each of which is reasonably probable. Single-step selection, by contrast, must overcome a vast improbability in one step, and thus cannot mitigate any probabilistic hurdles placed in its way. For Dawkins, what gives evolution its power is the ability to overcome vast improbability by replacing single-step selection with cumulative selection.

Dawkins illustrates the difference between these two forms of selection through a linguistic example that attempts to evolve the phrase METHINKS IT IS LIKE A WEASEL (a phrase taken from Shakespeare's *Hamlet*). Dawkins asks us to imagine evolving this phrase, which consists of 28 letters and spaces and has 27 possibilities in each position (26 capital letters plus a space for separating letters). METHINKS IT IS LIKE A WEASEL is thus 1 of 27^{28} (approximately 1.197×10^{40}) possible phrases consisting of 28 characters.

Dawkins assigns to each such phrase the same probability (i.e., phrases consisting of 28 letters and spaces with 27 possibilities in each position). Given that letters and spaces don't care how they're arranged, this uniform probability assumption seems reasonable. And so, the probability by single-step selection of getting METHINKS IT IS LIKE A WEASEL will be 1 in 1.197×10^{40} for each generation, or roughly 1 in 10 thousand trillion trillion trillion. Single-step selection is thus, at the time of this writing in 2023, highly unlikely to find this phrase, even with a billion Cray supercomputers running for over a decade (recall Section 4.2).

But single-step selection, according to Dawkins, is not how evolution operates. Instead, it operates by cumulative selection, which takes a divide-and-conquer approach that breaks what would be a highly improbable single step into a sequence of probabi-listically manageable smaller steps. Each of these smaller steps then has high probability. Moreover, the entire sequence of steps leading to the final product will then have high probability as well.[6] In this way, METHINKS IT IS LIKE A WEASEL becomes achievable with high probability. By randomly varying a few letters at a time and by assigning higher fitness to phrases that, letter-for-letter, match METHINKS IT IS LIKE A WEASEL more closely, this phrase can be evolved with high probability in short order.

Thus, whereas single-step selection would, on average, evolve METHINKS IT IS LIKE A WEASEL in roughly 10^{40} generations (a very long waiting time corresponding to a very small probability), cumulative selection evolves it, on average, in around 40 generations (a very short waiting time corresponding to a large probability).[7] This vast drop in waiting times results from cumulative selection, and it corresponds to a vast increase in probabilities. For Dawkins, cumulative selection epitomizes how the Darwinian mechanism empowers the evolutionary process, enabling it to render highly probable what for single-step selection would remain highly improbable.[8] For more on waiting times and their connection to probability, see Appendix B.5.

Mathematician Jason Rosenhouse agrees that Dawkins' METHINKS IT IS LIKE A WEASEL example captures how Darwinian evolution overcomes the vast improbabilities that would otherwise help to justify a design inference. He writes: "Dawkins' ... simulation captured enough of the important aspects of evolution to show that cumulative selection will very quickly achieve what blind search will never achieve at all."[9] To leave no doubt about how Darwinian processes overcome apparent improbabilities, Rosenhouse offers a coin-tossing example that is simpler than Dawkins' WEASEL:

> There is nothing analogous to natural selection when you are tossing coins. Natural selection is a non-random process, and this fundamentally affects the probability of evolving a particular gene. To see why, suppose we toss 100 coins in the hopes of obtaining 100 heads. One approach is to throw all 100 coins at once, repeatedly, until all 100 happen to land heads at the same time. Of course, this is exceedingly unlikely to occur. An alternative

approach is to flip all 100 coins, leave the ones that landed heads as they are, and then toss again only those that landed tails. We continue in this manner until all 100 coins show heads, which, under this procedure, will happen before too long.[10]

Tossing all coins simultaneously would yield, even with an enormous number of tries, a small probability of getting 100 heads. But re-tossing only the coins that landed tails yields, in very few tries, a large probability of getting 100 heads. This latter approach, for Rosenhouse, corresponds to Darwinian natural selection in making probable for evolution what at first blush would seem improbable.

Variations on the WEASEL

What should disturb readers about these examples by Dawkins and Rosenhouse is not just the total absence of any real biology but the tacit insertion of teleology in the form of an explicitly specified target. METHINKS IT IS LIKE A WEASEL is a specified sequence of letters and spaces. Moreover, Dawkins specifically arranged his fitness landscape to converge to this sequence. Similar considerations apply to Rosenhouse's example.

But why choose METHINKS IT IS LIKE A WEASEL? Why not a random sequence of 28 letters and spaces such as OLBFXG CDUOHVKMQV EVUGASYEWZ? Dawkins' algorithm could have been contrived to converge, with equal speed, to this sequence as well. What's so special about METHINKS IT IS LIKE A WEASEL (other than that it is a coherent English text and calls to mind the famous trope about monkeys given enough time typing the works of Shakespeare)?

Would Dawkins' example have been nearly as persuasive in underwriting the power of cumulative selection if the random sequence OLBFXG CDUOHVKMQV EVUGASYEWZ had instead been made the target? Of course not. But in fact, the evolution of neither sequence lends support to Darwinian evolution. This is a consequence of conservation of information for search, which is the topic of the sequel to this book and is briefly addressed in the epilogue.

The example by Dawkins, in different guises, has been a prime intuition pump for natural selection. Variants of his simulation have appeared over decades and continue to be touted as slam-dunk evidence for the power of Darwinian evolution. Preceding Dawkins by several years, Manfred Eigen and Peter Schuster performed a similar simulation, whose target phrase was TAKE ADVANTAGE OF MISTAKE.[11]

David Berlinski, in *Black Mischief*, called out the illicit introduction of teleology in Eigen and Schuster's simulation (illicit from the point of view of a stricter Darwinism that attempts to dispense with teleology).[12] Despite Berlinski's critique, Bernd-Olaf Küppers, a student of Eigen, continued to advance this simulation in the early 1990s. Küppers' target phrase was EVOLUTION THEORY.[13]

In exactly the same vein, Jeffrey Satinover attempted to demonstrate the power of evolutionary algorithms with a simulation that generated the target phrase MONKEYS WROTE SHAKESPEARE.[14] RNA-worlds researcher Michael Yarus reprised this simulation with the target phrase NOTHING IN BIOLOGY MAKES SENSE EXCEPT IN THE LIGHT OF EVOLUTION.[15]

No doubt, the history of this simulation as sketched here is incomplete, with the evolutionary literature containing many more such instances. The never-ending riffs on Dawkins' WEASEL suggest either a profound insight into evolution or a hopeless delusion about its working and power. Design theorists are convinced it's the latter.

Examples like this purport to show that the Darwinian mechanism of cumulative (or natural) selection functions as a probability amplifier, lifting small probabilities to the probable and even certain. Thus do mainstream evolutionists resist the design inference, seeing it as inapplicable to the history of life. Perhaps design theorists should therefore shift their focus to the origin of life. There, at least at some crucial stages where lifeless matter shades into the first self-replicating biological entity, no process of random variation plus natural selection could be operating, given the Darwinian mechanism's need for entities capable of self-replication and variation. Might the origin of life, then, be a more fertile ground for small probabilities than the subsequent evolution of life?

The evolutionary biologist Theodosius Dobzhansky famously remarked that "prebiological natural selection is a contradiction in terms."[16] His point was that natural selection presupposes the self-replication of what is being selected, and self-replication doesn't exist before life itself comes to exist. So the Darwinian evolutionary mechanism, which is said to undo the design inference in biological evolution, would therefore seem to be unavailable at the origin of life. Given natural selection's putative role as a probability amplifier once life is already here, its unavailability would seem to leave the design inference some breathing room at the very start of life.

The key question then becomes whether the Darwinian mechanism applies only to biological evolution or could also apply to chemical evolution. Chemical evolution is the evolution of life from non-life via chemical interactions among non-living materials. For naturalistic scientists, chemical evolution is thus the key to life's origin. Yet it also seems like a place within the study of biological origins where small probabilities might hold greater sway and where a design inference might be more forthcoming given that natural selection seems barred there from acting as a probability amplifier. All the same, mainstream origin-of-life researchers appear just as committed as mainstream evolutionary biologists to resisting design inferences and invoking some version of the Darwinian mechanism.

Origin-of-life researchers are especially interested in finding a self-replicating molecular assembly, such as a self-catalyzing RNA. Suppose such an assembly is so simple that it could arise through a process of self-organization.[17] Such a self-organizational process would operate by physical necessity, and thus with large probability, so it wouldn't trigger a design inference. Might it be possible for the Darwinian mechanism, once given such a simple initial replicator, then to do the rest by evolving it into a full-blown cell? Why shouldn't "molecular Darwinism," in which natural selection and random variation act at the molecular level, bootstrap an initial replicator all the way up to a full-blown cell?

There is good reason to doubt the effectiveness of natural selection before the advent of cellular life. The focus of origin-of-life research has mainly been on explaining the origin of a simple molecular replicator, such as a self-catalyzing ribozyme. In that vein, the research has also focused on the chemical precursors to such a replicator. But even with such a simple replicator, there's no reason to think that natural selection can somehow bootstrap it into a self-replicating system consisting of numerous interacting biomacromolecules, as typical of cellular life—which is the only life we know.

Such a system needs to perform the specific functions universally associated with life, such as material transport, metabolism, energy conversion, signal transduction, information processing, and sequestration from the environment. Many origin-of-life researchers think that once a primitive replicator comes to exist, Darwinian selection will take it the rest of the way to a complex self-replicating system of many interacting biomacromolecules. But that's wishful thinking with no evidence to back it.

Origin-of-life researchers have been guilty of lowering the bar for what qualifies as first life. The reason for this move is that full-fledged life is so complex and difficult to account for that it's easier to substitute a simpler problem on which they can realistically hope to make progress. Life traditionally has been defined by enumerating essential characteristics exhibited across all cells, as in the list just given. But these days we find life redefined, with redefinitions ranging from anything that undergoes Darwinian evolution to anything that exhibits memory.[18] No such redefinitions suffice to define life. Nor is replication per se enough to define life. A salt crystal could be said to result from self-replication in a medium of sodium and chloride. But when it comes to actual life, what needs to be explained is the origin of complex self-replicating systems that consist of many biomacromolecules embedded in many diverse subsystems and performing a wide range of coordinated functions of the kind found in actual cells.[19] In other words, what needs to be explained is the origin of the full cell as we know it. Molecular Darwinism offers no insight here.

In the end, the question facing biological origins—both chemical and biological evolution—is whether naturalistic causal mechanisms always exist to overcome the improbabilities that seem to block their path. This is a substantive scientific question, so the answer to it cannot simply follow from armchair speculations about what, for instance, natural selection is or is not supposed to be able to do. Examples such as Dawkins' METHINKS IT IS LIKE A WEASEL are suggestive but misleading, so removed from biological reality that they cannot decide the matter. Practically, this means that scientists must be able to form reasonable probability estimates for the evolvability of various biological systems even when factoring in naturalistic causal mechanisms such as natural selection. Moreover, such probability estimates must be formed without prejudice, allowed to be small or large, whichever way the evidence points.

Yet for naturalistic evolutionary biologists, natural selection and other naturalistic mechanisms[20] are supposed to constitute a universal solvent for ridding the study of biological origins of small probabilities (unless specification is absent). If they are right, then the design inference cannot get off the ground because it depends fundamentally on small probability. But what if some probabilities remain extremely small, and demonstrably so, despite such mechanisms? The challenge for intelligent design researchers intent on drawing design inferences in biology will then be to show that for at least some clearly specified biological systems, natural selection and other naturalistic mechanisms do nothing to alleviate their vast improbability. The rest of this chapter takes up that challenge.

7.2 Resetting Darwinian Evolution's Bayesian Prior

To demonstrate design in biology, it's not necessary to show that all aspects of biological systems are designed. Even one unequivocal case of design in biology would be enough. Naturalistic biologists, by contrast, maintain that every aspect of every biological system gives no evidence of actual design. To refute this claim, logic only requires showing that some biological system, even just one, gives solid evidence of actual design.

Given the design inferential apparatus laid out in this book, it would therefore seem straightforward to demonstrate that some clearly identified biological system cannot be produced by any naturalistic evolutionary process that brings it about with anything other than small probability. And since Darwinian natural selection is the only mechanism within naturalistic evolution known to act as a probability amplifier and thereby mitigate small probabilities (other mechanisms, such as genetic drift, don't presume to have that capability), it would be enough to find some biological system whose probability can be calculated precisely, whose probability is small, and whose probability remains small even with the help of the Darwinian mechanism.

But carrying out such a demonstration is easier said than done. In biology, we do not have the convenience of calculating probabilities for toy problems, such as tossing a coin a thousand times and getting all heads. Nor do we have the convenience of calculating probabilities for more complicated but still readily manageable problems, such as inferring design in the case of data falsification, financial fraud, or the search for extraterrestrial intelligence (recall Chapter 2). Real-life biological systems are far less tractable probabilistically. In trying to account for their origin, we often have little idea what their evolutionary precursors might have been and what evolutionary pathways they might have taken.

In fact, the complexities of biological systems can become so overwhelming that forming reliable probability estimates for their origin may not be possible. The challenge for design theorists looking to apply the design inferential method to actual biological systems is to find a *Goldilocks zone* in which a system is complex enough to yield a small probability if the probabilities can be calculated but also simple enough for probabilities actually to be calculated. Vertebrate eyes seem too complex. Gene fragments of a few codons seem not complex enough.

But before calculating probabilities for the emergence of actual biological systems with the aim of drawing a design inference, we need to bring some sense of proportion to the Darwinian mechanism of natural selection. Currently, Darwinian theory is held in such high regard among mainstream evolutionary biologists that any merits of intelligent design are simply ruled out of court. No theory is that well confirmed or deserves that kind of unmitigated acceptance. To establish a fair playing field for the design inference, we therefore take a Bayesian approach. Specifically, we'll knock down to size the prior probability for the Darwinian mechanism so that it cannot automatically invalidate design.

The natural selection hypothesis, denoted by *NS*, is within the biological mainstream considered to be so much better established than the design hypothesis, denoted by *DS*, that it's impossible for *DS* to receive fair consideration in relation to *NS*. To be clear, by the natural selection hypothesis we are not referring to the wholly uncontroversial idea that natural selection exerts some influence on the differential survival and reproduction of organisms. A healthy fast cheetah is more likely to survive and reproduce than one crippled from birth. A bacterium resistant to antibiotics will, in an environment with the antibiotic, be selectively favored over one that has no such resistance. Instead, we mean the view that natural selection can conquer just about any evolutionary challenge, whether it's building the first wings, engineering the vertebrate eye, or constructing entirely new animal body plans. It's this grander sense of natural selection that we intend by *NS*.

In Bayesian terms, the prior probability of *NS* is presently so large and so overwhelmingly close to 1 that for any event E signifying the emergence of some biological system, it doesn't matter what the likelihoods $P(E|NS)$ and $P(E|DS)$ are because the prior probability $P(NS)$ is so large that it swamps everything else.

Thus, in any likelihood ratio and Bayesian updating of probabi-
lities, the posterior probability of NS, namely $P(NS|E)$, will con-
tinue to stay large and far bigger than the posterior probability of
DS, namely $P(DS|E)$ (see Appendices A.3 and A.4 for the Bayesian
terminology and logic underlying this discussion).

It is a fact about Bayesian probability that when the prior
probability of a hypothesis is sufficiently close to 1, it becomes
impossible to dislodge the hypothesis. This can be seen from the
update rule for Bayesian evidence in Appendix A.4, only in this
case substitute NS for H_1 and DS for H_2:

$$\frac{P(NS|E)}{P(DS|E)} = \frac{P(E|NS)}{P(E|DS)} \times \frac{P(NS)}{P(DS)}.$$

The likelihood ratio $P(E|NS)/P(E|DS)$ controls how the event E acts
as evidence for NS and DS, strengthening support for one and
weakening support for the other.

So, let's say that E is an event giving rise to some complex
biochemical machine of the sort that design theorists take as
providing evidence of design. Even if this likelihood ratio favors
design (DS) by being substantially less than 1 (i.e., $P(E|DS) \gg$
$P(E|NS)$), if $P(NS)$ vastly exceeds $P(DS)$, with $P(NS)$ very close to
1 and $P(DS)$ very close to 0, then the ratio of posterior probabilities,
$P(NS|E)/P(DS|E)$, will be much larger than 1, and thus still favor
NS. $P(NS|E)$ and $P(DS|E)$, as posterior probabilities, reflect the
degree to which the event E, the evidence, has recalibrated our
probabilities of NS and DS respectively. The bottom line is that by
making $P(NS)$ too close to 1, we insulate NS from evidential
challenge by DS, which is the point we were making in the last
section, though without the Bayesian formalism. Insulated in this
way, NS functions more as a dogma than as a scientific framework
open to testing and refutation.

Our approach, then, is to reset the Bayesian prior for NS,
lowering it sufficiently so that its head can go on the empirical
chopping block (as required of all scientific theories), thereby
giving DS a fighting chance. Right now, it's hard to imagine how
the prior probability of NS, as understood by Darwinian biologists

and their supporters, could be less close to 1. Daniel Dennett, surveying the grand accomplishments of modern science, finds nothing greater than natural selection: "If I were to give an award for the single best idea anyone has ever had, I'd give it to Darwin, ahead of Newton and Einstein and everyone else. In a single stroke, the idea of evolution by natural selection unifies the realm of life, meaning, and purpose with the realm of space and time, cause and effect, mechanism and physical law."[21]

Not to be outdone in his praise for natural selection, Richard Dawkins goes even further by crediting Darwinian evolution with making atheism far more rationally acceptable than theism:

> What Darwin showed was the staggeringly counter-intuitive fact that life can be explained by an undirected process, natural selection, which is the very opposite of chance. That's the essence of what Darwin discovered... That is possibly the greatest achievement that any human mind has ever accomplished. Not only did he show that it could be done. I believe that we can argue that the alternative is so unparsimonious and so counter to the laws of common sense that reluctant as we might be, because it might be unpleasant for us to admit it, although we can't disprove that there is a God, it is very very unlikely indeed.[22]

This unflappable confidence in Darwinian natural selection has worked its way into the wider culture. Thus, novelist Barbara Kingsolver has described Darwin's idea of natural selection as "the greatest, simplest, most elegant logical construct ever to dawn across our curiosity about the workings of natural life. It is inarguable, and it explains everything."[23] Discipline after discipline has now been Darwinized. Cosmology has self-reproducing black holes. Ethics and psychology are now evolutionary ethics and evolutionary psychology. Professional schools now entertain evolutionary medicine, management of the human animal, economics as an evolutionary science, and evolutionary jurisprudence—all following Darwin's lead by embracing natural selection. Even the field of religious studies celebrates genes that predispose us to believe in God regardless of God's actual existence, tracing such genes to their adaptive value via natural selection.

Given this level of exuberance for natural selection, not only will the prior probability of *NS* be huge and the prior probability of *DS* be minuscule, but it's hard to imagine how any biological design inference can get a hearing in its face. Our task, therefore, is to give cogent grounds for lowering the prior probability of *NS* (i.e., *P(NS)*). Accomplishing this task is not difficult once we look past natural selection's imagined strengths, as well as the hype of its enthusiastic advocates such as Dawkins and Dennett, and instead come to terms with its actual limitations, which are serious and damning. These limitations will become evident in subsequent sections. Only after we have shown, on general grounds, that natural selection deserves not unflagging acceptance but skeptical scrutiny will we examine some actual biological systems and sketch how their probabilistic analysis could yield small probabilities and warrant design inferences.

As we've characterized design inferences, they are chance elimination arguments that *sweep the field clear of all relevant chance hypotheses*. Thus, in line with Chapter 5, when confronted with a collection of chance hypotheses $\mathcal{H}$, we apply the Generic Chance Elimination Argument to each chance hypothesis H in $\mathcal{H}$ and, if all are eliminated, we thereby underwrite a design inference. But mainstream evolutionary biologists resist any such design inference. They claim that we never have an adequate handle on the relevant chance hypotheses $\mathcal{H}$. At best, we have a handle on a subset of $\mathcal{H}$, call it $\mathcal{H}'$, for which we may be successful in eliminating all its chance hypotheses. But there will always be, so they contend, a further subset of $\mathcal{H}$, call it $\mathcal{H}''$, chance hypotheses non-overlapping with $\mathcal{H}'$, that we have not identified, that we may not be able to identify, and that we may not even suspect to exist. And yet there may be chance hypotheses in $\mathcal{H}''$ that characterize how natural selection can give rise, with high probability, to the biological systems in question.

But note what is actually happening here. It's not that evolutionary biologists have identified and exhibited $\mathcal{H}''$ and its hypotheses. Rather, they simply posit $\mathcal{H}''$ as a set of hypotheses about how natural selection might, with high probability, *possibly* bring about the biological system in question. No detailed Darwinian pathway of how the system could actually have arisen is on offer. The argument here is one of *sheer possibility*, with no evidential backing, and supported only by an overweening confidence in the power of natural selection. The thinking is, "We know that natural selection is extraordinarily powerful, so we are confident it can do the job, even if at the moment we have no idea how."

But once $P(NS)$ is downgraded (in the same way that the creditworthiness of a bond might be downgraded), it is no longer necessary to insist that sweeping the field clear of chance hypotheses means ruling out every hypothesis that might exist in such a collection as $\mathcal{H}''$. The chance hypotheses in this set reside in a fantasy world. They are merely posited but never actually identified or displayed. Nor are they used to trace clearly articulated and evidentially supported Darwinian pathways. Instead, such hypotheses are merely invoked, and the Darwinian pathways they are supposed to engender are merely gestured at.

Design theorists, in attempting to draw a design inference in biology, eliminate chance with respect to those chance hypotheses that have actually been identified and may rightly be deemed as relevant. If they've missed any relevant chance hypothesis, they are ready to include it in $\mathcal{H}$ and rerun the analysis. What they are unwilling to do is to negate their analysis because of the mere possibility that they might have omitted some relevant chance hypothesis that no one has yet considered. This is just ordinary scientific practice, namely, doing the best with what we know and not hamstringing ourselves with worries over what we don't know. That's not to say we may not reserve judgment when we recognize our knowledge to be spotty. But biological design inferences depend on deep study and thorough knowledge of the biological systems whose design is in question. They are not hasty inferences based on ignorance.

Who Is Arguing from Ignorance?

Critics of intelligent design often mistakenly charge the design inference with being an argument from ignorance (i.e., arguing for design based on our ignorance of chance alternatives). In fact, a design inference, by ruling out relevant chance hypotheses, engages in an eliminative induction, whose logic is sound and differs from an argument from ignorance. Eliminative induction is a method of reasoning used in science and philosophy to support a hypothesis by systematically eliminating competing hypotheses. The principle underlying eliminative induction is that if all alternative hypotheses can be falsified or shown to be less likely, then the remaining hypothesis (a design hypothesis in this case) gains credibility and support.

Thus, as alternative hypotheses are tested and rejected, the likelihood of the remaining hypothesis being correct increases. This approach is especially useful in situations where direct evidence for a hypothesis is unavailable (as with a design hypothesis in a typical design inference). While this method of reasoning is able to strengthen support for the remaining hypothesis, it cannot guarantee its truth, as there could always be other hypotheses or explanations that have not yet been considered or discovered. Leaving aside strict deductive reasoning, fallibility of this sort is a feature of all human reasoning.

Eliminative inductions rely on the successful falsification of competing hypotheses. Their strength therefore increases to the degree that they effectively eliminate such competition. The problem with eliminative inductions in practice, however, is that we don't have a neat way of organizing competitors so that they can be eliminated with a few manageable blows. As philosopher of science John Earman puts it,

> The eliminative inductivist [seems to be] in a position analogous to that of Zeno's archer whose arrow can never reach the target, for faced with an infinite number of hypotheses, he can eliminate one, then two, then three, etc.,

but no matter how long he labors, he will never get down to just one. Indeed, it is as if the arrow never gets half way, or a quarter way, etc. to the target, since however long the eliminativist labors, he will always be faced with an infinite list [of remaining hypotheses to eliminate].[24]

Earman then immediately answers this objection:

My response on behalf of the eliminativist has two parts. (1) Elimination need not proceed in such a plodding fashion, for the alternatives may be so ordered that an infinite number can be eliminated in one blow. (2) Even if we never get down to a single hypothesis, progress occurs if we succeed in eliminating finite or infinite chunks of the possibility space.[25]

The key word in Earman's remarks here about eliminative induction is *progress*. Design inferences, as eliminative inductions, make progress in advancing our understanding of biological origins. To deny design inferences is to stymie progress, leaving biology in a holding pattern that reflexively invokes non-design naturalistic explanations even though these explanations come up short time after time.

Much more can be said in defense against the argument-from-ignorance objection, but we will not belabor the issue here. Suffice it to say that design theorists have responded cogently and at length to this objection.[26] There is, however, an argument from ignorance in play in the Darwinism/design controversy. It lies not with design theorists, but with Darwinian evolutionists who struggle in vain to explain how natural selection could have produced the biological systems that lead design theorists to infer design.

For example, cell biologist Franklin Harold, who is not a supporter of Michael Behe, states, "We should reject, as a matter of principle, the substitution of intelligent design for the dialogue of chance and necessity." However, the basis of this principle and why it should be followed he leaves unanswered. Indeed, the principle becomes implausible given what Harold says next: "But we must concede that there are presently no detailed Darwinian accounts of the evolution of any biochemical or cellular system, only a variety of wishful speculations."[27] Harold made this remark back in 2001 and in doing so explicitly cited Behe. As we will discuss later in this chapter, such wishful speculations by Darwinists persist, and the design-related challenges facing biology continue to grow.

   In inflating natural selection's prior probability and in dismissing design inferences as arguments from ignorance, mainstream evolutionary biologists assume no burden of proof. Instead, whenever a design inference for the emergence of a biological system threatens to be drawn, they invoke unidentified and indeed unidentifiable chance hypotheses that render the probability of its emergence high enough to avoid a design inference. They do this not by actually exhibiting such chance hypotheses (that's why they are unidentified and even unidentifiable) but by an act of faith in the wonder-working power of natural selection. Lowering the Bayesian prior in natural selection blocks this maneuver and opens the way for design inferences in biology that would otherwise get blocked.

   Before turning to actual limitations and shortcomings of the Darwinian selection mechanism with the aim of establishing a more sensible Bayesian prior, we need to make a general point about Bayesian probabilities. What is relevant to determining the prior probability of a hypothesis in general, and of natural selection in particular, is the evidence supporting it, which includes everything from its conceptual soundness to its empirical adequacy. But it does not include the mere feeling of confidence in a hypothesis or the intensity of that feeling. Biologist and Nobel laureate Peter Medawar put it best: "*I cannot give any scientist of any age better advice than this: the intensity of the conviction that a hypothesis is true has no bearing on whether it is true or not.* The importance of the strength of our conviction is only to provide a proportionately strong incentive to find out if the hypothesis will stand up to critical evaluation."[28]

7.3 John Stuart Mill's Method of Difference

In resetting natural selection's Bayesian prior, let's start with a point of logic articulated by John Stuart Mill. Mill (1806–1873) was a close contemporary of Charles Darwin (1809–1882). As a philosopher, Mill is best remembered in our day for his writings on liberty and utilitarianism. Yet what interests us here is his work on logic. Biology's reception of Darwinism might have been far less favorable had scientists heeded Mill's work on logic and applied it to natural selection.

In 1843, sixteen years before the publication of Darwin's *Origin of Species*, Mill published the first edition of his *System of Logic* (which by the 1880s had gone through eight editions).[29] In that work, Mill laid out various methods of induction. The one that interests us here is his *method of difference*. Mill described this method as follows: "If an instance in which the phenomenon under investigation occurs, and an instance in which it does not occur, have every circumstance in common save one, that one occurring only in the former; the circumstance in which alone the two instances differ is the effect, or the cause, or an indispensable part of the cause, of the phenomenon."[30]

According to this method, to discover which of a set of circumstances is responsible for an observed difference in effects requires identifying a circumstance that is present when the effect occurs and absent when it doesn't occur. An immediate corollary of this method is that *common circumstances cannot explain a difference in effects*. Note this last point well in the following discussion.

Suppose you and a friend have been watching television, eating popcorn, and lounging on a couch. Yet your friend is now staggering about bleary-eyed while you are walking around no problem and feeling just fine. Precisely because the television, popcorn, and couch are common to both your experiences, they do not explain why your friend is having difficulties and you are doing fine. To explain the difference, you need to find not what's common to your circumstances but what's different. And then you discover that your friend also consumed considerable quantities of alcohol, mixing it into his lemonade, whereas you only drank plain lemonade. In citing alcohol to explain the difference, you are applying Mill's method.

Mill's method of difference, so widely used in everyday life, is crucial for assessing the power and applicability of natural selection in biological evolution. The fact is, natural selection never operates in isolation. Accordingly, Kenneth Miller will underscore the need for more in evolution than just natural selection, though not a lot more. He claims that what's needed to drive Darwinian evolution and its increases in biological information is "just three things: selection, replication, and mutation."[31] Miller is certainly right that evolution by natural selection requires random variation (or mutation) and replication (or heredity). But, as we'll see by using Mill's method, Miller's list of three items is actually missing quite a lot.

For evolution to work, natural selection must act on things that are replicating, and those things that are replicating must also vary, preferably randomly but also with the possibility of beneficial variations. Or, as Darwin put it in the *Origin*: "Unless profitable variations do occur, natural selection can do nothing."[32] Moreover, all three of these (selection, variation, and replication) must operate in an environment capable of contributing information through the evolutionary process to its newly emerging products. Thus, Jason Rosenhouse will remark, "Natural selection serves as a conduit for transmitting environmental information into the genomes of organisms."[33]

Darwinian evolution therefore depends on natural selection, random (sometimes beneficial) variation, and replication working together within an environment. But what renders one environment acting by selection, variation, and replication a source of evolutionary change that's interesting but another that's uninteresting? The words "interesting" and "uninteresting" in this context pertain to why we should care about biological evolution at all. The reason evolutionary biology attracts so much attention is that real-life biological systems, such as are said to have evolved, are so cool and interesting. Yet many environments operating with selection, variation, and replication don't lead to any interesting evolutionary products. The products are boring. For instance, constrain variation enough, and the products of evolution will not form a diverse and thriving evolutionary tree but instead consist of items that are all very similar, with nothing interesting to see.

For an illustration of Mill's method with real biological significance, consider Sol Spiegelman's work on the evolution of polynucleotides in a replicase environment. One thing that makes real-world biological evolution interesting, assuming it happens, is that it increases complexity in the things that are undergoing evolution. Yet Spiegelman demonstrated that even with selection, variation, and replication in play, complexity steadily decreased over the course of his experiment. The environment Spiegelman set up therefore put a premium on simplicity over complexity. Brian Goodwin, in his summary of Spiegelman's work, underscored this point:

> In a classic experiment, Spiegelman in 1967 showed what happens to a molecular replicating system in a test tube, without any cellular organization around it. The replicating molecules (the nucleic acid templates) require an energy source, building blocks (i.e., nucleotide bases), and an enzyme to help the polymerization process that is involved in self-copying of the templates. Then away it goes, making more copies of the specific nucleotide sequences that define the initial templates. But the interesting result was that these initial templates did not stay the same; they were not accurately copied. They got shorter and shorter until they reached the minimal size compatible with the sequence retaining self-copying properties. And as they got shorter, the copying process went faster. So what happened with natural selection in a test tube: the shorter templates that copied themselves faster became more numerous, while the larger ones were gradually eliminated. This looks like Darwinian evolution in a test tube. But the interesting result was that this evolution went one way: toward greater simplicity.[34]

Examples like this one cited by Goodwin, in which evolution driven by natural selection occurs in an environment where nothing interesting ever evolves, are not hard to find. Certainly, they are common with computer simulations of evolution that go absolutely nowhere. At the same time, it's also possible to write such simulations that solve interesting problems and produce salient patterns, such as evolutionary computing programs that solve intriguing engineering problems.[35] But because selection, variation, and replication acting within an environment are common to both such simulations, they cannot, as Mill's method makes clear, account for the difference.

The best known recent biological example where natural selection comes up short despite being given overwhelming opportunity to succeed is Richard Lenski's extended experiment with *E. coli*. Even with 75,000 generations and counting, Lenski's experiment found that the bacterial genomes of his *E. coli* "rapidly decayed."[36] Such a result is hardly a positive showcase for Darwinian evolution, whose claim to fame is building complexity, not removing it. Lenski attributes this rapid decay to an inauspicious balancing of high mutation rate with weak selective pressure.

Here we see again Mill's method of difference: it's not just mutation (or random variation) and selection working together in some generic sense that are at issue. What's at issue, rather, is the way selection and mutation are calibrated in relation to each other so that the resulting evolutionary process leads, as here, to uninteresting evolutionary products but might otherwise, with a *different* calibration, lead to interesting evolutionary products. Lenski presumes that in the wild *E. coli* has evolved in interesting ways, with net genomic gains over time. Yet in his experiment, which corresponds to over a million years of human generations, *E. coli* experienced rapid genomic loss, doing nothing to display the putative power of Darwinian evolution. So yes, natural selection is operating in the Lenski experiment, but no, it is not the difference maker that explains why evolution works and is worth studying.

Problems for biological evolution like those identified in the research of Spiegelman and Lenski are the tip of the iceberg. Yes, evolution had better be complexity-increasing if it is to deserve the attention it receives. But complexity better not just be complexity for the sake of complexity. In the history of life, increasing complexity has been in the service of building magnificent structures of incredible sophistication and elegance (everything from the genetic code in the first bacteria to the cerebral cortex in humans). How could evolution accomplish such feats simply by means of selection, variation, and replication? Something more is needed. Biologist Stuart Kauffman, who has long had an uneasy truce with Darwinism, understands the challenge:

> Life uses mutation, recombination, and selection. These search procedures seem to be working quite well. Your typical bat or butterfly has managed to get itself evolved and seems a rather impressive entity... Mutation, recombination, and selection only work well on certain kinds of fitness landscapes, yet most organisms are sexual, and hence use recombination, and all organisms use mutation as a search mechanism... Where did these well-wrought fitness landscapes come from, such that evolution manages to produce the fancy stuff around us?[37]

According to Kauffman, "No one knows."[38]

Facile invocations of natural selection, even when combined with further invocations of random variation, replication, and information-generating environments, therefore don't even begin to address the real problems confronting biological evolution. Mill's method confirms that all these factors, even with pride of place going to natural selection, are just stage-setting. Yes, they must all be in place—just as a play needs props. But the real action of a play resides elsewhere—in the actors. Likewise, something else besides natural selection and its accoutrements must be going on if evolution is going to produce anything interesting. And that "something else" is not just peripheral. According to Mill's method of difference, it is the difference that makes the difference. It is the secret sauce.

Selection, variation, and replication operating within an environment can produce wildly different types of evolution:

interesting, uninteresting, complexity increasing, complexity de-creasing, elegantly engineered, kludgily engineered, etc. It follows from Mill's method that something else besides these factors must be in play. What is the difference that makes the difference? Design theorists would contend that the key difference maker here is design or intelligent activity. Thus, by infusing (complex specified) information into the environment—information that could be empirically ascertainable—a designing intelligence could make itself evident in biology, giving researchers grounds to draw a design inference.

Of course, design theorists have a burden of evidence here in attempting to show that design in biology is real (a burden that this book is meant to redress by laying out the statistical and empirical features by which evidence can count for design). But evolutionists have no less a burden in contending for the power of natural selection to act as a designer substitute. Part of that burden is meeting the challenge of Mill's method of difference—a burden that evolutionists, instead of carrying, have tended to sidestep.

Mill's method of difference shows that the information needed to round out a robust theory of evolution cannot be reducible to selection, variation, and replication operating within an environ-ment. All these elements are common to all types of evolution. So something else must be going on to explain the difference between different forms of evolution. The logic here is airtight. It means that Darwin's theory is radically incomplete. And this in turn means that the centerpiece of Darwin's theory, natural selection, cannot bear the weight that Darwinists place on it.[39]

7.4 The Challenge of Multiple Simultaneous Changes

Darwinian evolution, as understood in our day, is neo-Darwinian, a wedding of classical Darwinism with genetics, thereby making genetic mutations the primary source of variation. Neo-Darwinian evolution therefore happens gradually, one small step at a time, by single mutational changes. There's a sound probabilistic rationale for this allegiance to gradualism, underwritten by—or one might say, in reaction to—the design inference and its key criterion for design, specified complexity.

The alternative to single mutational changes is multiple simultaneous mutational changes. If simultaneous changes were required of evolution, then the steps along which Darwinian processes move would become improbable, so much so that Darwinian evolution itself would no longer be tenable. Darwin himself grasped the need for gradual evolutionary change, long before genetics was folded into his theory. And the need for gradualism became all the more obvious after genetics was folded in, due to the mathematical tractability of genetics. Our focus in this section, therefore, is on the gradualism inherent in Darwinian theory. This gradualism renders Darwinian evolution unable to accommodate multiple simultaneous and coordinated mutational changes, and that inability, as we will show, undercuts natural selection's efficacy and thus its Bayesian prior.

Darwin made his commitment to evolutionary gradualism clear in his *Origin of Species*: "If it could be demonstrated that any complex organ existed, which could not possibly have been formed by numerous, successive, slight modifications, my theory would absolutely break down. But I can find out no such case."[40] Of course, Darwin didn't just mean any sort of numerous, successive, slight modifications. What, after all, can't be formed gradually in the absence of any constraints whatsoever? Indeed, any system of parts can, in principle, be built up one part at a time, and thus gradually.

For Darwin, the constraint on gradual evolutionary pathways was, obviously, natural selection. The slight modifications acceptable to Darwin were those where each modification confers a selective advantage. Darwin was, after all, in the quotation just given, defending his theory. Like his modern-day followers, he was convinced that all evolutionary change happens gradually, with natural selection approving every step of the process. In our day, those changes are seen as genetic mutations, and the most common, and most gradual, of these is the single mutational change of a DNA base pair.

Jason Rosenhouse, whom we have met earlier in this chapter, is a contemporary defender of Darwinian gradualism who opposes design inferences in biology. As a mathematician, he therefore serves as a helpful foil to our case for biological design inferences. For Rosenhouse, every adaptation in organisms is the result of a gradual step-by-step evolutionary process in which natural selection ensures the avoidance of missteps along the way. Writing specifically about the evolution of "complex biological adaptations," he notes, "Either the adaptation can be broken down into small mutational steps or it cannot. Evolutionists say that all adaptations studied to date can be so broken down while anti-evolutionists deny this."[41]

In consequence, Rosenhouse rejects that existing biological adaptations could ever require multiple coordinated mutational steps: "Evolution will not move a population from point A to point B if multiple, simultaneous mutations are required. No one disagrees with this, but in practice there is no way of showing that multiple, simultaneous mutations are actually required."[42] This claim is remarkable. Rosenhouse is claiming not only that multiple simultaneous mutations never happen but also that we couldn't know if they did. But if their falsification is indemonstrable, whence his confidence that only single mutational changes ever happen? In science, if there can be evidence for a proposition, it must be possible also for there to be evidence against it. But for Rosenhouse, all the evidence can point in only one direction.

We differ with Rosenhouse in his claim that multiple simultaneous mutations are never required and will, shortly, present evidence to the contrary. But for the moment, let's focus on why he should strictly prohibit multiple simultaneous mutations. The answer is that they would render life's evolution too improbable. Simultaneous mutations throw a wrench in the Darwinian gearbox. If they played a significant role in evolution, Darwinian gradualism would become untenable. Accordingly, Rosenhouse maintains that such large-scale mutational changes never happen and, conveniently, that they are indemonstrable even if they do happen. Rosenhouse presents this point of view not with a compelling argument, but as a consequence of his commitment to Darwinian evolution.

To overturn Rosenhouse's view that all biological adaptations can occur through single mutational changes, one baby step after another, it is therefore enough to show that some biological systems resist gradual formation in this way and instead require multiple simultaneous changes. We shall turn to such systems later in this chapter when we consider, for instance, irreducibly complex biochemical machines. But in this section, our aim is to understand, in broader conceptual terms, the challenge that multiple simultaneous mutational changes pose to Darwinian evolution insofar as they would raise probabilistic barriers for natural selection, barriers that take the form of small probabilities.

To that end, consider a model of Darwinian evolution on a hypercube. What this model loses in biological realism it gains in conceptual clarity. Consider, therefore, a 100-dimensional discrete hypercube of 100-tuples of the form $(a_1, a_2, \ldots, a_{100})$, where every a_i is a natural number, inclusive, between 0 and 100. (In mathematics, an n-tuple is an ordered arrangement of n items. The simplest n-tuple is an ordered pair, which is a 2-tuple.) Consider, now, the following path in the hypercube starting at $(0, 0, \ldots, 0)$ and ending at $(100, 100, \ldots, 100)$. New path elements are now defined by adding 1s to each position of any existing path element, starting at the left and moving to the right, and then starting over at the left again. Thus, the entire path takes the form

$$0: \quad (0, 0, \ldots, 0)$$
$$1: \quad (1, 0, \ldots, 0)$$
$$2: \quad (1, 1, \ldots, 0)$$
$$\cdots$$
$$100: \quad (1, 1, \ldots, 1)$$
$$101: \quad (2, 1, \ldots, 1)$$
$$102: \quad (2, 2, \ldots, 1)$$
$$\cdots$$
$$200: \quad (2, 2, \ldots, 2)$$
$$\cdots$$
$$300: \quad (3, 3, \ldots, 3)$$
$$\cdots$$

1,000: (10, 10, ..., 10)

...

2,000: (20, 20, ..., 20)

...

10,000: (100, 100, ..., 100)

The hypercube consists of 101^{100} (or approximately 2.7×10^{200}) elements, but the path itself has only 10,001 path elements connected by 10,000 implicit path edges. For simplicity, let's put this discrete hypercube under a uniform probability distribution (we don't have to, and design inferences don't require this assumption, but it's convenient for the purposes of illustration). Given a uniform probability on the discrete hypercube, the path elements, all 10,001 of them considered together, have probability roughly 1 in 2.7×10^{196} (10,001 divided by the total number of elements making up the hypercube). That's very small, indeed smaller than our universal probability bound of 1 in 10^{150}.

Each path element of the hypercube has 200 immediate neighbors. Note that in one dimension there would be two neighbors, left and right; in two dimensions there would be four neighbors, left and right as well as up and down; in three dimensions there would be six neighbors, left and right, up and down, forward and backward; etc. Note also for path elements on the boundary of the hypercube, we can simply extend the hypercube into the ambient discrete hyperspace (consisting of the 100-fold Cartesian product of all integers), bringing in neighbors there that never actually end up getting used.[43]

Next, let's define a fitness function f that assigns to path elements of the form $(a_1, a_2, ..., a_{100})$ the sum $a_1 + a_2 + ... + a_{100}$. The starting point $(0, 0, ..., 0)$ then has minimal fitness and the end point $(100, 100, ..., 100)$ then has maximal fitness. Moreover, we assign negative fitness to all points off this path. Each successive path element, as illustrated above, therefore has higher fitness, by 1, than its immediate predecessor on the path. If we now use a uniform probability, and thus sample uniformly from the adjoining 200 neighbors, then the probability p of getting to the next element on the path, as judged by the fitness function f, is 1 in 200 for any given sample query. We can think of and describe this next element up the path as a *mutational step*.

The underlying probability distribution for moving between adjacent path elements is the geometric distribution.[44] Traversing the entire path from starting point to end point can thus be represented by a sum of independent and identically distributed (with geometric distribution) random variables. Thus, on average, it takes 200 evolutionary sample queries, or mutational steps, to move from one path element to the next, and it therefore takes on average 2,000,000 (= 200 × 10,000) evolutionary sample queries, or mutational steps, to move from the starting to the end point. Probabilists, as noted in Appendix B.5, call these numbers *waiting times*. Thus, the waiting time for getting from one path element to the next is, on average, 200; and for getting from the starting to the end point is, on average, 2,000,000.

This example illustrates that with just one mutational change at a time, hypercube evolution happens efficiently with high probability. Moreover, it doesn't matter if all functional states (10,001 in this case) are extremely sparse, and thus improbable (which they are), within the totality of possible states (101^{100} or 2.7×10^{200} in this case). Evolution depends on path-connectedness through (biological) configuration space, not on the density of those paths within configuration space.

Daniel Dennett made this point about path-connectedness repeatedly in *Darwin's Dangerous Idea*, stressing that for evolution to succeed it needs *isthmuses* (his word) to connect what otherwise would be isolated islands of functionality.[45] This is correct as far as it goes. But it all depends on evolution happening one beneficial mutational change at a time: an advantageous mutation by chance happens, selection agrees that it is advantageous and thus preserves it, and on to the next round of mutation and selection. Rinse and repeat.

To recap, evolution on the hypercube proceeds by going from one path element to the next, querying up to 200 neighbors, with the probability of finding the next path element being 1 in 200 for each query. This is a geometric progression, so the average number of queries per successful evolutionary step is 200, and since the total path has 10,000 steps, the average number of queries, or the average waiting time, to go from $(0, 0, ..., 0)$ to $(100, 100, ..., 100)$ is 2,000,000. As we saw in Section 4.5, the fastest that bacteria can replicate is every 4 minutes, which is the fastest replication rate of life as we know it. So with life on Earth lasting close to 4 billion years, that puts an upper limit of about 526 trillion generations on any evolutionary lineage on planet Earth. So 2,000,000 would be doable.

But what if hypercube evolution required two simultaneous successful queries—of one neighbor and then the next—for the next successful evolutionary step? Because the queries must be successful simultaneously, they are probabilistically independent, and the probabilities multiply. So, the probability of two successful simultaneous queries is $1/200 \times 1/200$, or 1 in 40,000. Granted, each step now traverses two neighbors, so the total number of steps needed to get from $(0, 0, ..., 0)$ to $(100, 100, ..., 100)$ drops in half to 5,000. But the total average waiting time to get from $(0, 0, ..., 0)$ to $(100, 100, ..., 100)$ is now $40,000 \times 5,000$, or 200,000,000. That's a hundred-fold increase over non-simultaneous queries, but still less than 526 trillion, the maximum number of generations in any evolutionary lineage on Earth.

But let's now ramp things up. What if hypercube evolution required five simultaneous successful queries—of one neighbor, then another, and another, and another, and still one more—for the next successful evolutionary step? Because the queries must be successful simultaneously, they are probabilistically independent, and the probabilities multiply. So, the probability of five successful simultaneous queries is $1/200 \times 1/200 \times 1/200 \times 1/200 \times 1/200$, or 1 in 320 billion. Each step now traverses five neighbors, so the total number of steps needed to get from $(0, 0, ..., 0)$ to $(100, 100, ..., 100)$ drops to 2,000. But the total average waiting time to get from $(0, 0, ..., 0)$ to $(100, 100, ..., 100)$ is now 2,000 times 320 billion, or 640 trillion. That's a 320-million-fold increase over non-simultaneous queries. Moreover, 640 trillion now exceeds 526 trillion, the maximum number of generations in an evolutionary lineage on Earth.

As is evident from this example, if evolution requires simultaneous successful queries (i.e., simultaneous mutational changes), then natural selection is rendered powerless and Darwinian evolution is dead in the water. In that case, the waiting times simply become too great for evolution to do anything interesting. Waiting times are inversely correlated with probabilities—the greater the waiting time, the smaller the probability. And so, for evolution to require multiple simultaneous changes would provide a clear opening to small probabilities and thus to the design inference. Now granted, the hypercube is an artificial, stylized example. Yet real-life examples of evolution exist that seem to require not successive, but numerous simultaneous mutational changes. We'll return to this point later in this chapter.

The lesson of this section is that natural selection loses any edge it has in amplifying probabilities (and thus overturning design inferences) as soon as it must reward large multiple simultaneous mutational changes in order to progress. Provided that biological reality is crisscrossed with evolutionary pathways that connect all organisms and their adaptations via small single-step mutations, natural selection can navigate those pathways. But as soon as small single-step mutations lose this path-connectivity, then large multiple simultaneous mutational changes become necessary, and natural selection can no longer navigate those paths.

Whether multiple simultaneous mutational changes pose a practical obstacle to real-world Darwinian biological evolution is a question to which we will return. But the very prospect that such changes could pose an obstacle in itself constitutes a challenge to natural selection. It indicates that natural selection lacks robustness, only being able to carry out certain types of evolutionary change but not others. That's a limitation. And it's a limitation that intelligent agency does not face, which is well able to conceive and implement systems that require the multiple simultaneous coordination of parts, as is exemplified in most engineering designs.

7.5 What to Make of Bad Design?

Bayesian priors should, ideally, be evaluated on rational and objective grounds. Unfortunately, that doesn't always happen. Psychological factors, some no better than ideology, wish-fulfillment, or matters of taste, often play a decisive role in setting Bayesian priors. One objection to intelligent design that has been particularly effective at derailing the idea for many, and from a Bayesian vantage in sharply elevating the prior probability of natural selection above that of design (i.e., $P(NS) \gg P(DS)$), is the charge of bad design in biology. Biology, we are told, is littered with so much pathetic and substandard design that no self-respecting designer would have invented and built things that way. Those who make this charge are called *dysteleologists*.[46] Dysteleologists see themselves not so much as disproving design as simply hauling it off stage, akin to yanking a performer who's flopping.

Dysteleologists often suggest that design proponents have invited this treatment. The adjective *intelligent* in front of the noun *design*, after all, connotes that the design is intelligent, which is to imply that it is well-conceived and artfully executed. Any substandard design that is then found in biology is thereby thought to invalidate intelligent design. Such critics of intelligent design are looking for an unfair advantage. The reason for putting *intelligent* in front of *design* is to stress that the design in question is indeed the product of a real intelligence, a mind. Richard Dawkins writes on the first page of *The Blind Watchmaker*, "Biology is the study of complicated things that give the appearance of having been designed for a purpose."[47] Dawkins then takes three hundred additional pages to convince readers that such design is only an appearance and not real. The word *intelligent* in front of *design* is there to underscore that the design is real.

Design, however, can be real even if it is poorly conceived and executed. Companies have faced significant financial disappointments precisely because of their faulty—but still real—designs: the anti-inflammatory drug Vioxx was supposed to be an effective pain remedy, but it caused heart attacks and strokes; the Firestone 500 tire was supposed to perform well at high speeds but had a disproportionate number of blowouts; Microsoft's Windows Vista was heavily criticized for its high system requirements and incompatibility with legacy applications, and was therefore quickly replaced by Windows 7.[48]

In trying to make sense of the quality of design in biology, we need to keep in mind that all actual design involves tradeoffs, and so no perfect design exists that is optimal in every conceivable way. As Henry Petroski, an engineering professor at Duke University, has aptly pointed out, "All design involves conflicting objectives and hence compromise, and the best designs will always be those that come up with the best compromise."[49] Real design cannot avoid tradeoffs among conflicting objectives. Perfection of design is therefore a fantasy. In consequence, all optimization of design is constrained optimization, which means that making one thing better may necessitate making another thing worse.[50]

So, what sort of design, regardless of whether it really is design, do we find in biology? Opinions differ. Some naturalistic thinkers, unconvinced that biology contains actual design, nevertheless regard the design they find in biology as impressive and even praiseworthy. Here is a brief sampling that can readily be expanded:

1. Nobel laureate and CRISPR co-discoverer Jennifer Doudna on DNA: "The double helix beautifully reveals the molecular basis of heredity." She then adds that the double helix is "this simple and beautiful molecular structure."[51]

2. Nobel laureate geneticist Frank Nurse on the cell: "The cell is made up of lots and lots and lots of little compartments, and it is compartmentation that allows all these different chemistries to occur simultaneously in such a small space. I want to emphasize how fantastic that is!"[52]

3. Biochemist Alonso Ricardo and Nobel laureate geneticist Jack Szostak on the cell: "Every living cell, even the simplest bacterium, teems with molecular contraptions that would be the envy of any nanotech-

nologist. As they incessantly shake or spin or crawl around the cell, these machines cut, paste and copy genetic molecules, shuttle nutrients around or turn them into energy, build and repair cellular membranes, relay mechanical, chemical or electrical messages—the list goes on and on, and new discoveries add to it all the time."[53]

4. University of Chicago molecular biologist James Shapiro on natural genetic engineering: "Living cells and organisms are cognitive (sentient) entities that act and interact purposefully to ensure survival, growth, and proliferation... They have the ability to alter their hereditary characteristics rapidly through well-described natural genetic engineering and epigenetic processes as well as cell mergers."[54]

5. Biomimicry expert Janine Benyus on the engineering capabilities of organisms: "Nature, imaginative by necessity, has already solved many of the problems we are grappling with. Animals, plants, and microbes are the consummate engineers. They have found what works, what is appropriate, and most important, what lasts here on Earth."[55]

Some evolutionists, however, make a cottage industry of highlighting what they regard as poor designs in biology. In fact, a few have turned sneering at biological design into an art form. MIT bioengineer Erika Benedictis has even added a twist to the sneer. She calls herself an advocate of intelligent design, but only in the sense that biology needs intelligent designers like herself and other scientists to play god and correct all the wrong designs that they find in biology. Biology unimproved by her handiwork and that of her colleagues therefore displays no actual intelligent design, at least as far as she is concerned.[56]

A game of ping-pong exists between those who claim biology is chock-full of bad design and those who claim, to the contrary, that attributions of bad design tend to be overblown, and that what is said to constitute bad design is often in fact good or even outstanding design. Each side tries to deflect the balls sent to their end of the table and preferably smash the ball onto the other side's end of the table. The best-known player on the side of bad design these days is biologist Nathan Lents, through his 2018 book *Human Errors*. More recently, in 2022, engineer Steve Laufmann and medical doctor Howard Glicksman published *Your Designed Body*, which takes the side of good design.

Readers of this book interested in following this ping-pong game can start by looking at these books, though they are only the tip of the iceberg in the literature on good and bad design. The fact is, however, that logically speaking, this ping-pong game is irrelevant to the question of actual design in biology. Regardless of whether the design is bad or good, if it is actual and ascertainable—as by a design inference—then Darwinism is dead in the water.

Nonetheless, to admit that the design in biology, even if real, is substandard or embarrassing scores rhetorical points against intelligent design irrespective of whether such an admission logically refutes intelligent design. Consider what is probably the most widely cited example of bad design in biology—indeed, the one with which Nathan Lents begins his book—namely, the human eye's inverted retina, which situates the photoreceptors behind nerves and blood vessels and thus would seem to obstruct the incoming light. The inverted retina in humans—and vertebrates more generally—has become the locus classicus of poor biological design.

As it is, good functional reasons exist for this construction of the eye. A visual system needs speed, resolution, and sensitivity. Speed is unaffected by the inverse wiring. Resolution is unaffected as well (except for a tiny blind spot, which the brain works around without difficulty). For comparison, the cephalopod retina of squids and octopuses, which is said to be "correctly wired" by having photoreceptors in front of nerves and blood supply, is no better at resolving objects in its visual field.

One reason the "incorrect wiring" doesn't affect resolution is that the nerve cells leading from the retina to the brain are surrounded by Müller glial cells that serve double duty, not just as insulation for nerve signals but also as optical fibers that transmit light with minimal distortion to the retina (some might regard this feature of the glial cells as good design!).[57]

As for sensitivity, the inverted retina enhances it. Retinal cells need more oxygen when the incident light is minimal. Placing the blood supply in front of photoreceptors ensures that retinal cells will have the oxygen they need to be as sensitive as possible when incident light is minimal. Some vertebrate eyes with inverted retinas, such as in frogs, are so sensitive that they can respond to single photons.[58]

Now the point of these musings is not that the human eye or any other features of humans cannot be improved or are in some clear and incontrovertible sense optimal. The point, rather, is that simply drawing attention to the inverted retina or any other supposedly suboptimal features is no reason to think that they actually are suboptimal, inferior, or otherwise embarrassing, and that no self-respecting designer would have done it that way. Most of the time we don't understand these systems well enough even to judge the quality of their design (imagine a Stone Age artist of cave paintings criticizing Rubens or Velasquez). We may simply be blind to functionalities of design that are hidden to us because of our ignorance, as was the case even recently with "junk DNA" (more about that in a moment).

Dysteleologists who criticize biological design are all blame and no game. That's true even of those like Erika Benedictis and Nathan Lents, who promise to bring "real intelligent design" into biology by improving it through their recommendations or handiwork. Invariably, they offer no detailed concrete proposals for how the human eye or other biological systems might be improved while also ensuring no loss of function there or elsewhere. Lents, for instance, charges the human ankle with needless complexity for having "too many bones," most of which he regards as "pointless" (his choice of words).[59] Accordingly, he recommends that some of those bones would be better off fused. But as mechanical engineer Stuart Burgess notes:

> If the ankle-foot complex were badly designed, it should not be difficult to define a better design. Lents has attempt-

ed to define a better design by claiming that a fused ankle joint would be better. Lents states: "Because many of the bones of the ankle do not move relative to one another, they would function better as a single, fused structure, their ligaments replaced with solid bone. Thus simplified, the ankle would be much stronger." It is wrong to say that many of the bones of the ankle do not move relative to each other because it is well known that the bones have significant relative movements. It is also wrong to state that a fusion would be better and stronger. It is well known in the medical field that ankle fusions lead to a degradation of ankle performance. One hospital report states: "Walking on rough ground is difficult after [an ankle]... fusion. Most people cannot play vigorous sports such as squash... after a... fusion."[60]

Vague, ill-defined, or simply mistaken recommendations for improving supposedly bad biological designs are common coin among dysteleologists. Detailed, well-considered recommendations, by contrast, are universally absent. Moreover, dysteleologists are light-years away from actually implementing any proposed improvements and confirming experimentally that they actually are improvements. At the same time, confirming that their proposed improvements fail may be feasible, as when ankle bones are medically fused and lead to degradation of performance, as pointed out by Burgess in response to Lents. A further wrench in the dysteleologists' gears is this: structures like the eye and ankle have to be built in embryological development, and this fact alone constrains how their design can be implemented and thus what potential form their design can take.

Design, as noted, is a matter of tradeoffs. There's no question that we would like to add to or improve existing functionalities. Yet when dysteleologists raise the suboptimality objection, invariably they merely imagine additional functionalities or improvements, but without offering any details of their design, construction, or implementation (as in embryological development), and with no thought about unforeseen deficits or impairments that the proposed improvements might create. With design, and especially with redesign—as these are imagined rather than delineated by dysteleologists—the devil is in the details.

As impressive as whole organisms are, most of the action in drawing design inferences in biology is not at the level of full multi- or even unicellular organisms, such as humans or bacteria, but at the level of biochemical systems inside the cell, which are easier to study and analyze. Is the design there, if it really is design, any better respected among biologists than, say, for the human body? Take the bacterial flagellum, a bidirectional motor-driven propeller used by certain bacteria to navigate their watery environments. Whatever biologists may have thought of its ultimate origins, they tended, in the past, to regard it with awe. Howard Berg discovered that flagellar filaments rotate to propel bacteria through their watery environments.[61] In public lectures he would refer to the flagellum as "the most efficient machine in the universe."[62] And indeed, its proton-powered motor is almost 100 percent efficient, well beyond any engines humans have constructed.

The bacterial flagellum contains many standard engineering components, including a rotor, a stator, bushings, and a drive shaft. It typically requires the coordinated interaction of thirty or more complex proteins. Moreover, the bacterial flagellum is hardwired into a signal transduction circuitry that enables a bacterium to sense its environment and engage in chemotaxis, moving up positive nutrient gradients but also reversing course out of negative nutrient gradients. It can spin as fast as 100,000 rpm and change direction in a quarter turn. In 2009, it elicited praise such as the following in the mainstream biological literature:

> Since the flagellum is so well designed and beautifully constructed by an ordered assembly pathway, even I, who am not a creationist, get an awe-inspiring feeling from its "divine" beauty. However, if the flagellum has evolved from a primitive form, where are the remnants of its ancestor? Why don't we see any intermediate or simpler forms of flagella than what they are today? How was it possible that the flagella have evolved without leaving traces in history?[63]

Writing about such systems in general back in 1998 for a special issue of *Cell*, the National Academy of Sciences president at the time, Bruce Alberts, remarked:

> We have always underestimated cells... The entire cell can be viewed as a factory that contains an elaborate network of interlocking assembly lines, each of which is composed of a set of large protein machines... Why do we call the large protein assemblies that underlie cell function protein *machines*? Precisely because, like machines invented by humans to deal efficiently with the macroscopic world, these protein assemblies contain highly coordinated moving parts.[64]

A few years later, in 2003, Adam Wilkins introduced a special issue of *BioEssays* devoted to nanomachines:

> The articles included in this issue demonstrate some striking parallels between artifactual and biological/molecular machines. In the first place, molecular machines, like man-made machines, perform highly specific functions. Second, the macromolecular machine complexes feature multiple parts that interact in distinct and precise ways, with defined inputs and outputs. Third, many of these machines have parts that can be used in other molecular machines (at least, with slight modification), comparable to the interchangeable parts of artificial machines. Finally, and not least, they have the cardinal attribute of machines: they all convert energy into some form of "work."[65]

Neither of these special issues offered step-by-step Darwinian pathways for how these machine-like biochemical systems might have evolved, but they did talk up their design characteristics. Despite the special treatment that these systems received in these journals, none of the mystery surrounding their origin has in the intervening years been dispelled. Nonetheless, the admiration that they used to inspire has diminished. This is not to say that some researchers have lost all awe for this system. As recently as 2015, the bacterial flagellar motor (BFM) came in for the following praise, albeit on evolutionary grounds: "The bacterial flagellar motor is the pinnacle of evolutionary bionanotechnology." It "informs and inspires the design of novel nanotechnology in the new era of synthetic biology." It is "a self-assembling nanoscale electric rotary motor that performs at higher speed and with greater efficiency than any man-made device."[66]

But a change in attitude among biologists toward the bacterial flagellum seems to be happening—from favorable to unfavorable. Consider the following comment about the flagellum from a 2020 review article on propulsive nanomachines. Jason Rosenhouse cites it approvingly, prefacing the quotation by claiming that the flagellum is "not the handiwork of a master engineer, but is more like a cobbled-together mess of kludges":

> Many functions of the three propulsive nanomachines are precarious, over-engineered contraptions, such as the flagellar switch to filament assembly when the hook reaches a pre-determined length, requiring secretion of proteins that inhibit transcription of filament components. Other examples of absurd complexity include crude attachment of part of an ancestral ATPase for secretion gate maturation, and the assembly of flagellar filaments at their distal end. All cases are absurd, and yet it is challenging to (intelligently) imagine another solution given the tools (proteins) to hand. Indeed, absurd (or irrational) design appears a hallmark of the evolutionary process of co-option and exaptation that drove evolution of the three propulsive nanomachines, where successive steps into the adjacent possible function space cannot anticipate the subsequent adaptations and exaptations that would then become possible.[67]

The shift in tone from earlier to the present is remarkable. What happened to the awe these systems used to inspire? Have investigators learned so much in the intervening years to say, with any confidence, that these systems are indeed absurdly engineered kludges? Or are they merely countering the favorable press that these systems have received from design-friendly scientists, who see in them evidence of design? Consider the charge of over-engineering. To say that something is over-engineered is to say that it could be simplified without loss of function (as with a Rube Goldberg device). But what justifies that claim here, especially since these systems have to assemble themselves, and thus cannot be put together piecemeal in the same way as a Johnson or Mercury outboard motor might be put together for a speedboat?

Have scientists invented simpler systems that in all potential environments perform as well as or better than the systems in question? Are they able to go into existing flagellar systems, for instance, and swap out the over-engineered parts with more efficient systems and subsystems that they've invented? Have they in the intervening years gained any real insight into the step-by-step evolution of these systems? Or are they merely engaged in rhetoric to make flagellar motors seem less impressive and thus less plausibly the product of design? To pose these questions is to answer them.

Such downgrading of design in nature and thus of any designer responsible for the design can be found as far back as David Hume's *Dialogues Concerning Natural Religion* (and millennia earlier in the work of Lucretius and Epicurus). There Hume, in a remarkably Darwinian passage appearing eighty years before the *Origin*, asks whether "many worlds might have been botched and bungled" with "much labour lost, many fruitless trials made, and a slow, but continued improvement carried on during infinite ages in the art of world-making."[68] No real intelligent design need apply. A bungling designer that creates through trial and error suffices. Darwin is supposed to have gone Hume one better by providing a designer also bereft of mind—this designer being natural selection.

In same spirit, Rosenhouse offers his own anti-design argument. Humans are able to build things like automobiles, but not things like organisms, he notes. Accordingly, ascribing design to organisms is an "extravagant extrapolation" from "causes now in operation." Rosenhouse's punchline: "Based on our experience, or on comparisons of human engineering to the natural world, the obvious conclusion is that intelligence cannot at all do what they [i.e., ID proponents] claim it can do. Not even close. Their argument is no better than saying that since moles are seen to make molehills, mountains must be evidence for giant moles."[69]

This argument is ridiculous. So, primitive humans living with Stone Age technology, if they were suddenly transported to Dubai, would be unable to get up to speed and recognize design in the technologies on display there? Likewise we, confronted with space aliens whose technologies can build organisms using ultra-advanced 3D printers, would be unable to recognize that they were building designed objects? The obvious answer to such questions is that of course we would appreciate the design in such cases. What underwrites our causal explanations is our exposure to and understanding of the *types* of causes now in operation, not the *idiosyncrasies* of their operation. Because we are designers, we can appreciate design even if we are unable to replicate the design ourselves. Lost arts are lost because we are unable to replicate the design, not because we are unable to recognize the design.

In making a charge of poor design in biology, dysteleologists typically depend on a guilty-until-proven-innocent strategy. In effect they say, "Prove us wrong by showing that the design really is good and cannot be made palpably better." And what cannot be criticized by suggesting that it leaves "room for improvement," especially when critics face no obligation to describe the improvement in any detail? But in the case of junk DNA, dysteleologists went too far. Their rhetorical strategy backfired here because their charge of dysteleology was, in the end, shown to be false.

Until about 2010, many mainstream biologists interpreted the non-coding regions of DNA (those that don't code directly for proteins) as junk. Junk DNA was said to result from a sloppy evolutionary process that put a premium on survival but not on editing and excising DNA that had lost its use (thus offering no benefit but also posing no threat to survival). The challenge of junk DNA to intelligent design emerged in the late 1990s. In response, I [WmD] predicted in 1998 that function would be discovered for what was being called junk DNA:

[Intelligent] design is not a science stopper. Indeed, design can foster inquiry where traditional evolutionary approaches obstruct it. Consider the term "junk DNA." Implicit in this term is the view that because the genome of an organism has been cobbled together through a long, undirected evolutionary process, the genome is a patchwork of which only limited portions are essential to the organism. Thus, on an evolutionary view we expect a lot of useless DNA. If, on the other hand, organisms are designed, we expect DNA, as much as possible, to exhibit function... [D]esignating DNA as "junk" merely cloaks our current lack of knowledge about function... Design encourages scientists to look for function where evolution discourages it.[70]

Other intelligent design proponents around that time sounded similar notes, including Richard Sternberg and Jonathan Wells.[71]

The first decade of the new millennium saw the heyday of using junk DNA to criticize intelligent design. One of the more memorable criticisms came in 2006 from Francis Collins. According to him, our genome is bursting with "genetic flotsam and jetsam." Any designer would therefore be a jokester who put junk DNA into the genome "to confuse and mislead us" into thinking that "these are actually functional elements placed there... for a good reason." It was vain, in Collins' view, to hope that junk DNA might turn out to be functional after all and that the term "just betrays our current level of ignorance." For Collins, the evidence was overwhelming and the verdict was in: junk DNA really was junk, and any view to the contrary strained credulity. He concluded that junk DNA provides compelling evidence for Darwinian evolution.[72]

Collins was in good company. Richard Dawkins had likewise heralded junk DNA as a vindication of Darwinian evolution. For decades he had written of surplus DNA existing in cells—DNA that performed no function and that was parasitic.[73] Even as late as 2009, he affirmed what was becoming an increasingly wobbly position, claiming that "the greater part (95 percent in the case of humans) of the genome might as well not be there, for all the difference it makes."[74] By that point, the evidence had been accumulating that junk DNA was not nearly as junky as it had once seemed. And then, in 2012, the results of the ENCODE project were announced, and Dawkins rewrote history. He would now claim:

> There are some creationists who are jumping on [ENCODE] because they think it's awkward for Darwinism. Quite the contrary, of course, it is exactly what a Darwinist would hope for—to find usefulness in the living world... We thought that only a minority of the genome was doing something, namely that minority which actually codes for protein. And now we find that actually the majority of it is doing something... The rest [of the genome] which had previously been written off as junk [is now understood as] the program [that's] calling into action the protein coding genes.[75]

So, what exactly was behind Dawkins' about-face? In 2012, the premier journal *Nature* published definitive results from the ENCODE (Encyclopedia of DNA Elements) Project Consortium.[76] Begun in 2003, this consortium's goal was "to build a comprehensive parts list of functional elements in the human genome, including elements that act at the protein and RNA levels, and regulatory elements that control cells and circumstances in which a gene is active."[77] This ambitious research project enlisted 442 scientists from across the globe to investigate noncoding DNA in the human genome. Its key findings radically undermined the junk-DNA theory.

The 2012 *Nature* article concluded that the "vast majority" of the human genome exhibits biochemical function: "These data enabled us to assign biochemical functions for 80 percent of the genome, in particular outside of the well-studied protein-coding regions."[78] Ewan Birney, one of ENCODE's star researchers, foresaw that this number would go up to 100 percent once more cells in the human body were examined.[79] The results were so compelling that a *Science* headline read, "ENCODE Project Writes Eulogy for Junk DNA."[80]

Darwinists who had claimed that junk DNA was real now had to backpedal. Some tried to minimize the ENCODE results: yes, most of the genome is biochemically active, but no, most of it doesn't carry out any important biological function.[81] Like extras in a Hollywood film, they serve a role but are peripheral to the real action. In any case, as Dawkins illustrated, it was easy for Darwinists to reclaim, or at least appear to reclaim, their credibility. They simply had to ascribe to natural selection greater efficiency than at first suspected, thus making it more plausible that the genome should largely be functional.[82] Yet in accurately predicting that what was being called junk DNA would, on closer scrutiny, display function, intelligent design clearly came out on top. But entrenched paradigms like Darwinism are able to weather many such storms, as the history of science attests.[83]

A less obvious but perhaps even more threatening storm for Darwinism than the demise of junk DNA is the high informational density that is increasingly being uncovered in genomes. It's one thing to find that the genome is largely functional. It's quite another to find that the genome can contain multiple overlapping layers of functional information. It's like the information in a crossword puzzle. Normally we read left to right and that's all the information we expect to find. But with crossword puzzles, we also read top to bottom. To the degree that letters in a crossword puzzle do double duty, its information density will be greater than one. A perfect crossword puzzle, where every letter does double duty, gives twice the information for the price of one—an information density of two.

Overlapping genes facilitate the compact organization of genetic information. They have been found in the genomes of viruses, bacteria, and eukaryotes.[84] They occur when two or more genes share a portion or all of their nucleotide sequences but are transcribed or translated in different reading frames or in different directions (regarding the latter, think of a hetero-palindrome, which has different meanings when read forward and backward, such as the word *stressed*, which read backward is *desserts*). This phenomenon allows for the efficient use of genetic information, as multiple proteins or functional RNA molecules can be generated from a single stretch of DNA (a fact appreciated by viruses and other small-genome organisms that have limited genomic space).

The usual story to explain the emergence of overlapping genes invokes mechanisms for manipulating genes, such as gene duplication or fusion.[85] But these mechanisms simply address the storage capacity of genomes, not the high density of biologically significant information. The point of high-density information is how to pack increasing amounts of information into the same amount of storage capacity. Double the size of your hard drive, and you have more storage capacity. Double the amount of information you can pack into an existing hard drive (for instance, through data compression), and you have more information density. Because existing evolutionary mechanisms merely manipulate storage capacity, they offer no help or insight with information density. In any case, if evolutionary biologists committed to a theory of blind trial-and-error evolution were understandably surprised to find that "junk DNA" is in fact functional, much more should they be surprised to find the high informational density of overlapping genes.

What is the probability of getting highly dense biological information in a genome, and is such a probability small? The probability models for evaluating such probabilities have yet to be spelled out (here's a dissertation topic). But on its face, this high informational density speaks of actual intelligent design rather than unguided naturalistic processes, and indeed, of ingenious design. High-density information packing and layering are common not just in crossword puzzles but also in the field of digital data embedding technologies (DDET), which includes steganography, watermarking, and digital data forensics. Humans have long put a premium on nesting information within information, and of course they accomplish such feats by design—indeed, ingenious design. There's no good naturalistic mechanism for doing the same.

In sum, the move to eliminate design from biology by portraying it as substandard cannot be honestly maintained. Some of the structures found in biology are clearly remarkable, even inspiring the work of engineers (in the field of biomimetics). Some structures that appeared substandard or even nonfunctional have turned out to have remarkable functionality, as with "junk DNA." Other structures in biology may continue to raise concerns about the quality or excellence of their design. But even in such cases, it is not clear if the design can be improved without incurring deficits that make the final product, on balance, worse rather than better.

If life is actually designed, it must work with existing materials, it must reproduce, it must self-assemble, it must maintain homeostasis, it must take in nutrients and expel wastes, etc. Given such constraints, biologists are a long way from being in a position to lecture any putative designer of life on how exactly the designer might have acted more intelligently to produce better overall designs. But even if biology contains botched and bungled designs, to the degree that biology also contains designs that are truly magnificent, they will call for explanation. Moreover, it may well be that botched and bungled designs result from the effects of history in degrading designs that initially were excellent. In any case, ostensibly bad designs in biology, whatever their ultimate cause or status, ought to play no role in maintaining a large Bayesian prior for natural selection.

7.6 Doing the Calculation

In this section, we're going to run through some calculations of specified complexity for a few biological systems. The calculations are correct in the sense that they commit no mathematical errors. What's wrong with them, however, as far as Darwinian evolutionists are concerned, is that none of them may legitimately be put forward to underwrite a biological design inference. In particular, once the Darwinian mechanism of natural selection is properly factored in, the small probabilities inherent in specified complexity and needed to warrant a design inference are said simply to evaporate.

It's a story we've seen before. A biological system appears to be designed. The system is clearly specified. But what about its probability? Design theorists attempt to lay out a plausible scenario in which the system might have evolved, but then they find that evolving the system within this scenario would require a very small probability event. Instead of drawing a design inference, evolutionary biologists turn the tables, charging design theorists with mistakenly laying out a scenario that required too large an evolutionary change resulting in too small of a probability. If only the scenario had properly factored in the small incremental changes by which Darwinian evolution can and must proceed, a larger probability would have been calculated, and a design inference would no longer be tenable. So the story goes.

We'll focus in this section on calculations involving proteins. In Section 7.2 we noted that biological design research needs to focus on a Goldilocks zone in which biological systems are simple enough to be thoroughly analyzed and yet complex enough to merit a design inference. Proteins reside in such a Goldilocks zone. They are large molecules formed of amino acids. Each amino acid is a small molecule, and a protein is assembled by forming a sequence of amino acids linked by peptide bonds (amino acids have other ways of being linked, but in that case, they can't form proteins). There are twenty standard amino acids across all of life. This means that there would be $20^{300} \approx 10^{390}$ possible proteins for proteins made up of 300 amino acids, which is just under the average size of a bacterial protein, 320 amino acids. Most possible amino acid sequences would not do anything useful, but a few will fold into the right shapes needed to perform useful functions.

Proteins assume a three-dimensional configuration that allows them to perform a host of functions inside cells. Any job that needs to get done in the cell is done by or in conjunction with proteins. To construct any cellular system thus comes down, in large part, to having an appropriate suite of proteins. A given protein is typically divided into several domains, which are self-stabilizing regions of the protein that fold independently into specific shapes. The ways these domains fold and the shapes they assume determine how proteins function. Almost any biological design inference for the design of proteins will focus on either entire proteins or on specific domains of proteins.

Given a particular protein or domain, what about it would entitle us to infer that it was designed? To answer that question, we need a chance-based hypothesis (or collection of such hypotheses) that might explain it apart from design. As a starting point, we can consider the hypothesis that the protein or domain arose by stringing together amino acids completely at random. A Darwinian evolutionist will object to this hypothesis, insisting that this isn't how evolution works. A design theorist will likewise object that this hypothesis is too restrictive on chance. Even so, let's start here with this simple hypothesis and then consider more complicated hypotheses.

How likely is it that a purely random sequence of amino acids will perform some biologically significant function inside the cell? It depends on the function in question. A simple function would be binding to adenosine triphosphate (ATP). ATP is the primary energy source for numerous biological functions. Many proteins have to bind to ATP in order to access the energy stored in the ATP and use it to perform biological functions. How likely is a completely random sequence to bind to ATP?

It turns out that this question admits an answer: construct a large number of random proteins and find out what proportion of them bind to ATP. Anthony Keefe and Jack Szostak did just that. They created 6 trillion random proteins and then estimated that approximately 1 in 100 billion random proteins bound to ATP.[86] Given these numbers, we can estimate the specified complexity of a protein that binds to ATP. For the description that specifies this function, we will take the phrase "binds ATP," which is two words long and thus we estimate would take around forty bits to describe (if we are generous in assigning 20 bits per word). Using the formula for specified complexity (see Section 6.4), we then calculate:

$$
\begin{aligned}
SC(X|H) &= I(X|H) - D(X) \\
&\approx -\log_2\left(\frac{1}{10^{11}}\right) - 40 \\
&\approx 36 - 40 \\
&= -4 \text{ bits.}
\end{aligned}
$$

This calculation gives a negative amount of specified complexity. It indicates that the existence of a protein that binds to ATP does not give good reason to reject the chance hypothesis. It is plausible that such an event could be explained by chance. This result follows because it's relatively common for a random sequence to bind to ATP. Even so, simply binding to ATP is not all that biologically useful. Biological usefulness typically requires something more sophisticated and therefore rarer. As such, estimating the probability of rarer functionality will require more sophisticated analytic and experimental techniques.

The 6 trillion distinct random proteins created by Keefe and Szostak is impressive and may seem like a lot, but not at the scale of protein spaces. Spaces of possible proteins are so huge that, by comparison, 6 trillion ends up being a negligible number. What if Keefe and Szostak needed to create not 6 trillion but 6 trillion trillion trillion ($= 6 \times 10^{36}$) possible random proteins? This would not be an unreasonably large number, as we'll see momentarily, for estimating the probability of forming some proteins by chance. Indeed, $6 \times 10^{36} \leq 20^{29}$, suggesting an exhaustive search of possible protein sequences not more than 29 amino acids in length—a length far smaller than most actual proteins.

An average amino acid weighs 110 daltons, or 1.826×10^{-25} kg. An average bacterial protein consists of 320 amino acids, and thus weighs 5.843×10^{-23} kg. Six trillion trillion trillion times this amount thus equals roughly 351 trillion kilograms, or 351 billion metric tons. And that's for exactly one of each of these possible proteins being created. Keefe and Szostak needed to create multiple copies of each protein in their experiment. For perspective, annual steel production worldwide is 2 billion metric tons. Keefe and Szostak's approach could not even get started if they needed to generate considerably more than 6 trillion possible bacterial proteins.

An alternative approach is therefore to take a known protein with a particular function and, through a combination of theory and experiment, attempt to determine how much variation is possible in the amino acid sequence without losing the function. Hubert Yockey took this approach in the 1970s for the protein cytochrome c.[87]

Cytochrome *c* has about 100 amino acids (104 for many multicellular organisms). Thus, any particular cytochrome *c* molecule will have well over 400 bits of information. Yet given the variation that this protein allows, the number of bits of information associated with its function is estimated as falling below 300.

Cytochrome *c* is important for electron transport in the mitochondria of eukaryotic cells. Yockey examined the known variations of cytochrome *c* in different species. He also drew from theoretical knowledge about which amino acids were likely to admit substitution. If we take "electron transport" (two words, estimated at 40 bits total) as the description that specifies the function of cytochrome *c*, and if we use Yockey's estimates of the variation that this molecule allows, the formula for specified complexity then associates the following number of bits with cytochrome *c*:

$$SC(X|H) = I(X|H) - D(X) \approx 298 - 40 = 248 \text{ bits.}$$

In contrast to the case of ATP binding, Yockey's analysis provides evidence that cytochrome *c* is not the result of randomly stringing together amino acids. The probability of the corresponding rejection region then comes to 2^{-248}, or less than 1 in 10^{74}.

Is cytochrome *c* therefore designed? Yockey's analysis shows that on an assumption of uniform probability, cytochrome *c* is very sparsely distributed within protein space. His analysis is suggestive of design. But by not controlling for its evolvability, Yockey's analysis is less than dispositive. Of course, evolutionists have no detailed Darwinian pathways to cytochrome *c* from some protein with a different structure performing a different function. But to draw a compelling design inference for cytochrome *c* would require a more precise probabilistic analysis of its evolvability.

Molecular biologist Douglas Axe was sensitive to this loophole in Yockey's work. In the 2000s, he analyzed one of the domains making up beta-lactamase, and published the findings in the *Journal of Molecular Biology*.[88] Beta-lactamase is an enzyme that bacteria use to break down penicillin-like antibiotics. It thus confers antibiotic resistance. Axe, through a combination of experimental work and theoretical analysis, estimated the probability of obtaining a functional domain of this protein to be around 1 in 10^{77}. If we take the description of the domain to be "resists antibiotics," with a descriptive complexity of 40 bits, we can then estimate the specified complexity of this domain to be:

$$SC(X|H) = I(X|H) - D(X) \approx -\log_2 10^{77} - 40 = 215 \text{ bits.}$$

The probability of the corresponding rejection region then comes to 2^{-215}, or less than 1 in 10^{64}. This probability is greater than the universal probability bound of 1 in 10^{150} that we give in Chapter 4. Nonetheless, it is still incredibly small, and amounts to far more luck than scientists may legitimately wish upon themselves in their scientific theorizing. So, with Axe, as with Yockey, we find that this protein and its function are exceedingly rare and yet easily described. We therefore have good reason to reject the hypothesis that it was generated by randomly stringing together amino acids.

But Axe's work was not just aimed at refuting a "random stringing" hypothesis. Axe was testing the evolvability of beta-lactamase by Darwinian means. Thus, his small probability of 10^{-77} for generating beta-lactamase antibiotic-resistant functionality was not just the probability of finding such a functional protein by purely random means. Rather, the point of Axe's research in calculating a 10^{-77} improbability was to characterize how a protein with a different fold could change folds and thereby evolve into one with a beta-lactamase fold. This is precisely what Darwinian evolution needs in order to explain the evolution of protein folds: the structure must change without losing function, and at some point, the structure and the function must both change to achieve a new function.

What Darwinian evolution doesn't allow is for the beta-lactamase fold to just magically materialize. It had to evolve from somewhere. For any product of evolution, there's always a prior product, namely, a product from which it evolved. Go back far enough, and the prior product had to do something different from what the current product does. Note that Axe's assumption of uniform probability here, in treating as probabilistically equivalent all the possible variants of beta-lactamase, was justified: any random forces transforming these proteins would, according to neo-Darwinian theory, be mediated through genetic point mutations that operate uniformly (or at least roughly so) across the genome.

So, did Axe's work with beta-lactamase win the day, establishing that the domain he analyzed was indeed highly improbable, not just under some chance hypothesis or other, but under a Darwinian chance hypothesis, thereby ruling out the evolvability of that system by Darwinian means? And, to take this question to its logical conclusion, was the field thereby swept clear of relevant chance hypotheses, warranting a design inference and thus demonstrating that beta-lactamase is indeed the product of design?

Design theorists and Darwinists answer this question differently. Design theorists would say that Axe's work provides the best probabilistic analysis to date of the (un)evolvability of beta-lactamase. The specified complexity calculated is large. And even though the probabilities calculated are not on the order of the 1 in 10^{150} needed to satisfy a stringent universal probability bound, they are small enough for most scientific purposes to rule out chance and infer design. His work was published in a reputable, peer-reviewed science journal. If Axe's analysis of the underlying probabilities here leaves something to be desired, the burden is on his critics to show what, if anything, he missed and how the probabilities can be improved.

To that end, Darwinists argue that Axe's probabilistic analysis omitted some crucial Darwinian evolutionary pathways to beta-lactamase from some prior protein/domain having a different structure and function. Factor in these as yet unknown pathways, and the small probability Axe calculated will disappear. This, of course, falls far short of actually showing what Axe missed, but for Darwinists, such pathways can and must always exist, where structures and functions coevolve and where at each step on the evolutionary pathway selective advantage is preserved.

For Darwinists, because such pathways, even if unidentified, assuredly exist, the probabilities cannot be as small as Axe made them out to be. It's just that Axe's analysis somehow missed them. The sparsity of functional folds is thus, for them, largely irrelevant. Yes, if functional folds were scattered purely randomly through the space of possible protein configurations, then getting a functional path through that space would be highly improbable. But once on a path, evolution can proceed with high probability, a point illustrated in Section 7.4 with evolution on the hypercube.

Two criticisms of this Darwinian position need now to be considered. The first is that the Darwinist assumes here no burden of evidence. All the work of analyzing probabilities and assessing evolvability is on the shoulders of the design theorist. The Darwinist simply assumes that Darwinian pathways exist to account for the system whose evolution is in question, but without offering any details about the pathway or any alternative probability calculations. In consequence, no probability calculation will ever convince the determined Darwinist that a design inference is warranted.

This immunity to disconfirmation is not to Darwinism's credit. Darwinists will cite experimental support for natural selection, and when they do, it's because the probabilities for natural selection to bring about some results are high. Kenneth Miller, for instance, will cite an experiment by Barry Hall in which the *lac* operon in *E. coli* is disabled and then, with high probability, is restored through selection pressure.[89] In fact, as Michael Behe notes, when the experiment is properly analyzed, it's clear that "the system was being artificially supported by intelligent intervention."[90]

But the point to note is that Miller regards Hall's experiment as a "glittering example" of natural selection's power precisely because of the experiment's high probability of restoring the *lac* operon. Parity of reasoning, however, requires that if large probabilities confirm Darwinian theory, then small probabilities should disconfirm it. To refuse parity of reasoning in such cases, as Darwinists typically do, commits the fallacy of special pleading. It is special pleading to use a mode of reasoning when it serves one's purposes, but then to reject it when it yields results that one would rather avoid or ignore.

The second criticism of Darwinists rejecting small probabilities—such as those Axe uncovered as evidence of design—is this: there is a definite fact of the matter that underlies whether Darwinists or design theorists are right about the evolvability of such systems as the beta-lactamase enzyme. This is not a historical question about whether beta-lactamase over the course of time did indeed evolve. It is, rather, a biophysical question about the structure and function of the proteins and amino-acid sequences in question, and the biological configuration spaces in which they reside and are said to evolve.

An analogy will help to illustrate the point at issue. Chess is a finite game for which it could, in principle, be determined whether, with optimal play, white is guaranteed to win, black is guaranteed to win, or a draw is guaranteed to occur. One of these options must hold.[91] The problem is that the totality of different possible moves in chess is so immense that we simply don't have the computational firepower to determine which of these options holds. But it is mathematically certain that one of these options holds.

Similarly, with Douglas Axe's beta-lactamase, there is a definite fact of the matter regarding its evolvability. However one wants to interpret Axe's 1 in 10^{77} improbability of forming functional beta-lactamase domain variants, this improbability involves a huge denominator of possible beta-lactamase variants, with the totality of such variants exceeding 10^{77}. There is thus no way for the present state of science to examine all such possible variants and determine their functionality as proteins. But imagine that we did have enough experimental firepower, say, by moving the research about these variants to a Star Trek experimental station where the investigation could be conducted at "warp speed" and thereby run through all the possible variants. We could then determine experimentally which variants conferred antibiotic resistance, which could serve alternative functions, and which simply broke down by refusing to fold, thereby precluding any possibility of biological function.

Such an analysis might, for the beta-lactamase that Axe considered, reveal an island of functionality surrounded by an ocean of non-function. In consequence, the probability of traversing that ocean by Darwinian means would be very small. The probability in this case would not merely be estimated—it could be exactly calculated. Indeed, there is a real objective probability for the evolvability by Darwinian (or other) means of beta-lactamase (and of many other biological systems). It's just that in most circumstances (the case of Keefe and Szostak being an exception), we do not have the experimental means to determine such probabilities exactly.

Because of the sheer size of the possibilities at play in Axe's work, we cannot simply inspect these possibilities one by one. Brute force doesn't work in such cases. Instead, we must use more indirect theoretical and experimental means to estimate the probabilities. That's what Axe did. By contrast, it is merely an article of faith to say, on Darwinian grounds, that the probabilities must be large. Whether they are large or small must be established not by armchair speculation and not by reflexively invoking a conventional orthodoxy, but by rolling up one's sleeves and doing what's necessary to understand the underlying probabilities. Accordingly, when a researcher does this, as Axe has, those calculations need to be taken seriously rather than dismissed on a priori grounds. And if the results lead to a design inference, so be it.

In this section, we have focused on the evolvability of proteins by Darwinian means as well as on the challenges to their evolvability from specified complexity. In closing this section, we make an obvious point about transitivity of design: if individual proteins (or domains) exhibit specified complexity, then so do functional systems that are composed of them. Proteins are widely regarded as the building blocks of life. As building blocks, they typically function not in isolation but with other proteins and constituents of life (such as lipids). Systems composed of many proteins working together to perform a function are the norm. Such systems will then exhibit at least as much specified complexity as their individual protein parts. It follows that if the individual proteins are sufficient to trigger a design inference, then so is the functional system consisting of these proteins.

But the case for significant amounts of specified complexity in biology can be made still stronger. Systems of proteins such as the bacterial flagellum or the electron transport chain often have short descriptions. As such, when we consider whole systems rather than individual proteins, the description length tends not to increase. However, the complexity (improbability) will increase as the number of different parts that have to be accounted for increases. As such, the specified complexity of a system as a whole can be greater than the sum of the specified complexity of the parts. Thus, when we consider a large system of proteins, the evidence against chance and for design can accumulate rapidly.

The Breakdown of Evolvability

Darwinian evolution holds to common descent. It thus regards every living form as connected to every other living form by gradual evolutionary pathways in which each step along a path is selectively advantageous or at least selectively neutral. But whence the confidence that Darwinian evolution should allow untrammeled interconnectivity from any one living form to any other, whether by direct evolution or by indirect evolution from a common evolutionary precursor? This question about evolutionary interconnectivity applies not just to evolutionary biology but to any evolutionary processes whatsoever. It's always possible that the very structure of the evolutionary configuration space hampers evolvability, imposing isolated islands of function that are unevolvable from one to the other whereas full-scale evolution requires bridges or isthmuses to connect them.

Because protein evolution faces challenging empirical and theoretical problems, its limits remain controversial. It can, however, be simply illustrated by comparing proteins with English words. Both utilize a fixed alphabet: for proteins, twenty L-amino acids; for English words, twenty-six letters. In order for a protein to evolve into another through a Darwinian mechanism, minor amino acid alterations must progressively change one protein into another, ensuring that each intermediate stage maintains a similar amino acid sequence to its immediate predecessor and successor while also folding properly to exhibit a biological function. Preferably, the function would also improve (as in Darwinian evolution). But at least it must not degrade (as in neutral evolution).

Similarly, the evolution of one English word into another through a (quasi-) Darwinian process requires minor letter alterations that gradually convert one word to another, making sure that each transitional stage has a closely similar sequence of letters to any adjacent stage and also constitutes an English word. To be specific, let's institute the following transformation rule: each evolutionary stage allows a single letter to be added, removed, or altered within any given letter sequence. This rule ensures the gradual evolution of English letter sequences. Note that the evolution of words via this transformation rule would be selectively neutral—all evolution proceeds unabated among actual English words, not English words that are in some sense selectively more fit than others. That said, in this scenario, arbitrary letter sequences that correspond to no English word would be selected against.

As an example of such English-word evolution, let's evolve the word A (the indefinite article). The following is an example of an evolutionary path originating from A that uses the transformation rule of add, remove, or insert a single letter: A → AN → TAN → TIN → TIP → TIPS. Note that this evolutionary path to TIPS is not unique. We could instead, given our transformation rule, have evolved A into TIPS as follows: A → AT → BAT → BATS → BASS → BOSS → LOSS → GLOSS → GROSS → DROSS → DROPS → DRIPS → DIPS → TIPS.

How should we make sense of this example? Clearly, the word A is evolving into other English words quite different from it. But the real question is how much can it evolve and are there words into which it cannot evolve. Evolvability can come in degrees and have limits. Typically, there are no first principles for deciding the extent of evolvability within a given evolutionary scenario. Rather, we must get down to brass tacks with the configuration space in which evolution is occurring and with the rules governing the evolutionary process over it. Through theoretical insight and empirical inquiry, we will then need to assess the power of evolution—in whatever guise it happens to take, Darwinian or otherwise—to traverse the configuration space.

Given the changes allowed at each stage in the evolution of English words (i.e., adding, removing, or modifying a single letter), it is clear that words diverging significantly from the original—in this case from the word A—can evolve. Even so, the big question is whether it's possible to evolve every English word from this original word through a series of gradually changing English words. In other words, is A a universal common ancestor for all other English words via a (quasi-) Darwinian evolutionary process that puts a premium on legitimate English words rather than, as in biology, differential survival and reproduction?

The claim that A could, based on our transformation rule, serve as a universal common ancestor to every word in the English language seems implausible. Do we really think that, for instance, a word like CONSTANTINOPLE is evolvable in this way from A? But plausibility is not a good guide for scientific inquiry. What if the range of letter sequences taken to be legitimate English words is extremely large? Make it large enough, and everything is a word, and all words are evolvable into each other. Certainly, the more letter sequences we treat as legitimate English words, the greater their evolvability. Fortunately, what can be taken as legitimate English words is reasonably circumscribed. To avoid placing undue restrictions on evolvability, we therefore use the largest dataset of English words that we know, namely, the Unix file consisting of 200,000 English words. This file is denoted simply as words (written in a monospace font, as is typical among computer scientists).[92]

Given our transformation rule of add, subtract, or insert a letter, and given this Unix file, evolvability within this configuration space of letter sequences is ascertainable. The analysis to confirm or disconfirm the universal common ancestry of A is easily performed by a computer program taking the word A as starting point and exploring potential evolutionary paths from it in accord with the allowable changes at each evolutionary step. We've done this and found that none of the following four-letter words is evolvable from A via our transformation rule: IOTA, ECRU, FUJI, HYMN, SEMI, KIWI, ENVY, AMMO, TOFU, EXPO, ANKH, OVUM, IMAM, AHOY, ONYX, OBEY, IBEX, IFFY, KOHL, IDOL, NAZI, and ULNA. As we discovered, CONSTANTINOPLE is also unevolvable in this way.

Have we therefore disproven the power of natural selection to underwrite the vast evolutionary interconnectedness of all organisms and biological structures? Not quite. Maybe such interconnectedness exists and Darwinists are right, though we have no convincing evidence of any complex system evolving by single random point mutations into a fundamentally different system. Unlike our computer simulation, which allows for an exhaustive and definitive analysis, real-life biology is so complex that the corresponding analysis has not been and, for now, cannot be performed. But even though we haven't shown that such evolutionary interconnectedness does not exist, we have provided a convincing rationale for why it need not exist.

The takeaway, then, is that the total evolutionary interconnectedness of biological forms cannot be taken as a foregone conclusion. Any such conclusion must be based on evidence. For now, however, we have no detailed step-by-step path-based evidence that such interconnectedness exists—a case in point being the lack of intermediates between distinct phyla in the Cambrian.[93] Moreover, our analogy of English-sentence evolution raises the intriguing possibility that there could be compelling reasons to think that such interconnectedness breaks down completely. The point of this chapter is to expand on the lesson of this excursus, arguing that there are indeed good reasons to think that a gradualist Darwinian form of evolution is very limited in what it can accomplish in biology, and that some features of biological systems not only resist Darwinian explanations but also invite design inferences.

7.7 Where to Look for Small Probabilities in Biology?

The first edition of this book didn't challenge Darwinian evolution. Biology figured only tangentially in the first edition, in a section on the origin of life.[94] The point of that section was to show how the logic of the design inference could lead to different conclusions about the origin of life. Depending on which premises were regarded as true or false, design inferential reasoning could lead either to the conclusion that the origin of life was the product of design, or that design was absent from the origin of life.

The design inference was, in the first edition, presented as a neutral method whose conclusions for biology could go either way. What's more, since naturalistic views of life's origin and subsequent evolution were widespread and firmly held, early readers of the first edition—even those who were thoroughly Darwinian—saw nothing particularly controversial or threatening about the book. It was only when members of the intelligent design community applied the logic of the design inference to biological systems and asserted it confirmed that biology contained actual design that the first edition became controversial.[95]

The work of applying the design inferential logic to biological systems with the aim of displacing Darwinian evolution and making room for actual design in biology is an ongoing project. Published peer-reviewed work to that end is increasing, though often without explicitly drawing a design inference so much as taking readers to the edge of the precipice (by demonstrating small probabilities) and then letting them take the final leap (by drawing the inherent design inference). The reason for this approach is prudential: Darwinian evolution is the reigning orthodoxy, and to get published in the mainstream literature requires not taking it on too directly.[96]

The design inference is a method. Whether and to what degree this method applies to biological systems depends on biologists successfully using it to demonstrate design. One of us is a pure mathematician. The other is a computer scientist. Neither of us is a biologist. Our task therefore is not so much to apply this method to biology as to hand it off to biologists, who can then apply it properly to their discipline, nailing down the design inference for particular biological systems.

Once small probabilities can be nailed down for the emergence of biological systems, the path is clear to drawing full-blown biological design inferences. In the rest of this section, we will describe what we regard as a particularly promising place to look for identifying small probabilities, and therewith for inferring design, in the field of biology. This is not to rule out other places to look, such as individual protein folds (Section 7.6) or high genomic information density (Section 7.5). But the prime place we would suggest looking is *irreducible complexity*, and especially two of its variants, *minimal irreducible complexity* and *bio-imperative irreducible complexity*.[97]

Michael Behe introduced the concept of irreducible complexity in his 1996 book *Darwin's Black Box*. His original definition has undergone refinement over the years, yet the basic concept still stands. An up-to-date understanding of the concept can be found in his 2020 book *A Mousetrap for Darwin*. The idea behind irreducible complexity is that a system exhibits this property if it consists of multiple parts each of which is indispensable to its function. Remove any part or parts, and the function is lost. So defined, irreducible complexity is an all-or-nothing proposition in the sense that function is kept or lost completely. But it can also allow for degrees, as when the degree of function is so diminished that the retained function is biologically irrelevant and thus effectively lost.

To determine experimentally if a system is irreducibly complex, knock out its parts (both individually and in combination) and determine if function is lost. If in each case function is lost, the system is irreducibly complex. Irreducible complexity can also be determined theoretically, through a conceptual analysis of the system and what is required to maintain its function. And of course, experiment and theory can work in tandem to determine irreducible complexity. Thus, conceptual analysis might reveal that a few parts are indispensable for function, but knock-out experiments might reveal that other parts are indispensable, especially if we don't know exactly what those parts do.

Several caveats are in order. Consider, for instance, a standard mousetrap. It consists of a platform, a hold-down bar, a spring, a catch (for the cheese), and a hammer (which crushes the mouse). It is irreducibly complex in the sense that the removal of any of these parts causes the trap to cease functioning *as a trap*. Remove the hammer, for instance, and what's left will no longer catch mice. But remove the hammer, and what's left might nonetheless serve as a perfectly good doorstop. The point with irreducibly complex systems is that they serve a basic or primary function, and that function is lost, and not merely attenuated, if any of the parts are removed. The ability to repurpose such a system once a part is lost doesn't count against the system's irreducible complexity.

Additional caveats hold. Imagine that two parts of a complex system are joined and that the removal of one of them causes the other to act as a wrench in a gearbox, making the system grind to a halt. But imagine further that if the two parts are jointly removed, the system keeps performing its function. The two parts, taken together, therefore don't do anything for the function of the system. Taken jointly, they are dispensable. Such a system should not be regarded as irreducibly complex. And yet it would be if irreducibly complex systems were defined as those that fail given the removal of any single part. What we want, then, in the definition of irreducible complexity is that all the parts, individually and collectively, are conducive to the function of the system, and where the removal of any combination of parts causes the system to stop functioning, the function in question being the basic or main function, and not some peripheral function. Systems like the one just described would therefore not be irreducibly complex, though they might contain an irreducibly complex core (see the next caveat).

Another caveat is that even though a complex system composed of numerous parts may not be irreducibly complex, it may contain an irreducibly complex core, all of whose parts are indispensable. For instance, a car has many components that are not strictly speaking necessary for its function. The roof, the windshield, the rear seats, etc. might be removed without preventing the car from propelling itself on roads, which is the car's basic function. But other components, such as the engine, the chassis, the wheels, and the driveshaft are indispensable. Together they are constitutive of the car's irreducible core. Thus, we can think of a system as irreducibly complex if it contains an irreducible core.

A final caveat is that the parts of an irreducibly complex system need to be specifically adapted (Behe used the term "well matched") for performing its basic function. Moreover, the function must be performed in a particular way. For instance, imagine a propeller plane powered by a gas engine. Suitably individuating the plane's parts will reveal an irreducible core. But we can also imagine a propeller plane powered by a rubber band. These are two separate irreducibly complex systems. Their parts are specifically adapted to each other in different ways, and the particular way each system performs its basic function is different (even though the basic function can be described similarly as "propeller-powered flight").

According to Behe, irreducibly complex systems constitute a barrier to Darwinian evolution. How so? Suppose a biological system possesses an irreducibly complex core. The parts of this core are then all necessary to the system's basic function. Remove any part or combination of parts, and the system ceases to function. So how is the Darwinian mechanism going to evolve such a system? Darwinian evolution proceeds gradually, step by small incremental step, at each step preserving or improving function. It would seem, however, that all the pieces of an irreducibly complex system need to be in place for such a system to function at all. Getting all these pieces to come together simultaneously is simply not an option for Darwinian evolution (it would be too improbable). And yet, such a simultaneous coming together of diverse parts, or multiple coordinated changes, is what these systems seem to require to bring them about.

Not so fast, say Darwinists. Take the bacterial flagellum, described in Section 7.5. In fact, many different bacterial flagella exist, belonging to many different bacteria. One of the bacteria most studied for its flagellar assembly is *Salmonella typhimurium*.

Its flagellum requires well over 30 proteins (many of which are repeated, as in the flagellar filament, which can contain thousands of subunits).[98] So how could this flagellum have evolved? The flagellum in *Salmonella typhimurium* seems to be more complicated than it needs to be. An analysis of different bacterial flagella suggests that only about 20 proteins are essential for a functioning flagellum.[99]

The flagellum in *Salmonella typhimurium* might therefore have evolved gradually under selection pressure by folding in new proteins individually (step by baby step), each of which conferred added function, but where over time, by a kind of ratcheting, those new proteins became indispensable. In this way, the complexity of an evolving flagellum could have increased over time. Thus, by first modifying one part and then removing another, an irreducibly complex system might retain its function where simply removing parts would break function. This suggests an add-and-modify approach to evolving irreducibly complex systems. It also suggests a way to run evolution backward by not just eliminating parts, but also by modifying some of them slightly so that function won't be lost when other parts are eliminated, because now their absence won't be felt on account of the modification.[100]

A problem with such an add-and-modify approach to the evolution of irreducibly complex systems is that it constitutes an argument from imagination. Indeed, we have no clear and compelling evidence that this form of evolution has ever accounted for any irreducibly complex biological system of any sophistication or interest. Most ID proponents therefore regard this approach as not so much implausible as simply unverified. Nonetheless, it is a theoretically possible route for evolving an irreducibly complex system from prior simpler irreducibly complex systems, and we can't be dogmatic that this approach will never be verified.[101]

Such an add-and-modify approach, however, cannot constitute a general solution to the Darwinian evolution of irreducibly complex systems. The problem is that irreducibly complex systems face a minimum complexity threshold. In other words, there's only so far we can go to simplify an irreducibly complex system while retaining its function. With the bacterial flagellum, for instance, flagella consisting of 30-plus proteins require more proteins than are absolutely necessary for a functioning flagellum. But the biological literature indicates that there's a lower limit to how simple a flagellum can be made. Twenty proteins seems to be a lower complexity threshold. And indeed, a conceptual analysis of the flagellum indicates that flagellar complexity can only be lowered so much. Stators and rotors are needed. A motor, whether proton or sodium powered, is needed. There needs to be a flagellar filament to act as a propeller. The whole assembly needs to be mounted. And for chemotaxis, there needs to be signal transduction circuitry.[102]

So, the challenge with irreducible complexity is not just irreducible complexity as such, but *minimal irreducible complexity*. And here the game changes. It's no longer possible to evolve a minimally irreducibly complex system by this add-and-modify technique, because if the system is minimally complex, any prior system that requires an addition won't be performing the function in question. At best, it will be performing a different function. And that's exactly what Darwinists admit:

> One could still question how... natural selection could lock on to an evolutionary trajectory leading to an organelle of motility [i.e., the flagellum] in the first place, when none of the components alone confer the organism with a selective advantage relevant to motility. The key missing concept here is that of exaptation, in which the function currently performed by a biological system is different from the function performed while the adaptation evolved under earlier pressures of natural selection. For example, a bird's feathers might have originally arisen in the context of selection for, say, heat control, and only later have been used to assist with flight. Under this argument, a number of slight but decisive functional shifts occurred in the evolution of the flagellum, the most recent of which was probably a shift from an organelle of adhesion or targeted secretion, such as the EspA filament, to a curved structure capable of generating a propulsive force.[103]

If you are a convinced Darwinist, then some scenario like this has to be true. On the other hand, if you are an evolutionist but not a Darwinist, then all bets are off. If evolution can proceed without selection pressure and if it can summon, as needed, multiple coordinated changes, then the evolution of even minimally irreducibly complex systems can happen.[104] Yet in the absence of intelligent design, it's unclear what the mechanism for such a form of evolution might be. Absent both selection pressure and intelligent design, non-Darwinian evolution seems able only to invoke neutral evolutionary processes, and the likelihood of finding a new function under such a regime can do no better than a random walk. That's not a promising mechanism for building complex machines.

A terminological point is worth making here. *Exaptation* as it appears in the above quotation is often also described as *co-option*, in which a structure with a given function is modified and then co-opted to serve a new function that allows the system to evolve in new directions, all with the blessing of Darwinian natural selection. *Co-optation* is less frequently used as a synonym for co-option. Another way to describe the folding in of an old structure for a new function is in terms of *repurposing*.

The problem that Darwinian evolution faces with minimal irreducible complexity is that we never seem to discover a clearly articulated gradual Darwinian pathway to these systems. This absence of clearly articulated Darwinian pathways holds not just for minimal irreducible complexity but for irreducible complexity in general. As we've seen, a Darwinian evolutionary path for irreducibly complex systems that are not also minimally irreducibly complex is at least plausible on Darwinian grounds by taking an add-and-modify approach. But with the bacterial flagellum, the closest known system from which it might have evolved is a type III secretion system, or T3SS, which performs a very different function (injecting toxins into multi-celled host organisms). A minimally irreducibly complex bacterial flagellum has at least 20 proteins. An arbitrary T3SS lacks at least ten proteins from a flagellum. So how does one explain the addition of those ten or more proteins in evolving from a T3SS into a bacterial flagellum?[105]

In trying realistically to assess how the bacterial flagellum arose, it doesn't help to invoke homologies of flagellar-like proteins in other bacteria or to note the modularity of these proteins in such systems. Darwinists will convince themselves that they have explained the evolution of these systems with rationalizations such as the following: "It is clear that designing an evolutionary model to account for the origin of the ancestral flagellum requires no great conceptual leap. Instead, one can envisage the ur-flagellum arising from mergers between several modular subsystems: a secretion system built from proteins accreted around an ancient ATPase, a filament built from variants of two initial proteins, a motor built from an ion channel and a chemotaxis apparatus built from pre-existing regulatory domains."[106] Such an argument will only seem compelling to true believers. Indeed, what can't be thought to evolve by "envisaging" such "mergers"?

It would be one thing if we could start with a T3SS, essentially a pump, and evolve it via a continuous, gently flowing stream of exaptations or co-options: Add or modify a single protein, and the system does something else. Add or modify another protein, and the system does still something else. Keep adding and modifying proteins so that step by gradual step we have a succession of new structures and new functions that finally terminate in a functioning bacterial flagellum.[107] In principle, this could work. But we never see anything like that. Instead, we are merely given a list of multiple proteins that would need somehow to be folded into a putative precursor to yield a system whose origin is in question, but without exhibiting a concrete co-optative or exaptative chain in which structures exhibiting new functions can be placed into a gradual succession.

So, how probable is a minimally irreducibly complex bacterial flagellum? If it had to evolve directly from a T3SS apart from intermediaries performing different functions, then it would need to fold in at least ten new proteins. Even if those proteins are homologous to existing proteins for use outside a bacterial flagellum (as such proteins would have to be if we are trying to explain the evolution of the first flagellum), getting them all to break free from existing systems and coalesce at the same time to transform a T3SS into a bacterial flagellum would be vastly improbable, even if we can't calculate the precise probability. If such a multiple coordinated addition of proteins into a T3SS is how a bacterial flagellum arose, then the process was non-Darwinian and a design inference would be warranted.

Of course, Darwinists will insist that this is not how the flagellum arose, and that instead it arose by a gradual chain of exaptations. Actual physical evidence, however, fails to support a gradual chain or succession of exaptations in accounting for a minimally irreducibly complex flagellum or any other minimally irreducibly complex biochemical system. Indeed, what was such an evolving system doing at each step of exaptation if not providing some targeted swimming or motility function? The physical evidence therefore supports a design inference. But to the degree that Darwinian evolution is accepted and even embraced, an exaptative chain, such as would lead from a T3SS to a flagellum, while lacking in empirical support and detail, will nonetheless seem plausible and even compelling, as in the last quote above about "envisaging mergers."

Does this mean that we are at an impasse? Will Darwinists simply think design theorists are being recalcitrant and incredulous, suffering from a failure of imagination and refusing to see that a system like the bacterial flagellum could very well have evolved from a T3SS? On the other hand, will design theorists simply think Darwinists are being dogmatic and credulous, indulging in fantasies and refusing to acknowledge the severe probabilistic hurdles that the minimal irreducible complexity of the bacterial flagellum imposes on the prospect of evolution from a T3SS? Yes, on both counts. Even so, we see no parity between Darwinian and design arguments for flagellar irreducible complexity (minimal or otherwise). Design has the vastly stronger position precisely because Darwinists only have Kiplingesque just-so stories to support their position. That said, there are biological systems where the design argument is logically stronger and tighter than for the bacterial flagellum.

To escape the impasse just described, we need a still stronger version of irreducible complexity, namely, *bio-imperative irreducible complexity*, and this will require systems other than the bacterial flagellum. Systems that exhibit bio-imperative irreducible complexity are, by definition, not just minimally irreducibly complex but also necessary for cellular life. The flagellar systems we've considered do not exhibit bio-imperative irreducible complexity. Flagella are not necessary for life. Indeed, many bacteria are stationary and thus get on quite well without having this motility device. In consequence, we can reasonably discuss flagellar evolution from living forms that do not have flagella and for which the evolutionary precursors to the flagellum function quite differently from the flagellum.

But if, hypothetically, the bacterial flagellum did constitute a bio-imperative, and thus was presupposed by cellular life as we know it, its evolution could not be explained as described earlier in this section. Clearly, neither an add-and-modify nor a modify-and-exapt approach to its evolution would fly. Such approaches assume that the flagellum evolved within a living form that at one point did not have that system and function. But for a system that is bio-imperatively irreducibly complex, the system must evolve from prior versions of the system that keep the evolving organism alive by always maintaining the biologically imperative function. And how does that happen when that system is also minimally irreducibly complex, as we are assuming with bio-imperative irreducible complexity?

There's an old joke about a mechanic and a heart surgeon that illustrates what's at issue. The mechanic complains that it's unfair for the heart surgeon to make so much more money working on a patient's heart than the mechanic makes working on a car's engine. In reply, the heart surgeon says, "Try working on the engine while it is running." That's the problem with bio-imperative irreducible complexity—the engine of life has to be kept running while a system that is making life possible is itself evolving. This requires not just an evolving system, but a system that in its evolution is at every point actively keeping the organism alive. Once you're dead, you're no longer a promising candidate for evolution.

Bio-imperatively irreducibly complex systems exist, as is obvious upon a moment's reflection. Take the genetic machinery that uses DNA to produce proteins. This machinery is ubiquitous in all cellular life and necessary for it. This machinery includes DNA for storing genetic information, enzymes for reading that information, RNA for transcribing that information, and ribosomes for translating that information into proteins. In fact, this entire genetic machinery contains not just a minimally irreducibly complex core, but consists of nested irreducibly complex systems. The ribosome, for instance, is an indispensable part of the entire genetic machinery, which is irreducibly complex when its parts are suitably individuated. But the ribosome is also irreducibly complex in its own right in the way it combines its protein and RNA parts to read messenger RNA and thereby produce proteins.

So how did this genetic machinery come to be? It is a bio-imperative. All cellular life as we know it has this machinery. Life is said to have begun about 4 billion years ago on planet Earth, and all evidence suggests that that machinery was present back then. In the origin-of-life literature, one will read speculations about how the present genetic machinery, which is based on codons of three nucleotides, may have evolved from a simpler genetic machinery based on codons of two nucleotides. But all such speculation is evidence-free. In the genetic machinery as we know it, we have an immense minimally irreducibly complex system that is necessary for life. We lack plausible precursors because those precursors—at least those so far identified—would not exist without that machinery. Any probability model for this machinery based on known parts composing it thus requires these parts to come together simultaneously and de novo. The probabilities will perforce be ludicrously small, thus mandating a design inference.

Of course, even with bio-imperative irreducible complexity, it's always possible (in some completely untethered-to-reality sense) to imagine that such systems are somehow amenable to a modify-and-exapt approach in which the exaptation includes the bio-imperative function of keeping the organism alive. But biology needs more than such sheer possibilities. It would be a coup indeed if origin-of-life biology could find a detailed Darwinian evolutionary account of how the genetic machinery in all cells might have evolved from a much simpler genetic machinery capable of sustaining life. This is not to require that life actually evolved that way. But it would need to provide a detailed enough evolutionary model to convince disinterested parties that it might reasonably have happened that way. As always, the Darwinian evolutionist here would be attempting to explain the more complicated in terms of the less complicated. Bio-imperative irreducible complexity effectively blocks such explanations (even if it cannot rule them out with deductive certainty). This form of complexity thus seems a particularly promising place to look for small probabilities and therewith to draw design inferences.

This brings us to the end of the road, the proper conclusion of this book. This last section has focused on where to look for small probabilities in biology for the purpose of drawing biological design inferences. The focus has been on irreducibly complex biological systems, and especially on minimally and bio-imperatively irreducibly complex biological systems. We've described *where to look*. In the appendices we also examine *how to look*. Thus, in Appendices B.6 and B.7, we describe some of the nuts-and-bolts probability considerations and methods for calculating small probabilities in biology. Biologists interested in applying the design inference to biological systems should study these appendices.

As a closing thought, let's put to rest a common misconception about biological design inferences, namely, that they are science killers. The worry among many biologists, even those who see some merit in intelligent design for spotlighting key unresolved problems in the study of biological origins, is that to embrace intelligent design would be to give up on science. Biologist Joana Xavier of University College London epitomizes this view. In an interview with Perry Marshall, she remarked:

> But about intelligent design, let me tell you, Perry, I read *Signature in the Cell* by Stephen Meyer... I must tell you, I found it one of the best books I've read, in terms of really putting the finger on the questions. What I didn't like was the final answer, of course... I think that we must have a more naturalistic answer to these processes. There must be. Otherwise, I'll be out of a job.[108]

But what if the job is not to find a naturalistic explanation for the origin of these systems, but, once a design inference is drawn for a biological system, to understand its engineering characteristics and how it might have arisen through a process of intelligent or technological evolution? Answering such questions is the focus of systems biology, TRIZ (theory of inventive problem solving), and an emerging synthesis of biology and engineering.[109] Real biological design will leave biologists with plenty to do and understand regardless of whether that design results from directed panspermia or a vitalistic teleology or a theistic creator God or any other purposive cause active in nature.

〰️ EPILOGUE

BEYOND THE DESIGN INFERENCE — CONSERVATION OF INFORMATION

The Basic Idea of Conservation of Information

CONSERVATION OF INFORMATION IS A NATURAL CONTINUATION OF the design inference. When we planned the second edition of this book, we intended to round out our discussion of the design inference with a thorough treatment of conservation of information. But on further reflection, it seemed to us that conservation of information is now sufficiently well developed to merit a book of its own. So, we decided to write a separate book on conservation of information and here touch on it only briefly.

Conservation of information was absent from the first edition, which, like this second edition, focused on specified complexity as a method for detecting and inferring design (though in the first edition it was called specified improbability rather than specified complexity). Moreover, detecting and inferring design via specified complexity can proceed even without conservation of information. Specified events of small probability trigger design inferences irrespective of conservation of information.

What, then, is conservation of information? Conservation of information is an accounting tool for tracking information and making sure it is not illicitly smuggled in or siphoned off. It arises not in dealing with small probabilities as such, but in dealing with processes that take seemingly small probabilities and make them so big that design inferences no longer apply. We therefore thought it better to let our case for the design inference stand on its own merits. If, for instance, certain irreducibly complex biochemical machines can be shown to be specified and sufficiently improbable *even on Darwinian grounds*, then a design inference would be warranted and directly disconfirm the power of natural selection and nonteleological evolutionary processes in general.

By contrast, conservation of information takes a more indirect approach. It becomes relevant in the debate over Darwinian evolution when natural selection, acting as a probability amplifier, is said to overturn what otherwise would be a small probability leading to a design inference. Conservation of information shifts the focus from the products of evolution to the processes of evolution. The design inference asks what about the products of evolution could lead us to think that they might be designed. Conservation of information asks what about the processes of evolution could lead us to think that they have been infused with information requiring intelligence. Products can be designed. But processes that build products can also be designed. The design inference tries to make sense of improbable products. Conservation of information tries to make sense of improbable processes that output probable products.

Biologist and Nobel laureate Peter Medawar introduced the term *conservation of information* in the 1980s, and the term gained currency in the following decade.[1] I [WmD] began using the term conservation of information in the late 1990s.[2] In those early days I also used the term *displacement* to mean the same thing. Conservation of information as we in the ID community developed the concept applies to search, so it could be more fully described as *conservation of information for search*. Search is a very general phenomenon, and it can characterize evolution. Indeed, the term *evolutionary search* has been widely used. Those who resist seeing evolution in terms of search stumble on the traditional view of search, which carries the sense of a purposive intelligence seeking something out. But search can be generalized so that it proceeds by natural means for naturally given targets. Evolutionary search, for instance, can proceed through biological configuration spaces to functional targets without reference to intelligence.[3]

What prompted my own use of the term conservation of information in the context of search was repeatedly seeing in the evolutionary literature (both the evolutionary computing and the biological evolution literature) that whereas purely random searches could not find desired targets, success in reaching those targets could be achieved by giving the searches additional information. These were needle-in-a-haystack problems that purely random or uninformed searches could not resolve. On the other hand, searches given additional information (in the form of fitness functions intended to mimic Darwinian natural selection) were much more readily able to find the desired targets.

Information accounted for the difference. There was always an information cost, typically unacknowledged, in substituting one of these better, more effective searches for a random or otherwise blind search. And yet, when that information cost was factored in, it seemed that nothing had in the end actually been explained or accounted for. Darwinian searches, in particular, always seemed to smuggle in the very information that they claimed to be producing for free. More precisely, the original problem of randomly finding a given search target (a needle-in-a-haystack problem) was *displaced* onto the new problem of finding a Darwinian search (in the form of a fitness function) capable of finding the target. Left unexplained in this move was how the Darwinian search gained the crucial information necessary for its success. Hence the term *displacement*— the information needed for successful search was displaced from one search to another, but never actually accounted for. The image here was that of filling one hole by digging another, but without accounting for how to get rid of the holes as such.

For an example of displacement in action, consider an exchange I [WmD] had with Eugenie Scott back in 2001 on the campus of Stanford University. Peter Robinson was interviewing us for his program *Uncommon Knowledge*. Robinson raised the trope about monkeys randomly typing Shakespeare if given enough time, and he then asked how it related to Darwin's theory. Scott, president at the time of the National Center for Science Education and an ardent opponent of intelligent design, responded by saying that in trying to account for how a monkey could type Shakespeare, natural selection's role would be that of a technician with whiteout standing behind the monkey where "every time the monkey types the wrong letter, [the technician] correct[s] it. That's what natural selection basically does. It's not just the random production of variation."[4]

Scott's error-correction approach to overcoming randomness and to thereby empowering the evolutionary process has wide currency in Darwinian circles. We saw a more embellished version of it in Section 7.1 with Richard Dawkins' METHINKS IT IS LIKE A WEASEL example, where this target phrase is achieved by reducing the errors in transitional phrases evolving to it. (It's perhaps not coincidental that this target phrase is drawn from Shakespeare's *Hamlet*!) In the case of Scott's error-correction approach, it loses plausibility as soon as we ask the obvious question: Given a monkey at a typewriter, what exactly are the qualifications of the technician standing at the monkey's shoulder who is doing the erasing? Simply put: How does the technician know what to erase?

The whole point of having monkeys at a typewriter in such examples is to render a chance-based evolutionary mechanism plausible. But in this example, that would mean accounting for the emergence of Shakespeare's works without the need to invoke an intelligence (like Shakespeare) that already knows Shakespeare's works. In other words, the whole point was to get Shakespeare's works without Shakespeare. But that's not what's happening here. Clearly, the only way to erase errors in typing Shakespeare's works is to know Shakespeare's works in the first place. Indeed, the very concept of error presupposes that there is a right way that things ought to be. That's the problem: Scott's technician, to be effective, needs already to know the works of Shakespeare.

This example captures, in a nutshell, conservation of information. Darwinian evolutionists look at what otherwise might seem to be designed and say, "Wait a minute, we have a process, driven by natural selection, that can produce design-like things with high probability, so no design inference is warranted." With the Shakespeare example, it's the lab tech at the monkey's shoulder. With a Behe-type example, such as an irreducibly complex molecular machine, a Darwinian process, assuming the role of the lab tech, is supposed to render it reasonably probable.

But Darwinian biologists stop there and don't ask the obvious follow-up question, which is: How do you get a search process that brings about results that in any other context we would regard as designed? When pressed on this matter, Darwinists will typically invoke *the environment*, as though all the information natural selection ever needs is somehow (magically) encoded there. But if it is encoded there, how exactly is it encoded there? Certainly, in computational settings, most ways of setting up environments lead to no interesting evolution. So how does the environment get the information it needs to bring about a successful form of evolution that then produces all the "neat stuff" we see in biology? Some environments work, some don't. Why do they work when they work? We're back to John Stuart Mill's method of difference (Section 7.3). Something other than the environment, taken generically, must be the difference maker.

Darwinian biologists, such as Dawkins with his WEASEL, may therefore seem to have resolved the problem of accounting for why their evolutionary searches concluded successfully. But in fact, they've merely displaced this problem. Indeed, the problem of finding the search target has now been shifted to finding a search that with high probability locates the target. Yes, the new search works much better than the original (random) search, but the question of where the new search comes from in the first place remains unanswered. Simply put, different searches can be better or worse at finding a search target, so the problem is to sort through those searches, finding those that are effective at finding the search target. Searches can themselves be the object of search—this is the crucial insight. So the problem at the heart of conservation of information can be succinctly restated as *how to successfully conduct a search for a search* (abbreviated S4S).

To illustrate what's at stake here with conservation of information (or displacement), consider an Easter egg hunt example. Suppose you have to find an Easter egg in a large, densely wooded area. You have until lunch to find the Easter egg, and you have been blindfolded for good measure. Finding the Easter egg is a needle-in-a-haystack problem. The field is so large and there are so many places where the Easter egg could be hidden that it is effectively impossible (read "small probability") to find it by random search in the time allotted. But just as you are losing hope of finding the egg, someone yells out to you "warmer, warmer, colder, warmer, warmer, hotter, hotter, red hot." And there you are, suddenly, standing over the egg.

What exactly happened here? The person yelling to you was giving you information to guide you to the egg. But where did that information come from? Conservation of information says that when a search does better than a baseline search (i.e., a search that requires no special information, such as a purely random or a brute-force search), then that better search required special information. Moreover, the amount of information required to make the search better had to be at least as much as the information that the improved search outputted.

Back in the late 1990s and early 2000s, conservation of information was more a qualitative than a quantitative notion. Information was clearly being smuggled in to make Darwinian-like searches successful. But the precise accounting of how much information had to be inputted and how much was being outputted, and whether an exact proportion between the two always held, was still unclear. Displacement always seemed to occur, shifting around but leaving unexplained the information that made a search successful. When a quantitative analysis was applied, it seemed that at least as much information always had to be inputted into the successful searches as they outputted (e.g., 20 bits out seemed to require at least 20 bits in). Conservation of information therefore seemed to hold in the sense that breaking even on the amount of information outputted versus the amount inputted was the best we could do. But conservation of information still fell short of being a full-fledged mathematical theory.

The Challenge of Evolutionary Computing

In the decade following the publication of the first edition of this book, the debate over the design inference and its applicability to evolution centered on the extent to which gradual evolutionary pathways existed and how their existence or non-existence would affect the probabilities of Darwinian processes originating living forms. The focus was therefore on small probabilities and design inferences rather than on information tracking and displacement, as in conservation of information.

Design theorists identified a variety of biological systems that resisted Darwinian explanations, and they argued that in some cases probabilities could be confidently estimated and shown to be small. Thus, they concluded that some of these systems were effectively unevolvable by natural selection, from which they drew a biological design inference. Darwinian evolutionists disagreed. Kenneth Miller's *Finding Darwin's God* (1999) and *Only a Theory* (2008) both sought to rebut this design-inferential challenge to Darwinism and neatly bookended the state of the debate during this period. By the end of that period, the debate had reached an impasse. Neither side was willing to budge.

But something else happened at the end of that period. In the late 2000s, the Evolutionary Informatics Lab, a Baylor University research group, began exploring a way out of this impasse. Headed by engineering professor Robert Marks, the lab included me [WmD] along with two of Bob's graduate students, George Montañez and Winston Ewert, who is the co-author of this second edition. The beginnings of our research were serendipitous. Repeatedly in those years we witnessed how computer programmers, inspired by Darwin, claimed that code they had written produced novel information by imitating Darwinian processes.

In particular, and often with the stated aim of unseating intelligent design, these Darwin-inspired programmers claimed that their programs were capable of creating novel information—notably specified complexity—without any special input or knowledge from the programmers. Or as biologist and programmer Thomas Schneider put it in reference to his own evolutionary simulation (which he dubbed "ev"), once some minimal evolutionary scaffolding was in place, "the necessary information should be able to evolve from scratch."[5]

These Darwinian programmers were in effect claiming to get out more information from their programs than they put in. Such claims intrigued us because computer programs are open to full scrutiny, so we can track the information flow through them step by step, confirming or disconfirming whether novel information is indeed being created. Such scrutiny is impossible with real-life biological evolution, where the data are always incomplete and the causal forces in play can at best be determined circumstantially. As the Evolutionary Informatics Lab explored such claims by these pro-Darwinian programmers, we invariably found that any information that they claimed to get out of their evolutionary programs had first to be built into the programs. How it was built in was not always obvious, though sometimes it was.

One physicist, David Thomas, offered a Darwinian program for constructing Steiner trees (a type of graph that in some optimal way connects various vertices) and posed this challenge: "If you contend that this algorithm works only by sneaking in the answer (the Steiner shape) into the fitness test, please identify the precise code snippet where this frontloading is being performed." We found several such code snippets, but the most glaring is one that included the incriminating comment "over-ride!!!":

```
x = (double)rand() / (double)RAND_MAX;

num = (int)((double)(m_varbnodes*x);

num = m_varbnodes; // over-ride!!![6]
```

Winston Ewert, Robert Marks, and I elaborated on this code snippet from Thomas's algorithm:

> The claim that no design was involved in the production of this algorithm is very hard to maintain given this section of code. The code picks a random count for the number of interchanges; however, immediately afterwards it throws away the randomly calculated value and replaces it with the maximum possible, in this case, 4. The code is marked with the comment "override!!!," indicating that this was the intent of Thomas. It is the equivalent of saying "go east" and a moment later changing your mind and saying "go west." The most likely occurrence is that Thomas was unhappy with the initial performance of his algorithm and thus had to tweak it.[7]

The precise details here need not detain us, though if you want them, they are available in the above cited paper by Ewert et al. The point to note, however, is that whenever we analyzed programs like those of David Thomas, we always found that the very information allegedly created from scratch or for free had in fact been inserted from the start, and often not very subtly. In short, the necessary information that made these programs work was always fully front-loaded, even if not fully obvious. We therefore set about deconstructing these programs to show precisely where the programmers were inserting the information they claimed to get out of the programs for free. It became like coin magic in which a coin (the information in question) is supposed to magically materialize in the magician's hand, but if you watch carefully from the right vantage, you can see where the coin is non-magically inserted (usually the coin is concealed through the technique of "palming," in which it is held so that the hands appear empty).

Darwinian programmers made no secret of constructing these programs as a way to refute intelligent design.[8] That's what motivated much of their work on these programs. To be successful in accomplishing this aim, however, these programs needed to output more information than they started with, just as real-life evolution needed to end with more information than it started with. Richard Dawkins put it this way: "The one thing that makes evolution such a neat theory is that it explains how organized complexity can arise out of primeval simplicity."[9] Darwinian evolution requires getting more from less. It requires a free lunch. These programs were supposed to show that. Our work showed, instead, that a free lunch never emerged from the programs we analyzed. But it pointed to much more, namely, that such programs were inherently incapable of producing a free lunch. To say that something can't happen is much stronger than to say that it didn't happen.

To this day, you can find at EvoInfo.org our Weasel Ware (analyzing Richard Dawkins' Weasel program), Ev Ware (analyzing Thomas Schneider's ev program), and Minivada (analyzing Christoph Adami's and Richard Lenski's Avida program, which was featured in *Nature* and was advertised at the time as dealing a deathblow to intelligent design). Ewert and Montañez did the heavy lifting in writing these software analyses. Through their work, the Evolutionary Informatics Lab in essence ran computational audits of the evolutionary programs by Dawkins, Schneider, Lenski, and others, tracking the informational flow through the programs from initial input to final output. What we found, in case after case, without exception, was that the amount of information outputted never exceeded the amount of information inputted.

This work of the Evolutionary Informatics Lab convincingly refuted the use of evolutionary computing to justify Darwinian evolution in biology. In 2010, the National Science Foundation spent $25 million to sponsor Michigan State University's BEACON (short for Bio/computational Evolution in Action CONsortium), whose primary aim at the time was studying and advancing such algorithms.[10] Using evolutionary computing had until then been a weapon of choice by Darwinists to defeat intelligent design. Yet by 2012, such efforts largely ground to a halt. The credit for disabling this weapon goes to the Evolutionary Informatics Lab. Unfortunately, some disreputable ideas are just too attractive to be permanently disowned, and so it seems that now in the 2020s, these algorithms are once again gaining popularity as a prop for Darwinian evolution.[11]

The Mathematics of Conservation of Information

The work of the Evolutionary Informatics Lab did not stop with simply analyzing individual evolutionary algorithms and showing that they never created information from scratch or for free. In science, when you take a particular approach to a problem (in this case, trying to generate novel information through evolutionary algorithms), and the approach keeps failing, there comes a point where you have to ask whether the approach may be inherently incapable of working. By cutting our teeth on these evolutionary algorithms and seeing firsthand how they never outputted more information than had been put into them, we were led to formulate a general principle, which we called the Law of Conservation of Information. Again: this generalization is important, because to say that something can't happen is much stronger than to say that it didn't happen. How did we get from the one to the other? While the impetus for our generalization was inductive (certain informational barriers kept arising for various searches), the principle we formulated found independent support in a class of mathematical theorems, which then provided a theoretical justification for the failures in search that we kept witnessing.

The Law of Conservation of Information is a proscriptive generalization—in other words, it is a general statement of what can't happen (and not simply of what doesn't happen). The best-known proscriptive generalization in science is the second law of thermodynamics, which effectively rules out perpetual motion machines. The USPTO, for instance, categorically refuses to consider patent applications for perpetual motion machines on the grounds of this law. But there are other known and accepted proscriptive generalizations: objects having mass cannot travel faster than the speed of light; temperatures cannot fall below zero degrees Kelvin; baseball pitchers cannot, simply with the use of their human physiology, accelerate baseballs to velocities in excess of 1,000 mph; etc.

Conservation of information, as we noted earlier in this epilogue, is a term that already had some history when around 2010

we at the Evolutionary Informatics Lab put it on a precise theoretical foundation. Our conception of it is more general and powerful
than what Peter Medawar and others meant by it in the 1980s and
1990s. As we use the term, conservation of information does not
refer to a quantity of information staying exactly unchanged as
information undergoes certain types of transformation. Our use of
the term "conservation" therefore differs from how it is used in the
phrase "conservation of energy," where it denotes that the quantity
of energy in a closed system stays exactly unchanged.

Rather, conservation of information, as we use it, signifies that
the best that can happen is that information is conserved, but that
the processing of information can also encounter inefficiencies,
where the quantity of information outputted is actually less (often
exponentially less) than the quantity inputted. We see analogous
inefficiencies with mechanical systems, where the effective energy
outputted is substantially less than the effective energy inputted
(the lost energy typically taking the form of heat from friction).
Similarly, with conservation of information, inputted information
tends to get underutilized, thereby leading to a loss in outputted
information.

Conservation of information therefore poses the following
challenge to conventional evolutionary theory: Darwinian evolution uses a natural-selection-based process to explain a seemingly
designed result, but then doesn't ask the natural follow-up question,
which is whether the process, in what it needs to produce something
design-like, might itself be designed and show evidence of design.
Darwinian evolutionists like Richard Dawkins reflexively assume
that the process itself is undesigned. Theistic evolutionists like
those at BioLogos, who are Darwinian in their science, may agree
that the process is designed (in some metaphysical or theological
sense) but also hold that it can give no evidence of design.[12] Both
camps thereby convince themselves that the design inference is
inapplicable to evolutionary biology because the process renders
the products of evolution reasonably probable and thus immune to
small-probability specification arguments of the sort required for a
design inference. Conservation of information shows that this
"passing of the buck" from the products of evolution to the

processes responsible for those products cannot invalidate design. In effect, it shows that processes that produce designed things must themselves be designed.

The Law of Conservation of Information, as we formulate it, is a serious scientific result. It derives not from a mere inductive generalization of past failures by Darwinian processes, or stochastic processes in general, to generate novel information needed for successful search. Rather, this law derives from a class of mathematical theorems first proven by researchers at the Evolutionary Informatics Lab. These theorems characterize and quantify how stochastic search processes produce information. The theorems then establish, with mathematical precision, that at least as much information has to be built into these processes as they output through successful search.

To convey a flavor of these theorems, without actually stating or proving any one of them, we offer the following example. Imagine that you are tossing a fair die. The probability of rolling a 6 is therefore 1/6. But now imagine that you have the power to load the die so that any two distinct numbers are bound to come up, and either of the numbers that could come up will then have probability 1/2. Suppose, therefore, that the die has been loaded to come up either 5 or 6, each with probability 1/2. You now have a "search" for the number 6 that, instead of having probability 1/6, has probability 1/2. That's a significant improvement in the probability of successfully searching for 6.

But how did you achieve this level of improvement? We assume you could load the die so that any two numbers come up. To increase your probability from 1/6 to 1/2 of rolling 6, you could have, instead of 5-or-6, loaded the die to come up 4-or-6, 3-or-6, 2-or-6, or 1-or-6. Thus, you had five ways to load the die so that with probability 1/2 you would roll a 6. But you also had ten other ways of loading the die that would have ensured that you would not roll a 6: 1-or-2, 1-or-3, 1-or-4, 1-or-5, 2-or-3, 2-or-4, 2-or-5, 3-or-4, 3-or-5, or 4-or-5. With any of these ways of loading the die, your probability of rolling a 6 would go down to 0.

Because of the symmetry of the die, any of these ways of loading the die seem equivalent, so the most reasonable probability distribution to put over these ways of loading the die would render them all equiprobable. In general, conservation of information theorems do not need uniform or equiprobability, but in this case it's appropriate. In consequence, your loading of the die on 5-or-6 is one among five ways of loading the die so that it comes up 6 with probability 1/2. But there are also ten other ways that you could have loaded the die so that it comes up 6 with probability less than 1/2 (in fact, with zero probability).

It follows that only 1/3 of the ways of loading the die raises to 1/2 your probability of rolling a 6. In other words, if you randomly selected from among these dice-loading/search methods, you would have a 1/3 chance of selecting one that boosted the chances of rolling a 6 to 1/2. Thus, at a probability "cost" of 1/3, you were able to load a die that with probability 1/2 rolls 6. These probabilities multiply not because of stochastic independence but because of conditional dependence, as when the probability of A and B is the probability of A times the probability of B given A (see Appendix A.2). In this example, when these probabilities multiply, they yield 1/6, which is the original probability of rolling a 6. It follows that this die-loading approach hasn't helped at all to increase our chances of rolling a 6. We initially had a probability of 1/6 of rolling a 6. And with our die-loading approach, we have a 1/3 probability of loading it in a way that then, with probability 1/2, rolls a 6, making the probability of getting a 6 again 1/6. So, procuring a better search has gained us nothing because of the probabilistic cost in procuring it.

In general, with conservation of information theorems, we are given a baseline probability p that is very small and for which success of the search would constitute finding a needle in a haystack. Next, we are given an improved search with probability q of search success, where q is much bigger than p ($q \gg p$). And finally, we are given a canonically associated search for a search (S4S) that accounts for the improved search increasing search performance from p to q. Conservation of information theorems then establish that the search for this improved search has probability no greater than p/q. If you now match up these numbers with the previous example, then $p = 1/6$, $q = 1/2$, and the probability of finding a successful search is $1/3 \leq p/q$. Granted, this die example is extremely simple and stylized. But the conservation of information theorems we proved establish these probabilistic relations with wide generality and applicability.

And so, to sum up, around the early 2010s, the Evolutionary Informatics Lab published a series of articles on conservation of information in the peer-reviewed engineering literature.[13] These articles established with precision that improving search performance entails an information cost that is at least as great as the information gained through improved search performance. So where did the information that makes up the difference in search performance come from? Conservation of information rules out that chance processes can pay the information cost needed to improve search performance. That leaves intelligence as the key to creating the information needed for improved search performance (an implication confirmed by engineering practice). This work on conservation of information logically completes the design inference. We will elaborate and validate this work in a sequel to this book.

⅏ APPENDIX A

A PRIMER ON PROBABILITY AND INFORMATION

THIS APPENDIX PRESENTS THE PROBABILISTIC AND INFORMATION-theoretic background needed to work with design inferences. The material is presented without frills. Readers with formal training in probability and information theory will know most of the material here but should still find a few useful nuggets to help in understanding design inferences. Readers with little technical facility will benefit from at least skimming the material here before reading the more technical chapters of this book, and then referring back to this appendix as needed.

Information and probability are closely allied notions, but they come with different intuitions. A bookie concerned with betting odds is in the business of probability. An accountant concerned with tracing and analyzing the flow of data is in the business of information. There's overlap here since probabilities depend for their determination on data and data emerge out of a probabilistic context.

A.1 The Bare Basics of Probability

In the mathematics of probability, probabilities are numbers between 0 and 1 that are assigned to *events* (and by extension to objects, descriptions, patterns, information, statements, hypotheses, and states of affairs, which can result from events or else operate logically in the same way as events). Events always occur with respect to a reference class of possibilities, which, mathematically, is a set. We will typically refer to such an underlying reference class with the Greek capital letter omega: Ω. Consider a die with faces 1 through 6. The reference class of possibilities in this case can be represented by the set $\{1, 2, 3, 4, 5, 6\}$ ($= \Omega$). Any subset of this reference class then represents an event. For instance, the event E_{odd}, namely, that an odd number was tossed, corresponds to $\{1, 3, 5\}$.

Such an event is said to occur if any one of its outcomes occurs. That's to say, in this example, that either a 1 or a 3 or a 5 is tossed. An *outcome* is thus any particular unitary thing that could happen. Outcomes can therefore be represented as singleton sets with respect to the reference class of possibilities. These are sets with only one element. Thus, the outcomes associated with $E_{odd} = \{1, 3, 5\}$ are $E_1 = \{1\}$, $E_3 = \{3\}$, and $E_5 = \{5\}$. It follows that outcomes are themselves events. Nevertheless, they are a special type of event, namely, events that, in a given inquiry, cannot be subdivided into still more elementary events. Outcomes are therefore sometimes also called *elementary events*. Events include not only outcomes but also *composite events* such as E_{odd}, which includes more than one outcome.

Events drawn from a reference class of possibilities Ω can be formed into new events via intersection, union, and complementation. $E \cap F$ is the intersection of E and F and denotes the event such that both E and F occur. $E \cup F$ is the union of E and F and denotes the event such that either E or F or both occur. E^c is the event complementary to E and denotes the event consisting of outcomes mutually exclusive of E.

Intersection and union, as defined here, are binary operations, but they are readily extended to finitely many (and even infinitely many) intersections and unions. Thus, for events $E_1, E_2, \ldots, E_n$, it makes sense to write $E_1 \cap E_2 \cap \ldots \cap E_n$ (the event in which all of these events occur) and $E_1 \cup E_2 \cup \ldots \cup E_n$ (the event in which at least one of these events occurs). Intersection and union are both commutative and associative, so there's no need to worry about the order of events or about grouping them with parentheses. Complementation, however, remains a unary operator, and is such that double complementation returns identity, i.e., $E^{cc} = E$.

Intersection, union, and complementation are set-theoretic operators that apply to events that, mathematically, are represented as sets. Yet these set-theoretic operators also mirror, respectively, conjunction, disjunction, and negation, which are logical operators that apply to descriptions or patterns and, more generally, to linguistic entities. To see the connection, suppose V and W are descriptions of events. Then, as linguistic entities, we can act on them conjunctively by forming $V \& W$ ("V and W"), disjunctively by forming $V \lor W$ ("V or W"), and by negation by forming $\sim V$ ("not V").

If we now let the asterisk * denote the event identified by a description, then V^* and W^* denote the events identified by V and W respectively. Moreover, the following connection then exists between conjunction and intersection, disjunction and union, and negation and complementation: $(V \& W)^* = V^* \cap W^*$, $(V \lor W)^* = V^* \cup W^*$, and $(\sim V)^* = (V^*)^c$. In this way the logic of events and the logic of descriptions mirror each other.[1] Because of this duality between events and the descriptions that identify them, we speak of *event-description duality* or *event-pattern duality* (thinking of descriptions also as patterns).

Probabilities of events from a reference class of possibilities Ω obey the following three axioms:

(1) The impossible event (i.e., an event that entails a physical or logical impossibility) is represented by the empty set ($\emptyset$) and has probability zero.

(2) The necessary event (i.e., an event that is guaranteed to happen) is represented by the entire reference class of possibilities and has probability one (e.g., with the die example, $E_{nec} = \{1, 2, 3, 4, 5, 6\} = \Omega$ has probability one).

(3) Events that are mutually exclusive have probabilities that sum together. Thus, in the example given in the

previous section, where distinct die tosses are mutually exclusive, $P(E_{odd}) = P(E_1) + P(E_3) + P(E_5)$ (i.e., $P(\{1, 3, 5\}) = P(\{1\}) + P(\{3\}) + P(\{5\})$.

From (2) and (3) if follows that mutually exclusive and exhaustive events always sum to one. A corollary of this last fact is that for E^c, the complement of E, $P(E^c) = 1 - P(E)$.

Note that the third axiom here only asserts the finite additivity of mutually exclusive events. Much, but not all, of probability theory is formulated in terms of σ-algebras that admit the countable additivity of mutually exclusive events, thus allowing for infinite summations of probabilities. For a case where finite additivity of probabilities holds but countable additivity does not, and where this failure of countable additivity makes a palpable difference for design inferences, see Appendix B.4 (specifically, for cosmological fine tuning).

Probabilities are interpreted in three principal ways:

(1) Frequentist approach—probability is a relative frequency (i.e., the number of occurrences of an event divided by the number of observed opportunities for the event to occur; relative frequencies are also called empirical probabilities). This approach is empirical in that one looks at the behavior of the system, records its relative frequencies, and then equates those relative frequencies with probabilities.

(2) Theoretical approach—probability derives from known properties of the system generating the events (e.g., dice are rigid, homogeneous cubes whose symmetry confers probability 1/6 on each face; quantum mechanical systems have probabilities derived from eigenvalues associated with the eigenstates of an observable).

(3) Degree-of-belief approach—probability measures strength of belief that an event will occur. Degree of belief is sometimes cashed out as degree of evidence or degree of confirmation.

These approaches are not mutually exclusive. For instance, one might conclude from the theoretical approach that each face of a die will appear with probability 1/6, but then look for confirmation from the frequentist approach—that is, by tossing the die repeatedly and recording the relative frequency of the various faces. The frequentist approach might discover that the die is unfair because it is unevenly weighted. But if the results matched up with the conclusion drawn from the theoretical approach, the frequentist approach would strengthen one's belief that the probability of each face is indeed 1/6.

Probability always reflects uncertainty in the mind of human inquirers. That raises the question whether this uncertainty is purely subjective or also has an objective dimension. The scientist Laplace, in the nineteenth century, thought that the world operated by Newtonian principles and thus was entirely deterministic. Probability for him was therefore simply a measure of human ignorance and thus reduced to pure subjectivity (if you knew exactly what nature was doing, all uncertainty would disappear).

With the rise of quantum mechanics, however, the idea that nature is irreducibly stochastic, or chance based, has gained currency. According to this interpretation of quantum mechanics, our uncertainty about various events in nature reflects not merely ignorance but an inherent propensity of the events to follow certain probability distributions. In any case, the logic and force of the design inference carries through whatever one's ultimate metaphysics of chance.[2]

A.2 Conditional Probability and Independence

Suppose an event E is known to have occurred, and we then ask what the probability of some other event F is given the knowledge of E's occurrence. In that case, the reference class of possibilities Ω contracts to E, and the probability of F is no longer simply $P(F)$ (i.e., the probability of F within the original reference class Ω), but the probability of that portion of F that resides within the new reference class E. This probability is called the conditional probability of F given E and is written $P(F|E)$, and it is defined as $P(F|E) = P(F \cap E)/P(E)$.

If E_1, E_2, ..., E_n are mutually exclusive and exhaustive, it follows that $E_1 \cup E_2 \cup \cdots \cup E_n = \Omega$ and that $1 = P(\Omega) = P(E_1 \cup E_2 \cup \cdots \cup E_n) = P(E_1) + P(E_2) + \cdots + P(E_n)$. In consequence, for any event F, $P(F) = P(F \cap (E_1 \cup E_2 \cup \cdots \cup E_n)) = P(F \cap E_1) + \cdots + P(F \cap E_n) = P(F|E_1)P(E_1) + \cdots + P(F|E_n)P(E_n)$ (by unpacking definitions). The first and last part of this equality, when juxtaposed, form what is known as the *law of total probability*:

$$P(F) = P(F|E_1)P(E_1) + \cdots + P(F|E_n)P(E_n).$$

Next, let us consider probabilistic independence. As noted in the last section, the axioms of probability require that for mutually exclusive events, probabilities add. Specifically, the probability of a union of mutually exclusive events is the sum of the probabilities of the individual events. Thus, if E_1, E_2, ..., E_n are mutually exclusive, $P(E_1 \cup E_2 \cup \cdots \cup E_n) = P(E_1) + P(E_2) + \cdots + P(E_n)$.

Does a corresponding relationship hold for intersection? For E_1, E_2, ..., E_n arbitrary events such that no intersection of any combination of them has zero probability, it follows from the definition of conditional probability that

$$P(E_1 \cap E_2 \cap \cdots \cap E_n) = P(E_1) \times P(E_2|E_1) \times P(E_3|E_1 \cap E_2) \times \\ \cdots \times P(E_n|E_1 \cap E_2 \cap \cdots \cap E_{n-1}).$$

To see this, in the case of just the two events E_1 and E_2, note that

$$
\begin{aligned}
P(E_1 \cap E_2) &= 1 \times P(E_1 \cap E_2) \\
&= [P(E_1)/P(E_1)] \times P(E_1 \cap E_2) \\
&= P(E_1) \times [P(E_1 \cap E_2)/P(E_1)] \\
&= P(E_1) \times P(E_2|E_1).
\end{aligned}
$$

If, now, $P(E_2|E_1) = P(E_2)$, it follows that

$$P(E_1 \cap E_2) = P(E_1) \times P(E_2).$$

In that case, we say that E_1 and E_2 are *probabilistically independent* or *stochastically independent*, or simply that they are *independent*. In general, we say that events E_1, E_2, ..., E_n are independent if for all distinct events taken from among them, i.e., $E_{i_1}, E_{i_2}, ..., E_{i_k}, 1 \leq k \leq n$,

$$P(E_{i_1} \cap E_{i_2} \cap \cdots \cap E_{i_k}) = P(E_{i_1}) \times P(E_{i_2}) \times \cdots \times P(E_{i_k}).$$

Events turn out to be probabilistically independent if they derive from causally independent processes. The converse, however, is not true—events can be probabilistically independent without being causally independent.

To see that probabilistic independence does not entail causal independence, suppose you are gambling in a game where you win (the event E) if two causally (and thus probabilistically) independent coin tosses turn up with the same face. Let H_1 and H_2 denote heads on the first and second tosses respectively; likewise, let T_1 and T_2 denote tails on the first and second tosses respectively. Then $E = (H_1 \cap H_2) \cup (T_1 \cap T_2)$.

Your victory (E) in this game thus clearly depends on how the coins land, and thus on H_1, H_2, T_1, and T_2. Moreover, your probability of victory is ½: $P(E) = P((H_1 \cap H_2) \cup (T_1 \cap T_2)) = P((H_1 \cap H_2)) + P((T_1 \cap T_2)) = (P(H_1) \times P(H_2)) + (P(T_1) \times P(T_2)) = (½ \times ½) + (½ \times ½) = ½$. That's because $H_1 \cap H_2$ is mutually exclusive of $T_1 \cap T_2$, and because the events being intersected here are probabilistically/causally independent.

And yet E, though not causally independent of H_1, H_2, T_1, and T_2, is probabilistically independent of them. Indeed, the calculation is straightforward to show that if F denotes any of H_1, H_2, T_1, and T_2, then $P(E \cap F) = P(E) \times P(F) = ½ \times ½ = ¼$, confirming probabilistic independence at the expense of causal independence.

A.3 Bayes' Theorem

Given events E and F, we saw in the last section that the conditional probability of F given E is, by definition, $P(F|E) = P(F \cap E)/P(E)$. From the definition of conditional probability, we can now rewrite the numerator of this fraction, namely, $P(F \cap E)$, as $P(E|F) \times P(F)$, and so

$$P(F|E) = \frac{P(E|F) \times P(F)}{P(E)}.$$

This last formula is the simple form of Bayes' theorem (named after the eighteenth-century Scottish clergyman Thomas Bayes).

Suppose, now, that instead of just E and F, we consider E and also the following n mutually exclusive and exhaustive events F_1, F_2, ..., F_n. Think of E as a given event known to have occurred ("evidence") and F_1, F_2, ..., F_n as possible mutually exclusive events of which one must occur. Then, given the axioms of probability and the definition of conditional independence,

$$\begin{aligned} P(E) &= P([E \cap F_1] \cup [E \cap F_2] \cup \cdots \cup [E \cap F_n]) \\ &= P(E \cap F_1) + P(E \cap F_2) + \cdots + P(E \cap F_n) \\ &= P(E|F_1) \times P(F_1) + P(E|F_2) \times P(F_2) + \\ &\quad \cdots + P(E|F_n) \times P(F_n). \end{aligned}$$

Substituting this last expression for the denominator in the simple form of Bayes' theorem now yields the standard form of Bayes' theorem for any of the events F_i $(1 \le i \le n)$:

$$P(F_i|E) = \frac{P(E|F_i) \times P(F_i)}{P(E|F_1) \times P(F_1) + \cdots + P(E|F_n) \times P(F_n)}.$$

This is the standard form of Bayes' theorem. It is commonly used to adjudicate between two alternatives, namely, between an event F and its complement F^c. These together then form mutually exclusive and exhaustive events. In this case, the standard form of Bayes' theorem becomes

$$P(F|E) = \frac{P(E|F) \times P(F)}{P(E|F) \times P(F) + P(E|F^c) \times P(F^c)}.$$

Bayes' theorem arises in situations such as the following: an event E is witnessed consisting of ten white balls drawn with replacement from an urn. But suppose there are two urns. Urn 1 has nine white balls and one black ball whereas urn 2 has one white ball and nine black balls. Suppose these urns are selected at random so that each has probability ½. Represent the selection of urn 1 by the event F_1 and the selection of urn 2 by the event F_2. F_1 and F_2 are therefore mutually exclusive and exhaustive ($F_1^c = F_2$ and $F_1 = F_2^c$).

Given that without additional information F_1 and F_2 are equally likely and given that we do have the additional information that E was observed and consisted of ten white balls drawn, it seems clear that F_1 rather than F_2 was much more likely to have been selected (i.e., that the ten white balls drawn were much more likely to be from urn 1 rather than urn 2). But how much more likely?

According to Bayes' theorem, given that ten white balls from urn 1 has probability $P(E|F_1) = (9/10)^{10} = .3486784401$ and given that ten white balls from urn 2 has probability $P(E|F_2) = (1/10)^{10} = .0000000001$, it follows that

$$P(F_1|E) = [P(E|F_1) \times P(F_1)]/[P(E|F_1) \times P(F_1) + P(E|F_2) \times P(F_2)]$$

$$= [.3486784401 \times .5]/[.3486784401 \times .5 + .0000000001 \times .5]$$

$$\approx .9999999997,$$

implying that $P(F_2|E)$ is approximately .000000003. Given E, F_1 is therefore much more probable than F_2.

Although Bayes' theorem can be formulated entirely in terms of events, as we have just done, it is more common to formulate it in terms of events and hypotheses, in which case it is common to substitute conjunction (&) for intersection (∩), disjunction (∨) for union (∪), and negation (~) for complementation (ᶜ). Thus, for events E and hypotheses H, we typically write E & H rather than $E \cap H$, $E \vee H$ rather than $E \cup H$, and $\sim E$ and $\sim H$ rather than E^c and H^c. For the underlying rationale, recall event-pattern duality from Appendix A.1.[3]

Consider, therefore, an event E and chance hypotheses H_1, H_2, ..., H_n that are mutually exclusive and exhaustive. What is the probability of any one of these hypotheses H_i given that E is assumed to have occurred? Given our use of logical rather than set-theoretic notation for combining events and hypotheses, and thinking of these hypotheses in terms of the events that produced them, our previous reasoning in formulating Bayes' theorem now shows that

$$
\begin{aligned}
P(E) &= P([E \,\&\, H_1] \vee [E \,\&\, H_2] \vee \cdots \vee [E \,\&\, H_n]) \\
&= P(E \,\&\, H_1) + P(E \,\&\, H_2) + \cdots + P(E \,\&\, H_n) \\
&= P(E|H_1) \times P(H_1) + P(E|H_2) \times P(H_2) + \cdots \\
&\quad + P(E|H_n) \times P(H_n).
\end{aligned}
$$

Substituting terms from this last equation in the standard form of Bayes' theorem given above, Bayes' theorem can now be restated as follows:

$$
P(H_i|E) = \frac{P(E|H_i) \times P(H_i)}{P(E|H_1) \times P(H_1) + \cdots + P(E|H_n) \times P(H_n)}.
$$

Given a chance hypothesis H, its negation is $\sim H$, and since H and $\sim H$ together are mutually exclusive and exhaustive, Bayes theorem then takes the following form:

$$
P(H|E) = \frac{P(E|H) \times P(H)}{P(E|H) \times P(H) + P(E|\sim H) \times P(\sim H)}.
$$

It's these two last formulations of Bayes' theorem that one typically sees in textbooks on Bayesian probability.

A.4 The Bayesian Approach to Statistical Inferences

In Bayesian statistical inference, one considers an event E and two competing hypotheses H_1 and H_2, which are mutually exclusive but not necessarily exhaustive. Think of H_1 as the hypothesis that a given coin is fair and H_2 as the hypothesis that a given coin has both sides heads. Moreover, think of the event E (say, ten coin flips) as evidence for either of these hypotheses. To decide whether the evidence E better supports either H_1 or H_2 therefore amounts to comparing the probabilities $P(H_1|E)$ and $P(H_2|E)$ and determining which is bigger. These probabilities are known as *posterior probabilities* and measure the probability of a hypothesis given the event/ evidence/data E.

Posterior probabilities like $P(H_1|E)$ and $P(H_2|E)$ cannot be calculated directly but must rather be calculated on the basis of Bayes' theorem. Using the simple form of Bayes' theorem in Appendix A.3, we find that the posterior probability $P(H_i|E)$ ($i = 1$ or 2) is expressed in terms of $P(E|H_i)$, known as the *likelihood* of H_i given E, and $P(H_i)$, known as the *prior probability* of H_i. Often prior probabilities cannot be calculated directly or even given a plausible rationale. Moreover, in calculating the posterior probability, we still need to compute the denominator in the simple form of Bayes' theorem, namely, $P(E)$.

Typically, this last term does not need to be calculated. If the aim is simply to determine which of these hypotheses is better supported by the evidence E, it is enough to form the *ratio of posterior probabilities* $P(H_1|E)/P(H_2|E)$ and determine whether it is greater than or less than 1. If this ratio is greater than 1, it supports the hypothesis in the numerator (H_1, which we are treating as the fair-coin hypothesis). If it is less than 1, it supports the hypothesis in the denominator (H_2, which we are treating as the double-heads hypothesis).

This ratio, using either the simple or standard form of Bayes' theorem, can now be rewritten as follows (note that the denominator for either form of Bayes' theorem simply cancels out in forming the product on the right):

$$\frac{P(H_1|E)}{P(H_2|E)} = \frac{P(E|H_1)}{P(E|H_2)} \times \frac{P(H_1)}{P(H_2)}.$$

The first factor on the right side of the equation is known as the *likelihood ratio* or *Bayes factor*; the second is the ratio of priors, which measures our relative degree of belief in these two hypotheses before E entered the picture as evidence. Since the ratio on the left side of this equation represents the relative degree of belief in these two hypotheses once E is taken into account, this equation shows that updating our prior relative degree of belief in these hypotheses (i.e., before the evidence E was factored in) is simply a matter of multiplying the ratio of prior probabilities times the likelihood ratio.

In this way, the likelihood ratio, namely, $P(E|H_1)/ P(E|H_2)$, is said to measure the strength of evidence that E provides for H_1 in relation to H_2. Thus, since we are treating H_1 as the fair-coin hypothesis and H_2 as the double-heads hypothesis, if this ratio is bigger than 1, E favors the former hypothesis. On the other hand, if this ratio is less than 1, it favors the latter hypothesis. Note that if E contains a tail, $P(E|H_2)$ will be zero, and the ratio will blow up to infinity, thus giving credence to H_1. But if E contains only heads, then how strongly E counts in favor of H_2 will depend on the prior probabilities.

How does all this apply to design inferences? A Bayesian approach to design inferences will make one of these hypotheses a design hypothesis and other a non-design hypothesis, and then pit them against each other over some event E responsible for a design-like object (such as a putative archeological artifact). If H_1 is a design hypothesis and H_2 a non-design hypothesis, then the evidence/event E favors design if $P(H_1|E) > P(H_2|E)$, non-design if $P(H_1|E) < P(H_2|E)$ (indifferent if they are equal).

A.5 The Fisherian Approach to Statistical Inferences

In Ronald Fisher's approach to hypothesis testing, one is justified in rejecting a chance hypothesis provided that a sample falls within a prespecified *rejection region* (also known as a *critical region*).[4] For example, suppose one's chance hypothesis is that a coin is fair. To test whether the coin is biased in favor of heads, and thus not fair, one can set a rejection region of ten heads in a row and then flip the coin ten times. In Fisher's approach, if the coin lands ten heads in a row, then one is justified in rejecting the chance hypothesis. But what does rejecting the chance hypothesis in that case really mean? And how confident can we be that the coin really wasn't fair?

Fisher's approach to hypothesis testing is the one most widely used in the applied statistics literature and the first one taught in introductory statistics courses. Nevertheless, as he formulated it, Fisher's approach failed to adequately answer an important question. For a rejection region to warrant rejecting a chance hypothesis, the rejection region must have a small enough probability. But what is a small enough probability? And why does small probability even matter?

To be more precise, given a chance hypothesis and a rejection region, how small does the probability of the rejection region have to be so that if a sample falls within it, the chance hypothesis can legitimately be rejected? Fisher never adequately answered this question. Sure, we all have intuitions that small probabilities can, under the right conditions, be used to justify eliminating a chance hypothesis. But Fisher was looking to do more than simply restate such intuitions.

The Fisherian approach to statistical inferences promised, by contrast, to give scientific rigor to such intuitions. The problem was therefore to rationally ground what, within the Fisherian approach, is called a *significance level*—the level of improbability at which chance no longer becomes a legitimate explanation. Fisher's solution was to say that whenever a sample falls within the rejection region and the probability of the rejection region given the chance hypothesis is less than the significance level, then the chance hypothesis can legitimately be rejected.

More formally, the Fisherian approach needed to justify a significance level α (always a positive real number less than one) such that whenever a sample (an event we will call E) falls within the rejection region (call it T) and the probability of the rejection region given the chance hypothesis (call it H) is less than α (i.e., $P(T|H) < \alpha$), then the chance hypothesis H can be rejected as the explanation of the sample. In the applied statistics literature, it is common to see significance levels of .05, .01, or even smaller. The problem with Fisher's approach as he originally formulated it is that any such proposed significance levels will seem arbitrary. Bayesian theorists Colin Howson and Peter Urbach therefore charged Fisher's approach with lacking "a rational foundation."[5]

In general, statistical theorists who take an inherently comparative approach to statistical inferences thus see the Fisherian approach as fatally flawed. The best known among the comparative approaches to statistical inferences are the Bayesian approach and the Neyman-Pearson approach. Fisher's approach is not comparative. It looks at one hypothesis at a time, and then uses small probabilities to eliminate hypotheses. Bayesians, by contrast, take a comparative approach to hypotheses, invariable pitting two or more hypotheses against each other, and then determining which is the more probable. Small probability thus ends up being irrelevant for Bayesians because all that matters is how the probabilities compare, not what they are in absolute terms.

Statistician Richard Royall takes a likelihood approach to statistical inferences that shares much in common with the Bayesians. (His main point of departure is in attempting to circumvent prior probabilities.) Royall's approach is comparative, and so he rejects Fisher. Nonetheless, Royall reluctantly admits that Fisher, and not Bayes, sets the standard for statistical inferences in science. Thus, Royall will write, "Statistical hypothesis tests, as they are most commonly used in analyzing and reporting the results of scientific studies, do not proceed... with a choice between two [or more] specified hypotheses being made... [but follow] a more common procedure." He then summarizes that "common procedure," which is exactly Fisher's approach.[6] All this suggests that Fisher's approach is not irretrievably broken but is instead on the right track, needing not to be jettisoned but merely amended.

The design inference, as developed in this book, provides such an amendment of Fisher. According to the design inference, significance levels cannot be set in isolation but must always be set in relation to the relevant opportunities for an event to occur (or what we call its *probabilistic resources*—see Chapter 4). The more opportunities there are for an event to occur, the more possibilities there are for it to land in the rejection region. And this in turn means that there is a greater probability that the chance hypothesis under consideration will be rejected. It follows that a seemingly improbable event can become quite probable once all the opportunities for its occurrence are factored in.

Chapter 5, on the logic of the design inference, clarifies how Fisher's approach to statistical inferences ought to be properly amended. There we need to address a logical point about what small probability events can tell us about the antecedents supposedly responsible for them (whether these antecedents be other events, hypotheses, or whatever). This is the focus of Section 5.7, on probabilistic modus tollens. Even though such a section may seem niche and boutique, in fact it gets at the core of why small probabilities can be used to eliminate chance but also why small probabilities by themselves are never enough to eliminate chance, since they require events that are not just improbable but also specified.

One final point: None of what's written in this chapter should be taken as an assault on Bayesian or other comparative approaches to statistical inferences. In particular, design inferences can be cashed out in Bayesian and other terms. And yet we would argue that when they are, they must incorporate small probability and specification. These are the key concepts of the design inference as developed in this book. Moreover, their logic follows as an extension of Fisherian rather than comparative approaches to statistical inferences. Hence the need to amend Fisher.

A.6 Random Variables and Search

Often, we are interested not so much in single events as in multiple events and in computing probabilities associated with their joint occurrence. In such cases, it is convenient to use *random variables*. A random variable associates probabilities with a range of possible outcomes, and multiple random variables can thus model the joint occurrence of complicated events, thereby making it easier to calculate their probability.

Take the occurrence of 100 heads in a row with a fair coin. Letting tails be denoted by 0 and heads by 1, we form the random variables X_n for n running from 1 to 100, where each X_n represents the tossing of a fair coin, and where all the X_n are independent of each other. The X_n are thus said to be *independent and identically distributed*—independent because no toss affects the probabilistic behavior of the others, and identically distributed because a fair coin assigns the same probability of ½ to each possible toss outcome. The probability of 100 heads in a row can therefore be represented as

$$P(X_1 \in \{1\} \, \& \, X_2 \in \{1\} \, \& \cdots \& \, X_{100} \in \{1\})$$

(or equivalently as $P(X_1 = 1 \, \& \, X_2 = 1 \, \& \cdots \& \, X_{100} = 1)$). Since all the events of the form $X_n \in \{1\}$ are independent of each other, probabilities multiply and the previous probability comes to

$$P(X_1 \in \{1\}) \times P(X_2 \in \{1\}) \times \cdots \times P(X_{100} \in \{1\}).$$

Moreover, since each event $X_n \in \{1\}$ has probability ½, this probability comes to $(½)^{100}$, or roughly 10^{-30}.

Random variables, however, need not be independent. Height and weight, for instance, tend, in humans, to be correlated. Let X_1 denote a human's height in feet and X_2 denote a human's weight in pounds. Then $P(X_1 \geq 7 \, \& \, X_2 \geq 220)$ (i.e., the probability that someone is simultaneously at least 7 feet tall and 220 pounds) will not equal $P(X_1 \geq 7) \times P(X_2 \geq 220)$ (i.e., the probability that someone is at least 7 feet tall times the probability that someone, possibly different, is at least 220 pounds). For instance, it might be the case that everyone 7 feet or taller is automatically 220 pounds or more. In that case, the first probability is just $P(X_1 \geq 7)$. But if the probability that there are people weighing less than 220 pounds is positive, then $P(X_2 \geq 220)$ is strictly less than one, and $P(X_1 \geq 7) \times P(X_2 \geq 220)$ will be strictly less than $P(X_1 \geq 7)$.

Sometimes, when random variables are indexed by numbers, the numbers denote times or trials at which the random variable registers a value. Such indexed random variables are often called *time series* or *stochastic processes*. Thus, for random variables X_i for i running from 1 to 100, each index number may denote the time of occurrence of an event at which the random variable with that index registers a value. X_{17} would thus represent the seventeenth occurrence of an event in question. Note that the numbers indexing random variables can be real numbers rather than natural numbers, thereby modeling continuous rather than discrete time.

It is convenient to use random variables indexed by natural numbers to represent search. Thus, we imagine a search space Ω and a target T that's an event/subset of Ω. We now consider random variables X_i for $i = 1$, 2, etc. defined on a probability space Ω' with probability measure P and with the X_is taking values in Ω. In general, all searches must be called off after a certain point because there are just not enough opportunities in the universe to carry them further. (See Chapter 4 on probabilistic resources.) Thus, any search must break off after a given natural number m (which may depend on the resources available for search given the universe as a whole, or on limitations specific to the search).

For instance, we might imagine that the X_is form independent and uniformly distributed random variables on Ω. So long as the X_is have a positive probability of finding T, however small, with enough such X_is, eventually one of them will indeed find T. This result follows from the law of large numbers.[7] Suppose, however, that m represents the outer limits of the number of X_is available to try to find T. If in m attempts the X_is will still be unlikely to find T, then the search will with high probability be unsuccessful.

Given a search X_i for $i = 1$, 2, etc. and where the object is to find T in Ω, we can ask what the probability is of the search successfully finding T in n steps, whatever n may be. Typically, what this means is forming X_1, X_2, ..., X_n into a single random variable Y_n taking values in Ω. Y_n then selects the first of the X_is among X_1, X_2, ..., X_n that falls inside T, or else takes an arbitrary value of these X_is if the search in n steps is unsuccessful. In this way, Y_n induces a probability measure P_n on Ω that assigns the probability of success of the random variables X_1, X_2, ..., X_n finding T. Such a probability can then be used to gauge the *waiting times* associated with the search for T—namely, how large n (the waiting time) needs to be before a certain probability of finding T is achieved.[8]

A.7 The Privileged Place of Uniform (or Equi-) Probability

A uniform probability assigns the same probability, or equiprobability, to each outcome.[9] Although most probabilities in practice are not uniform, there are exceptions. Games of chance played at a casino tend to follow uniform probabilities. Thus, for dice or cards, we assume any roll is as likely as any other and any hand is as likely as any other. Because the probabilities are all the same, they are uniform or equiprobable. But even with games of chance, we have to distinguish between underlying probabilities, which often are uniform, and actual probabilities relevant to betting outcomes, which often are not uniform.

Any die, for instance, has probability 1/6 of displaying a given face. Likewise, any pair of dice has probability 1/36 of displaying two given faces (that's because separate die rolls are probabilistically independent). These are uniform probabilities. But in the game of craps, the probability of interest is not the joint faces displayed by the pair of dice, but how the numbers displayed add up. In craps the number two comes up only when both dice show a one, and thus has probability 1/36. But the number seven has a much higher probability of coming up, 1/6, because multiple combinations sum to seven. The probabilities in craps are therefore not equiprobabilities. In blackjack, probabilities start out uniform, but once some cards have been dealt and noted, card counters, who are able to improve their odds of winning in light of cards already dealt, will update the probabilities of subsequent cards being dealt. These adjusted probabilities are conditional probabilities (see Appendix A.2), and they will no longer be uniform.

Even though probabilities in practice tend not to be uniform, uniform probabilities nonetheless tend to underlie many probabilities. We just saw this with the game of craps, where a pair of dice follows a uniform probability. Thus, if we let X_1 denote the number appearing on the roll of the first die, and X_2 for the second die (these are random variables—see Appendix A.6), then $P(X_1 = m \ \& X_2 = n) = 1/36$ for m and n natural numbers between 1 and 6. This is a uniform probability. At the same time, these probabilities enable us to determine the probability $P(X_1 + X_2 = k)$ for k a natural number between 2 and 12. This probability equals $(k–1)/36$ for $2 \leq k \leq 7$ and $(13–k)/36$ for $7 \leq k \leq 12$. This is not a uniform probability.

Even when a uniform probability does not accurately characterize a probabilistic phenomenon, it is often still relevant to the phenomenon. Letters that appear in English prose, for instance, are not uniformly distributed. The letter "e" occurs a lot more than the letter "z," and the letter "q" is almost invariably followed by the letter "u." Nonetheless, in trying to explain the role of chance in a sequence of letters, it is useful to calculate the uniform probability for the letter sequence. Letters don't care about their frequency or how they are arranged. Moreover, in trying to explain how letters might have been arranged by chance, uniform probabilities constitute the "worst case scenario" or baseline. In other words, any other chance process will impose more order or regularity than a uniform probability, so if we can't eliminate chance for a uniform probability, we can't eliminate it for probability distributions in general.

Or consider a specific gene or protein. The gene consists of a sequence of four types of nucleotide bases. The protein consists of a sequence of twenty types of amino acids. Both the nucleotide bases and the amino acids here are akin to the letters of an alphabet. And like the letters of the alphabet, nucleotide bases or amino acids do not have any preferential bonding affinities, so a good baseline for assessing the chance occurrence of a particular gene or protein is to consider a uniform probability as governing the order of nucleotide bases or amino acids. But when considering a gene or protein's evolvability, the probabilities of their constituent parts (nucleotide bases or amino acids) will typically not be uniform. For instance, hydrophobic and hydrophilic amino acids will need to be swapped out in kind rather than indiscriminately, as is the case with uniform probabilities.

In general, then, uniform probabilities provide a good starting point for any probabilistic analyses. If the point at issue is whether chance explains a particular probabilistic phenomenon, the first question is whether the phenomenon can plausibly be explained as the output of a uniform probability distribution. Uniform probabilities set a low bar for eliminating chance. If chance can't be eliminated in the face of uniform probabilities because the probabilities assigned aren't small, then chance becomes plausible.

On the other hand, if the phenomenon in question has small probability with respect to a uniform probability distribution, then the next question is whether the phenomenon can plausibly be explained as the output of some other probability distribution. Uniform probabilities help us get our bearings. In design inferences, they help us gauge what are the relevant probabilities that underlie the phenomena whose chance occurrence we are trying to assess. Obviously, in any probabilistic analysis, the point is to come to terms with the actual probabilities in operation.[10]

A.8 The Universality of Coin Tossing for Probability

Readers may wonder why such an abundance of coin-tossing examples exists in this book and whether the authors may be "returning to this well" one too many times. In fact, coin tossing is universal for probability. If we let 0 represent tails and 1 heads, then any probability with a natural number in the numerator and a power of two in the denominator can be represented with a designated sequence of coin tosses. For instance, $5/8$ ($8 = 2^3$) can be represented by 000, 001, 010, 011, and 100. Since probabilities with powers of two in the denominator can approximate any probabilities whatsoever, coin tossing can in this way be thought to capture all probabilities.

But in fact, the connection between probability and coin tossing goes deeper. Any probability that's a rational number of the form k/n where $0 < k < n$ can be captured precisely (and not just approximately) through a coin-tossing event as follows: Find a natural number m such that 2^m is as big or just bigger than n. Take all binary numbers between 0 and 2^m and divide them into three sets: those binary numbers with value strictly less than k, those greater than or equal to k and strictly less than n, and those remaining. Now flip the coin m times to determine an m-digit binary numeral. If the binary numeral is strictly less than k, the event happened. If greater than equal to k but strictly less than n, the event didn't happen. Otherwise, sample again, repeating the process.

This approach to capturing rational probabilities with coin tossing is known as *rejection sampling*, which means ignoring occurrences in the rejection region and repeating until you get an occurrence outside the rejection region. For instance, rejection sampling allows with a fair coin to allot an item to one of three people each with probability a third: 00 means give the item to person one, 01 means give the item to person two, 10 means give the item to person three, 11 means again flip the coin twice. The laws of probability guarantee that eventually you will land outside the rejection region and get an event with the probability that you want.

The connection between the mathematical theory of probability and coin tossing goes deeper still. Indeed, all the probability spaces of practical interest are Borel isomorphic to the Cantor cube.[11] The *Cantor cube*, denoted as $2^{\mathbb{N}}$ (where $\mathbb{N}$ is the natural numbers), is just the set of all infinite sequences of 0s and 1s. In other words, it is all infinite sequences of coin tosses. *Borel isomorphic* just means that the events to which probabilities are assigned can be put in one-to-one correspondence with each other, preserving both the event structures of the two spaces as well as their corresponding probabilities (though not necessarily any geometric properties of the underlying spaces). From a probabilistic perspective, probability spaces that are Borel isomorphic are equivalent. (This is the same Borel encountered in Chapter 1.)

One way to glimpse the universality of these infinite coin tosses for probability theory is to represent the real numbers between 0 and 1 by their infinite binary expansions. Doing so preserves the uniform probabilities on both spaces and places the spaces in one-to-one correspondence.[12] In fact, all the probability spaces we encounter in practice can be represented by this space of infinite coin tosses (they embed and have their probabilistic structure preserved inside this space). It's for this reason that the probabilist T. E. S. Raghavan used to quip: "If you understand coin tossing, you understand all of probability."[13] Of course, this insight applies only to those whose understanding of coin tossing is profound!

A.9 Information as Constraint on Contingency

Let's now shift gears and focus for the remainder of this primer on information. Most people have good intuitions about information. Yet they may be hard-pressed to give a good definition of it that covers most uses of the term. Is information a syntactic entity (comprising bit sequences or arrangements of symbols) or a semantic entity (comprising the meaning of syntactic entities)? It can be both, but it is also more than both. Information in its most general sense is a constraint on contingency. Let's now examine what that means. Once this contingency-constraint view of information becomes clear, it also will become clear how information relates to probability and also to computation.

Information arises from the interplay between *contingency* and *constraint*. Contingency refers to the different ways things might be. The plural here is important: way*s* things might be. If there's only one way a thing can be, there's no contingency and thus, as will become apparent, no information. Constraint refers to a narrowing of these contingencies, including some and excluding others. Suppose we tell you that it's raining outside. The contingencies here are that it's raining outside or that it's not raining outside (two contingencies). The constraint here is that by telling you it's raining outside (focusing on one contingency), we are excluding that it's not raining outside (ruling out the other contingency).

The constraint, if it is to be informative, must neither include all contingencies nor exclude all contingencies. If we tell you that it's raining or not raining outside, we haven't given you any information because you already knew that it had to be raining or not raining outside. We would thus have simply stated a tautology. Likewise, if we tell you that it's raining and not raining outside, we haven't given you any information because you already knew it couldn't be both raining and not raining outside. We would thus have simply stated a contradiction.

Writing about human communication, MIT philosopher Robert Stalnaker defined information by using the terms "possibility" and "exclusion" where we, respectively, use the terms "contingency" and "constraint." But his point was exactly the same. As Stalnaker defines information, "To learn something, to acquire information, is to rule out possibilities [contingencies]. To understand the information conveyed in a communication is to know what possibilities [contingencies] would be excluded [constrained] by its truth."[14] Going beyond human communication, philosopher Fred Dretske made the same point: "No information is associated with, or generated by, the occurrence... of events [constraint] for which there are no possible alternatives [contingency]."[15]

Information always presupposes a range of *contingencies* or *possibilities* or *alternatives*. And with these in place, it then requires a *focusing* or *narrowing* or *constraining* of some portion of that range to the exclusion of the rest. Information thus always presupposes both a *yes* and a *no*, a *yes* to what the constraint admits or includes, and a *no* to what the constraint dismisses or excludes. In fact, there's a sense in which *no* is more fundamental to information than *yes* because simply saying yes to everything guarantees that no information can be generated. The constraint that generates information is always about saying no to contingencies, ruling things out, excluding possibilities. Information demands negation.

In saying that information emerges through the interplay between contingency and constraint, it is helpful to think of information not merely as a *noun* (which it is grammatically) but also as a *verb* (as in an act of informing). Thus, information doesn't simply exist but it also happens—it informs. It is as much action as item. Because most of the information we deal with takes the form of alphanumeric characters arranged in sequence (this can be anything from bit strings on a computer to text resulting from ink on paper), we tend to think of information as *items of information*. But items of information are really the end product of an informational happening or activity. They result from an act of inputting information.

To see that information is as much a verb as a noun, consider how we get an item of information in the first place. Suppose you are reading Shakespeare's *Hamlet*. Shakespeare, in writing the play, applied ink to paper to form an item of information, namely, the first finished manuscript of the play. Simply looking at that manuscript, however, might not suggest information's defining interplay between contingency and constraint. Nevertheless, the activity of Shakespeare in writing the play does suggest this interplay. While composing *Hamlet*, Shakespeare might have written any number of other plays. In fact, he might have written any sequence of letters and spaces onto sheets of paper. Most such sequences would be gibberish. All these different texts that he might have written constitute the relevant range of contingencies. And then, by writing *Hamlet*, Shakespeare drastically constrained that range of contingencies.

That's how the information that constitutes *Hamlet* came to be, as a constraint on contingencies. Moreover, the text of *Hamlet* that you find in a bookstore comes through successive exchanges of information starting from the first finished manuscript and moving through various revisions, typesetting, and printing presses, and in our day through digital texts and eBooks. Each of these information exchanges occurs via constraints on contingencies. As the information is copied, it could be copied in any number of ways, only a small proportion of which are close enough to the original to count as *Hamlet*. The same principles apply to the MP3 music files you listen to or the MP4 video files you watch. In all such cases, there's a causal chain of information exchanges leading back to the author/artist/composer of the information from the point of origin.

Information is therefore not just dynamic rather than static but also inherently relational, connecting back to the originating source of the information. Accordingly, we can think of information not just as a verb but also as a *transitive verb*. As a transitive verb, information is thus about connecting a subject to an object. The subject is the giver or sender of the information, and the object is the receiver or target of the information. This implies a directionality to the information, going from the sender to the receiver, and explains why a synonym for *information theory* is *communication theory*.

In an exchange of information, information is communicated from a sender to a receiver. Yet we need to be careful here about identifying sender and receiver too rigidly and seeing the directionality of information too much as a one-way street. As a practical matter, we are often concerned with one direction in the information exchange (such as in sending an email from a writer to a reader). But we need to be sensitive to information flowing also in the other direction. Often there can be information blowback, where the receiver of information sends information back to the original sender (whether deliberately or inadvertently), thus interchanging their roles.

That said, information exchange is not like Newton's third law, in which for every action there is an equal and opposite reaction. In many inquiries, such as tracking the flow of information on the internet, we may be concerned with information flow in only one direction. But we always need to be aware that the exchange can be bidirectional. The case of an observer with a quantum mechanical system is a case in point: in the act of measurement, the system is sending the observer information, but likewise the observer is perturbing, and thus inputting information, into the system to elicit the measurement.

Quantum measurements are inherently bidirectional. But the same bidirectionality can occur with information generally. Thus, Shakespeare, as sender, gives *Hamlet* to the public, as receiver; yet the public's enthusiastic reception of *Hamlet*, now acting as sender, is not lost on Shakespeare, as receiver (the public's enthusiasm, communicated to Shakespeare as information, likely influenced subsequent plays that he wrote, such as *King Lear* and *Macbeth*).

A.10 Shannon Information

The contingency-constraint view of information generalizes Shannon information, the theory of information that Claude Shannon developed in the late 1940s and that to this day largely defines the field. Shannon's approach to information illustrates the contingency-constraint view of information. Shannon's master-work on information was his 1949 book *The Mathematical Theory of Communication*. Warren Weaver, in his introduction to that book, succinctly described Shannon's concept of information:

> The word *information*, in this theory, is used in a special sense that must not be confused with its ordinary usage. In particular, *information* must not be confused with meaning [i.e., semantics]. In fact, two messages, one of which is heavily loaded with meaning and the other of which is pure nonsense, can be exactly equivalent, from the present viewpoint, as regards information. It is this, undoubtedly, that Shannon means when he says that "the semantic aspects of communication are irrelevant to the engineering aspects." But this does not mean that the engineering aspects are necessarily irrelevant to the semantic aspects.[16]

It's therefore clear that Shannon's theory is, out of the gate, not a semantic theory of information. Indeed, Shannon's theory is at heart a syntactic theory. As such, Shannon's theory allows for semantic aspects because syntax can be a carrier for semantics (as noted in Weaver's introduction).

To say that Shannon's theory is a syntactic theory is simply to say that it is concerned with the linear arrangement of symbols (bits, alphanumeric characters, strings of these, etc.). Shannon was, after all, a communications engineer, and in formulating information theory, he was attempting to solve an engineering problem. The details of his theory need not detain us, but the gist is captured in what Shannon called the "schematic diagram of a general communication system":[17]

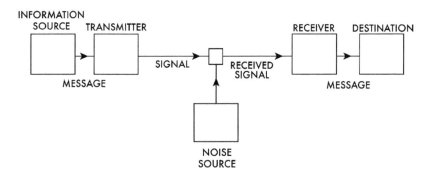

The engineering problem that Shannon faced was this: how to send a sequence of symbols drawn from a finite set of possible symbols across a communication channel so that the sequence received matched the sequence sent. Shannon worked for Bell Labs at the time, so he was thinking in particular about transmitting such symbols or character strings across phone lines (similar to sending them these days across the internet). But the phone lines, much as our internet lines (whether they be copper wires, fiber optics, or wireless), were all subject to noise that could corrupt the communications, altering the message received from the message sent.

Shannon's reference to "message" here can be confusing, a point underscored by Weaver in his introduction, because Shannon was, in the first instance, unconcerned with the meaning of a message but rather with its arrangement as a character string. We tend to think of messages as inherently meaningful, but Shannon, in referring to messages, had something else in mind. For Shannon, the important thing was to send character strings, whatever they might be, reliably across the communication channel so that sender and receiver would be looking at the same string ("message").

It didn't matter to his theory whether the strings were random or meaningful, gibberish or coherent, boorish or sophisticated. Shannon's theory decisively resolved the problem of getting character strings without alteration down communication channels despite the disrupting effects of noise. Given a strong enough signal and given effective mathematical techniques that he developed for overcoming noise (such as encoding schemes and error correction), Shannon showed that the character strings could be transmitted with extreme reliability (i.e., with arbitrarily high probability) from sender to receiver so that both could rest assured they were looking at the same string.

Focused as Shannon's theory is on character strings, it's clear that his theory is, in the first instance, a syntactic theory and that semantics is incidental to it. It's also clear that his theory exemplifies the contingency-constraint view of information, contingency signifying the different character strings that might be sent across the communication channel, constraint signifying the one string actually sent as well as those that with high probability would be received.

Even though information within Shannon's theory always starts in a syntactic space, it readily extends to a semantic space by focusing on the meaning of syntactical arrangements. Consider the symbol string "It's raining outside." Treated as a symbol string, it resides in a syntactic space with characters drawn from a word-processing keyboard and arranged linearly. But as a meaningful statement, it describes a fact about the weather, namely, that it's raining outside. As such, it resides in a semantic space, narrowing, and thus constraining, contingent weather patterns down to those that exhibit rain.

But consider instead the symbol string "It's raining outside or it's not raining outside." As a symbol string, it is longer than "It's raining outside" and thus has greater complexity and therefore, when considered syntactically, conveys more information than the shorter string (its representation requires more bits). But as a meaningful statement, it constitutes a tautology. As residing in a semantic space, it provides no constraint on contingent weather patterns, and therefore conveys no information. Thus, even though "It's raining outside or it's not raining outside" is more information rich when considered syntactically than "It's raining outside," the opposite is true when considered semantically.

Shannon's theory exemplifies the contingency-constraint view of information. Insofar as Shannon's theory applies directly to information sources in the actual world, it is largely confined to linguistic systems that convey meaning, engineering systems that send and receive alphanumeric characters, and living systems whose chemical constituents behave like alphanumeric characters (such as nucleotides and amino acids). Shannon's theory therefore applies directly to only limited portions of reality as we know it.

Yet when generalized to the contingency-constraint view of information, it applies much more widely. The contingency and constraint that are at the heart of information, and that occur outside human language, engineering, and biological contexts, can often be coded syntactically in ways that allow an application of Shannon's theory (as when a jpeg image of some physical object can be analyzed in terms of pixels). But the contingency-constraint view of information is not wedded to Shannon's theory and could be formulated without it. Still, Shannon's theory provides a particularly clear approach to information and, where applicable, turns the study of information into an exact science.

A.11 Connecting Information to Probability and Computation

What enables this contingency-constraint approach to information to become part of science? Two things: (1) quantification by means of probabilities, and (2) computability by representing information in bits.

Regarding point (1), the range of contingencies under consideration typically invites a probability distribution P, and the constraint, by identifying a subset of those contingencies, yields an event E with probability $P(E)$. So, for instance, there are 2,598,960 distinct poker hands. These constitute the relevant range of contingencies. One constraint would be hands with exactly one pair. These number 1,098,240. So, assuming all hands are equiprobable, the probability of getting exactly one pair is 1,098,240/2,598,960, or roughly 0.42. But consider instead a royal flush, of which there are only four. Assuming again equiprobability, the probability of getting a royal flush is 4/2,598,960, or very close to .00000154, a much smaller probability.[18]

Regarding point (2), in many situations, and all involving scientific measurement, information can be represented as character strings, which is to say, as linear sequences of characters drawn from a finite alphabet (or finite set of symbols). Because such characters can in turn always be represented by bits (think of the way UTF-8 or ASCII encodes the characters on a keyboard into bits), bit strings become the universal convention within science to represent information. Moreover, once this is done, it is straightforward to do computation on such bit strings, which can serve as stored memory as well as programs.

As soon as a probability distribution sits atop a range of contingencies, it makes sense to refer to the constraining of those contingencies as events. Thus, if P denotes a probability distribution over a range of contingencies, and if E denotes an event that coincides with a constraining of those contingencies, then $P(E)$ denotes the probability of that event and also constitutes a measure of the information inherent in that event (i.e., a measure of how contingencies were constrained). As an information measure, $P(E)$ therefore captures the information in E.

Notwithstanding, probabilities make for a counterintuitive measure of information, reversing that numerical order natural to information. We intuitively tend to think that information increases as possibilities get narrowed down, making the constraint on contingency more stringent. But this means that information increases as probabilities decrease. For instance, we learn more information about a poker hand if we learn that it is a royal flush as opposed to a hand other than a royal flush. There are a lot more ways for poker hands to be other than a royal flush than to be a royal flush. In consequence, we learn more information when we learn that a royal flush was dealt than when we learn that any hand other than a royal flush was dealt (i.e., the set-theoretic complement of royal flush).

Indeed, for a thoroughly shuffled card deck, the probability of being dealt a poker hand other than a royal flush is roughly

.999998, in contrast to the far smaller probability of being dealt a royal flush, which is roughly .000002. Smaller probability here signifies not less but more information. It is therefore customary to measure information by transforming probabilities logarithmically, and specifically by applying a negative logarithm to the base 2 to probabilities. This has the effect of reversing the numerical direction of probabilities so that less probability indicates more information. Moreover, by taking the logarithm to the base 2, probabilities can be reinterpreted as bits, thus drawing a clear connection with computation. In this way, a probability of getting five heads in a row, or $p = 2^{-5}$, yields an information measure of $-\log_2(p) = 5$ bits. For the probability $p = .00000154$ of getting a royal flush, the corresponding information measure is $-\log_2(p) \approx 19.3$ bits.

Information can therefore be defined as a logarithmic transformation of probability. Thus, for an arbitrary event E, we can define the information in E, or $I(E)$, as $-\log_2 P(E)$. It then follows that for events E and F, $P(E\&F) = P(E) \times P(F)$ if and only if $I(E\&F) = I(E) + I(F)$, so that events are probabilistically independent if and only if their probabilities multiply or their information measures add. In this vein, it also makes sense to define conditional information: if for events E and F, $P(F|E)$ denotes the conditional probability of F given E ($= P(F\&E)/P(E)$, as in Appendix A.2), we can define the conditional information of F given E as equal to $-\log_2 P(F|E)$ and denote it by $I(F|E)$. $I(F|E)$ is thus the information disclosed by the occurrence of F given that E has occurred or obtains. $I(E|H)$, for a chance hypothesis H, is crucial in the definition of specified complexity. (See Chapter 6.)

We close this section with two caveats. First, if you go to most mathematics of engineering books on information theory, the definition of information given there will look a bit different from what we've given here. Namely, instead of a straight logarithmic transformation of probability, as in $-\log_2(p)$, it will look like an average across different probabilities of this quantity, and thus something like $-\Sigma\, p \times \log_2(p)$, where the sum is across different values of p that partition the range of contingencies in question.[19] Information theorists refer to this term as *entropy*, and in communication contexts it tends to be more useful than the definition given here because communication theorists are concerned with the average performance of communication systems rather the precise information in any single communication. At the end of the day, however, the approach outlined here is more granular, coming to terms with information at its most basic (averaging always compresses and therefore diminishes information).

The other caveat is this: While information can be defined as a form or probability in which a logarithmic transformation takes a probability to a corresponding information measure (as in taking the probability p to $-\log_2(p)$), not all information measures are probabilistic. A branch of information theory known as algorithmic information theory (or AIT) applies a logarithmic transformation, but instead of being applied to a probability, it is applied to the length of the shortest computer program within a programming environment that generates a given bit string, thereby yielding a computational complexity measure.[20] Such an information measure still satisfies the contingency-constraint view of information (and often it has probabilistic aspects), but it does not constitute a straight-up logarithmic transformation of probability.

⌑⊐⊂⊐⊏ APPENDIX B

SELECT RELATED TOPICS

THE TOPICS ADDRESSED IN THIS APPENDIX, THOUGH INTERESTING and important in their own right, are not strictly speaking necessary for a basic understanding of the design inference. This appendix covers more peripheral or specialized topics related to specifications, probabilistic resources, and evolutionary biology. Parts of this appendix are more technically demanding than the rest of the book (especially B.5). Other parts are straightforward and accessible.

B.1 New Specifications from Old

From one or more existing specifications, it is often possible to generate new specifications. A specification is a pattern denoted by a linguistic entity that identifies an event. Linguistic entities, as well as any events they identify, can, however, be logically as well as syntactically manipulated. For instance, short descriptions (specifications) that identify events can be subjected to Boolean operators such as negation, conjunction, and disjunction. True, applying Boolean operators increases the description length of existing specifications. But if the existing specifications are short, then the Boolean operators will make them longer but still keep them reasonably short. Our use of "short" and "long" for description length remains, for now, intuitive. How to determine what descriptions may be regarded as sufficiently short to underwrite a rigorous chance-elimination argument is made clear in Chapter 6 on specified complexity, where we formalize the concept of description length.

Here's a simple illustration of how logical manipulations produce new specifications from old. If "all heads" is a specification, then so is "not all heads," albeit with a slightly longer description. Likewise, if "all heads" and "all tails" are specifications, then so are "all heads or all tails" and "all heads and all tails," albeit with description lengths greater than the sum of the description lengths of the specifications combined. "All heads and all tails" identifies a contradiction (no sequences of coin tosses can simultaneously be all heads and all tails) and will thus identify the null event, and thus have zero probability. "Not all tails," on the other hand, will identify an event of large probability, namely, $1 - 2^{-n}$ where n is the number of coin tosses under consideration. "All heads or all tails" identifies an event of probability $2 \times 2^{-n} = 2^{-(n-1)}$ (the probability of n heads plus the probability of n tails), which gets small very rapidly as n gets large. For design inferences, we want specifications of events that have small non-zero probability, as in "all heads or all tails."

The really interesting cases where new specifications are generated from old, however, arise when one probability space gets mapped, or transformed, into another probability space. Often such a transformation is from an underlying probability space Ω to itself, where Ω provides the backdrop for the events whose chance occurrence we wish to understand. Suppose, for instance, that Ω consists of sequences of ordinary letters, and consider the following sequence:

$$nfuijoltjujtmjlfbxfbtfm$$

We want to understand whether this sequence arose as a chance event in which the letters were generated randomly one after another. Given that specifications are critical to eliminating chance, we therefore try to determine whether this sequence of 23 letters is specified. A quick glance reveals no meaningful words. At the same time, the sequence seems a bit top-heavy in the letters "f" and "j." So perhaps it was not the result of sampling letters according to a uniform distribution. Still, from a cursory inspection, it seems random enough, even if some letters are more probable than others.

But in fact, the sequence is not random at all. It is the output of a Caesar cipher (named after Julius Caesar, who invented it). The Caesar cipher is perhaps the simplest cryptosystem that exists. In it, each letter in a cyphertext results from moving each corresponding letter in the plaintext a fixed number of steps. Think of the letters as arranged in a circle, with letters in the usual order adjacent to each other but with "z" next to "a." In the Caesar cipher, every letter is a fixed number of steps from its corresponding letter, whether one goes around clockwise or counter-clockwise. In the case at hand, each letter moves exactly one step up to encrypt, one step down to decrypt. Thus, to encrypt, "a" goes to "b," "b" to "c," etc., with "z" then going to "a." The preceding sequence is therefore a ciphertext, whose corresponding plaintext is:

methinksitislikeaweasel

If we add spaces for readability, this is just "methinks it is like a weasel," a sentence from Shakespeare's *Hamlet*. Clearly, this last sequence is specified, whether as a direct quote from English literature, or merely as a meaningful English sentence. But the crucial question for us is whether the original sequence—namely, nfuijoltjujtmjlfbxfbtfm—is also specified? In fact, it is. Given that Ω consists of sequences of ordinary letters, let f be the function or transformation from Ω to itself that encrypts plaintext messages by moving each letter one notch up the alphabet (circling back when the end of the alphabet is reached, thus moving "z" to "a"). Now let g be the inverse function on Ω that decrypts plaintext messages by moving each letter back a notch. Then

$$f(\text{methinksitislikeaweasel}) = \text{nfuijoltjujtmjlfbxfbtfm}$$

and

$$g(\text{nfuijoltjujtmjlfbxfbtfm}) = \text{methinksitislikeaweasel}$$

Given that methinksitislikeaweasel is specified, the specification of nfuijoltjujtmjlfbxfbtfm will consist simply of the specification of methinksitislikeaweasel along with a brief description of the function g that transforms the former into the latter. Schematically, but without full rigor, if the specification of methinksitislikeaweasel is σ, then the specification of nfuijoltjujtmjlfbxfbtfm is the combination $<\sigma,g>$. Note that because f denotes the encryption scheme for the Caesar cipher, it can be described very simply as "move every letter up one notch" (or in general for the Caesar cipher, "move every letter up m notches" where m is a number between 1 and 25). So, too, the decryption function g can be described very simply.

The sequences methinksitislikeaweasel and nfuijoltjujtmjlfbxfbtfm are short. Yet their shortness should not be taken as undercutting that they are specified. Given their shortness, it might be argued that any description of them is no shorter than the actual sequences, and so the sequences are not themselves specifications (we typically like to see a significant reduction in description length for sequences that are specified). The shortness of these sequences here is simply for ease of exposition. If these sequences consisted of thousands of characters, such as the whole of *Hamlet*, both σ and $<\sigma,g>$ would be vastly shorter than the sequences of letters they describe. In this case, σ could be the description "Shakespeare's play *Hamlet*," which is much shorter than the actual play.

The key takeaway from this example is that if a simple, easily described transformation yields something that is specified (as when a decryption function acts on a ciphertext to produce a meaningful plaintext), then folding the description of that transformation into the existing specification yields a new specification. As with Boolean operators, the description length of the new specification will be slightly longer than the original, but if the transformation can indeed be succinctly described, then the description length won't increase by much. Thus if σ is a specification and g is a succinctly described transformation, then so is $<\sigma,g>$, according to our schematic notation. In this case, σ specifies the plaintext whereas $<\sigma,g>$ specifies the ciphertext.

To summarize, if a simply described transformation outputs a specification, then its input is also a specification. Simply described transformations thus take us from specification to specification. This fact provides an illuminating perspective on cryptographic security. Suppose we have a cryptosystem in which f encrypts plaintext and g decrypts ciphertext. Suppose σ then is a specification of a plaintext message, which we'll simply call *text*. For definiteness, suppose that *text* is a meaningful message written in English. Then, $\langle\sigma,g\rangle$ is a specification of $f(text)$ provided that f and g are simply described. In the previous case, with the Caesar cipher, there are only 25 possible cryptosystems, and each can be described by simply associating the letter "a" with another letter in the Roman alphabet, after which the cryptosystem is entirely determined.

More complicated than a Caesar cipher is a substitution cipher, in which each letter is associated with another letter in the alphabet. By simply matching up letters with other letters, no more than 26! $(= 26 \times 25 \times \ldots \times 2 \times 1)$ substitution ciphers are possible. These ciphers are easily broken by analyzing letter frequencies of the ciphertext. Moreover, the encryption and decryption functions f and g do not depend for their description on the length of plaintext and/or ciphertext. Thus, for a specification σ of the plaintext, $\langle\sigma,g\rangle$ will be a specification of the corresponding ciphertext when g is the decryption function for a substitution cipher. But note, even though substitution ciphers are readily breakable, they are more secure than Caesar ciphers. That greater security consists in the added complexity of the decryption function g, which is more complex in the case of substitution ciphers than Caesar ciphers, and yet still simple enough to render $\langle\sigma,g\rangle$ a specification.

Not all cryptosystems have decryption functions that are easily described. The one-time pad is a case in point. The decryption functions of one-time pads have no short description. This fact ensures that the one-time pad is a secure cryptosystem. Indeed, it is the only cryptosystem that is provably secure. A one-time pad requires a random sequence of letters, call it *rand*, that is as long as *text*, the plaintext to be encrypted. Encryption occurs by then adding corresponding letters in *text* and *rand* according to modular (or circular) arithmetic: associate "a" with 1, "b" with 2, … "z" with 26. Now add the numbers associated with the letters and subtract 26 if you go over 26. Thus, a + b = 1 + 2 = 3 = c, etc. Decryption occurs similarly by subtracting *rand* modularly and then adding 26 to get back between 1 and 26: a − b = 1 − 2 = −1 = −1 + 26 = 25 = y, etc.

In a one-time pad, *rand* is used once and only once to encrypt *text*, after which it is retired (if it is reused, the cryptosystem is compromised and can be easily broken). Moreover, because *rand* is indeed random, it cannot be succinctly described. If we let +*rand* denote the encryption function *f*, and −*rand* denote the decryption function *g*, and σ denote the specification of the plaintext *text*, then unlike the two previous cryptographic examples, <σ,g> will be long and complicated because *g* = −*rand* is long and complicated. Indeed, if *rand* is truly random, it cannot be compressed. Thus, <σ,g> will not constitute a specification. *It is the failure of one-time pads to induce specifications that ensures their security.*

The case of public-key cryptography adds an interesting twist to cryptographic specifications. Unlike the Caesar, substitution, and one-time pad cryptosystems, where it is easy to reconstruct the decryption function from the encryption function and vice versa, with public-key cryptography it is not computationally feasible simply from the encryption function (without additional information) to construct the decryption function, or vice versa. Public-key cryptography exploits the computational difficulty of inverting certain mathematical functions. Encryption functions and decryption functions are inverse to each other in the sense that if *f* encrypts and *g* decrypts, then encrypting and decrypting in succession gets you back to where you started. In symbols, $f \circ g = g \circ f = id$, where *id* is the identity function, which for every input returns the value inputted ($id(x) = x$ for all *x*).

The advantage of public-key cryptography is that everyone can have access to the encryption function *f*, and therefore securely send encrypted messages, but only a select few have access to the decryption function *g* and can therefore read the encrypted messages. Suppose, then, you are confronted with a text that you can't make heads or tails of, call it *qtext*, for a "questionable" text that for all you can tell appears to be random. But perhaps it isn't random and instead is a ciphertext generated from the public-key (f,g)-based cryptosystem. Nonetheless, unless you have access to the decryption function g, you won't know whether *qtext* is indeed random or instead is an encrypted text, which, by means of f, encrypts some meaningful plaintext that we'll call *text*.

The problem is that the probability/search space Ω is so large that it won't be feasible to run through all its elements ω to determine which of those elements satisfy $f(\omega) = qtext$. Thus, even if $\omega = text$ satisfies $f(\omega) = f(text) = qtext$ where *text* is a meaningful English message and therefore is specified by the specification σ, we won't have any practical way of knowing σ unless we have the decryption function g to show that $g(qtext) = text$. Only then will we be able to identify the specification σ for *text* and the specification $<\sigma,g>$ for *qtext*.

If f is easily invertible, as in the previous examples, then g can be readily deduced from f, and the specification $<\sigma,g>$ of *qtext* will be easily formed. But for public-key cryptography, unless g is explicitly given, encrypted texts such as *qtext* will fail to register as specified (unless one gets extraordinarily lucky in guessing *text* and then showing that $f(text) = qtext$). They will be specified in the sense that short descriptions exist. But we won't know the short descriptions, and so won't know that they are specified.

With specifications, epistemology trumps ontology. It's not enough that a short description exists. We also need to know what it is. Thus with cryptography, the crucial question is not whether the decryption function g could or might be specified, but whether it can be known to be specified in the sense that a short description of it has been identified and made explicit.

Perhaps g has a very short description. In the case of public-key cryptography, it always does. In the RSA cryptosystem, for instance, the private key is known as soon as a several-hundred-digit number is factored into two primes. This several-hundred-digit number is never kept secret. Anyone who knows f knows it. But what makes it hard to find g by inverting f is that factoring several-hundred-digit numbers into primes is computationally difficult whereas determining whether a prime number is a factor of another number is computationally easy.

Once the prime factors in an RSA cryptosystem are known, so is g. The decryption function g can therefore be characterized typically by a few hundred digits (i.e., the prime factors). Such a characterization will be short compared to the long messages that the cryptosystem will be called on to encrypt (compare the one-time pad, where the description length of the decryption function is the same as the messages to be encrypted).

In general, to know that we are dealing with a specification, we need to explicitly identify a short characterization or description. Specifications, defined as they are in terms of description length, exist independently of their being known. But knowledge that something is a specification is often hard-won and in any case needs to be earned; it is not automatically conferred. That's true of cryptosystems in general and especially of public-key crypto-systems, where the decryption function is very hard to procure simply from the encryption function.

B.2 Transformations of Specifications

The lessons learned here from cryptography generalize to non-cryptographic contexts. Not all transformations that preserve specifications need be from a probability space Ω to itself. Often, the transformations will be between different spaces. Thus, we may have a transformation f from Ω to Ω' where Ω and Ω' are different probability spaces. Moreover, unlike cryptography, where we want encryption and decryption to be inverse functions, it may be that f is many-to-one, so that it has no well-defined inverse trans-formation, denoted by f^{-1} if it exists, from Ω' back to Ω. Still, f may be simply enough described to generate specifications on the spaces Ω and Ω'.

Consider, for instance, the genetic code. The genetic code is largely universal (we'll ignore exceptions for the sake of exposition). Thus, triplets of codons (three nucleotide base pairs, with four possibilities at each nucleotide, and thus sixty-four possibilities total) map to twenty possible amino acids. This coding, which we'll denote by the function f, associates sixty-four codons to twenty amino acids, and thus is many-to-one. For instance, six distinct codons map to the amino acid leucine, two map to tyrosine, and there are also a few codons that don't map to any amino acid and are known as stop codons. This mapping is well known, quite simple, and easily described (not much more complicated to describe than a substitution cipher).

Just by looking at a sequence of codons, it's not possible to tell whether it maps to a protein (i.e., a sequence of amino acids that performs a biological function).[1] But given a protein π, which is specified in virtue of its function (call its specification σ), it's typically possible to find the gene y in the cell's genome that maps to it so that $f(y) = \pi$. Then $<\sigma,f>$ specifies y. Note that this specification allows that there could be other genes, variants of y, that likewise map under f to π.[2] That's fine. Specifications describe an event but need not precisely identify an event, as in leaving no remainder. All the genes that map to π are in this way specified.

Such specification-preserving maps arise across the exact sciences. To take a stylized example from physics and cosmology, let Ω denote boundary conditions for the Wheeler-Dewitt field equation, and let Ω' denote solutions to it (these solutions are referred to as "universal wave functions"). Then we can let the mapping f from Ω to Ω' associate particular boundary conditions to particular solutions of the equation. Solutions that make for a life-permitting or otherwise life-conducive universe would then be specified, let us say by the specification σ, which applies to the space Ω'. A boundary condition β in Ω that maps to a σ-specified solution $f(\beta)$ of the Wheeler-Dewitt equation in Ω' will thus be specified by $<\sigma,f>$.

In this cosmological example, $<\sigma,f>$ will be a specification of β to the degree that f has a short description. Granted, a full mathematical description of f may be fairly involved, but it won't be long because the mathematical notation to characterize the Wheeler-Dewitt equation is concise. Moreover, given that the equation is common knowledge among cosmologists, simply referring to f as a mapping that associates boundary conditions with solutions to the Wheeler-Dewitt equation can be enough to specify it. This example provides a glimpse of conservation of information in cosmology, where the information in the solutions of cosmological equations gets smuggled in, but not adequately explained, at the boundary.[3]

Another way that mappings or transformations can generate new specifications from old is by forming two well-known set-theoretic variants of the original mappings or transformations. A mapping or transformation is just a function, call it f. Suppose then that f maps Ω and Ω'. It follows then that f also induces functions on the subsets of Ω and Ω' that are relevant to understanding how specifications relate between these two spaces. The events associated with Ω and Ω' can be represented by subsets of these spaces. All the subsets of Ω (resp. Ω') constitute what is known as the powerset of Ω (resp. Ω'), denoted by $pow(\Omega)$ (resp. $pow(\Omega')$).

Any function f from Ω to Ω' now induces two well-known set-theoretic variants on the powersets of Ω and Ω' respectively. These variants are well defined even if f is not invertible, as when it is many-to-one. The function f induces two functions on the powersets that (by abuse of notation) are denoted by f from $pow(\Omega)$ to $pow(\Omega')$, and by f^{-1} from $pow(\Omega')$ to $pow(\Omega)$. For each subset T of Ω, $f(T)$ is by definition $\{\, y \in \Omega' \mid$ there exists an x in T such that $f(x) = y \,\}$. Likewise, for each subset T' of Ω', $f^{-1}(T') = \{\, x \in \Omega \mid f(x) \in T' \,\}$.

If f is readily defined by a short description, then so is f as the induced function on the corresponding powersets. Thus, if T, as a subset of Ω, is specified by σ, then $f(T) = T'$ is specified by $<\sigma, f>$. On the other hand, if T' is specified by σ' and if the induced function f^{-1} on the corresponding powersets has a short description, then $T = f^{-1}(T')$ is specified via $<\sigma', f^{-1}>$. The logic here is a straightforward extension of the logic by which we generated new specifications from old in cryptography.

Of the two induced functions, f and f^{-1}, f^{-1} is, perhaps surprisingly, the better behaved. Mapping $pow(\Omega')$ to $pow(\Omega)$, f^{-1} is a homomorphism of Boolean algebras, preserving set-theoretic unions, intersections, and complements.[4] In consequence, f^{-1} pushes forward probabilities from Ω to Ω'. Thus, if P is a probability measure on Ω, then $P \circ f^{-1}$ is the natural induced probability measure on Ω'. (This fact underlies the change of variable formula in calculus.)

These facts about f^{-1} are relevant to specification-induced rejection regions (see Sections 3.6 and 3.7), ensuring that probabilistic properties of such regions are consistent under transformations of underlying probability spaces. In consequence, if f and f^{-1} both have a short description, if $T = f^{-1}(T')$ is specified, and if T has small probability under P in Ω, then T' likewise is specified and has small probability in Ω' (by the same arguments given earlier in this appendix). Moreover, it is invariably the case that the mathematical functions that characterize physical laws can be succinctly stated (otherwise, physicists and engineers could not begin to work with them). This confluence of specification and small probability under the induced mapping f^{-1} underlies conservation of specified complexity.

Conservation of specified complexity, also called conservation of complex specified information (or conservation of CSI for short), is real. Indeed, it is simply a fact about the way specifications and small probabilities behave and are preserved under transformations. Given a simply described function f (which is itself specified) that maps Ω and Ω', specifications on Ω push forward to specifications on Ω' under f, and specifications on Ω' pull back to specifications on Ω under f^{-1}. With proper bookkeeping that factors in the descriptive complexity of f and f^{-1}, both probability and descriptive complexity can be precisely preserved (compare Section 6.7).

It might therefore seem that conservation of specified complexity could be a silver bullet for refuting naturalistic theories of evolution, both biological and chemical, and thereby for supporting intelligent design. Given a case of specified complexity in an existing biological system, and given a simply described evolutionary process,[5] specified complexity in the present would imply specified complexity in the past and vice versa. The evolutionary process would thus be a conduit for specified complexity and therefore for design.

But such an argument cuts no ice with mainstream evolutionary biologists. On their accounting, the probabilities never get small enough and so no actual specified complexity exists in biological systems. What's needed, therefore, to refute Darwinian and other naturalistic forms of evolution is not conservation of specified complexity, but clear evidence that actual specified complexity exists in some actual biological systems.

Conservation of specified complexity is conditional. It says that *if* you've got specified complexity at one end of a simply described transformation, *then* you've also got it at the other end. But if no actual instance of specified complexity exists in biology, then conservation of specified complexity is an empty letter, underwriting no conclusions for evolutionary biology. The point therefore is to see whether there is at least one clear instance of specified complexity in biology that can stand on its own feet. This we argue for in Chapter 7.

B.3 Perturbation Neighborhoods

Philosopher John Leslie's famous fly-on-the-wall example illustrates *perturbation neighborhoods.*[6] Imagine a bullet hits an unlucky fly that's sitting on a wall. How can we rule out that the bullet happened by chance to hit the fly? Perhaps flies cover most of the wall. In that case, a random bullet might seem bound to hit one of the flies. But what if the local area surrounding the unlucky fly is empty (or extremely sparse) of other flies? What if hitting a fly by chance within that local area is highly improbable? In that case, it doesn't matter how large the wall is or how densely the flies are packed outside that local area. In that case, chance won't be a good explanation of the fly's demise.

Think of the unlucky fly as a tiny bull's-eye and the local area as a large target that surrounds it. Consequently, the global probability distribution of the flies on the wall as a whole is irrelevant to determining whether the unlucky fly was hit by chance. Instead, what's relevant to determining whether the bullet hit the fly by chance is the probability distribution of flies in the local area surrounding the unlucky fly as well as the number of bullets fired. For instance, with only one bullet fired and a large otherwise empty local area surrounding the unlucky fly, appealing to chance to explain the bullet hitting it becomes implausible. That local area constitutes a perturbation neighborhood.

Why refer to these local areas as perturbation neighborhoods? In mathematics, a perturbation is a small change where the smallness of the change is determined geometrically by a distance function or metric (small distance = small change). The idea with perturbation is to induce a small change in some input variable and see what degree of change is induced in an output variable. Sometimes the induced change is small, sometimes it is large. Often, when it is large, it is because of a phase transition. For instance, consider water temperature. A two-degree change from 72 to 74 degrees Fahrenheit doesn't cause any particularly noticeable change in the water. But a two-degree drop from 33 to 31 degrees Fahrenheit causes a considerable change in the water, turning it from liquid to solid. If a perturbation induces a small change, it's typical to keep perturbing until one observes a large change.

With perturbation neighborhoods the idea is to take a geometric point that answers to a prespecification and then perturb it, and keep perturbing the points generated by perturbing earlier perturbed points. Perturbation is a transitive process, where the results of prior perturbations can themselves be perturbed. In this way, a neighborhood gradually expands around the point of interest, usually uniformly in all directions or else uniformly in some preferred direction(s). Where does the perturbation process stop? Typically, one stops expanding the perturbation neighborhood when the probability of the prespecified portion of neighborhood become sufficiently small for the purposes at hand. Thus, in the case of a design inference, the perturbation neighborhood does not need to be made as large as possible, but just large enough to underwrite a small-probability chance-elimination argument.

With the fly-on-the-wall example, the underlying perturbation neighborhood now looks as follows. The flies are prespecified—it's assumed that we know in advance that there are flies on the wall. Next, we note that a fly on the wall gets hit by a bullet. That's the event. We now expand the neighborhood around the fly that was hit, perturbation by perturbation, calculating the probability of a bullet by chance hitting a fly (any fly) as the neighborhood expands. This approach allows that there could be more flies in the neighborhood than the one at the center. For purposes of illustration, we chose one unlucky fly at the center of a perturbation neighborhood and no others besides. But all that's really necessary is that the perturbation neighborhood expand around a salient point (in this case, the unlucky fly) and be extremely sparse of flies, thus making the hitting of a fly in the neighborhood extremely improbable.

This fly-on-the-wall example, because it focuses on circular areas surrounding individual flies, naturally suggests uniform probability distributions assigned to those areas. But in general, for perturbation neighborhoods, chance need not just mean a uniform probability distribution. With bullets hitting flies on walls, that's the most obvious probability to use. Nonetheless, perturbation neighborhoods allow for any probability distribution. For a successful chance-elimination argument with respect to a given probability distribution, it's enough that as the perturbation neighborhood expands, at least one of these neighborhoods assigns small probability to an event in the neighborhood that answers to the (pre)specification under consideration (in this case, a bullet hitting a fly by chance).

Let's now fill in the mathematical details for perturbation neighborhoods. A perturbation neighborhood starts with an underlying probability space Ω that has not only a probability measure P but also a metric d to characterize proximity relations and thus local areas. To say that d is a metric means that for points x and y in Ω, $d(x,y)$ denotes the distance between them. This distance is always a nonnegative real number that is equal to zero when x equals y and strictly greater than zero when x and y are distinct. The metric d reflects our ordinary understanding of distance, such as symmetry (the distance from x to y is the same as the distance from y to x) and the triangle inequality (the distance between two points is always no more than the sum of the distances of those two points from an intermediate point).[7]

With perturbation neighborhoods, there's always a prespeci-fication. In our running example, it's flies on the wall, which are assumed to be known about in advance of any event whose chance occurrence lies in question (here, a bullet randomly hitting a fly). Call this prespecification V and the event to which it corresponds V^* (in line with our asterisk notation). Note that V^* need not have small probability. In fact, $P(V^*)$ could be any probability short of 1. (If it is 1, perturbation neighborhoods cannot be made to have small probability.) Assume now that an event E is now observed and falls under V^* (i.e., $V^* \supset E$). How do we determine whether E happened by chance in line with the probability distribution P?

For a point c in Ω, form what topologists call a ball of radius r (r is a positive real number). This ball of radius r is typically denoted by $B_r(c) = \{\, x \in \Omega \mid d(c,x) < r \,\}$.[8] It corresponds to a local area of radius r surrounding the fly on the wall. Next choose c in Ω and r greater than zero so that this ball or radius r encompasses E, i.e., $B_r(c) \supset E$. Because E is contained in V^*, E is prespecified, and $P(E|B_r(c)) \leq P(V^*|B_r(c)) = P(V^* \cap B_r(c))/P(B_r(c))$. $P(V^*|B_r(c))$ is the conditional probability of V^* occurring given (or within) $B_r(c)$, which is the relevant probability of an event matching the specification V and simultaneously happening in the ball of radius r around c. Just how c was chosen is unimportant to the proba-bilistic argument we make here, though typically c is chosen as the precise outcome that was observed (in this case, the hitting of the unlucky fly). $P(V^*|B_r(c))$ may then be treated as a function of r.

Let, therefore, α denote a small probability cutoff, which is to say a probability value at or below which a probability is deemed small enough to underwrite a chance-elimination argument (recall how in Chapter 4 we defined probability bounds in terms of probabilistic resources). Given α, now let r increase to the smallest value of s so that $P(V^*|B_s(c)) \leq \alpha$. $B_s(c)$ is then a perturbation neighborhood that underwrites a small-probability chance-elimination argument. V here is an explicit prespecification, but in conjunction with the probability measure P, the metric d, and the probability cutoff α, it also prespecifies the event $V^* \cap B_s(c)$, which includes the observed event E.

The metric/probability space Ω can be discrete (finite or countably many points) or continuous (like the real line) or even a combination of the two (as when a discrete space and a continuous space are kludged). Often, in information-theoretic contexts, the space Ω consists of alpha-numeric strings from some abstract symbol convention (as with Shannon information) or polymers from interchangeable physical units (as with DNA, RNA, and proteins). Such informational spaces are discrete and the metric of choice here is typically the Hamming distance.[9]

The Hamming distance takes a string of alpha-numeric characters or string of repeatable physical constituents (as in a polymer) and calculates the number of positions in the string from which another string departs. For instance, the string of ordinary Roman alphabetic characters "boot" differs from "boat" by a Hamming distance of 1 whereas the string "boss" differs from "boat" by Hamming distance of 2. Perturbation neighborhoods based on the Hamming distance have been applied to polymers such as DNA and polypeptides to analyze the evolvability of biological systems (as in analyzing the ability of genetic information and protein folds to evolve).[10]

What makes perturbation neighborhoods work is that they make the very simplest use of geometry in setting up these neighborhoods. Imagine, instead, a fly on the wall that is hit by a bullet, but in this case the area around the fly is not a large disk otherwise empty of flies but a squid-shaped area with many tentacles reaching out over the wall, but all carefully avoiding other flies. For such a neighborhood, ascribing to something other than chance why the bullet hit the fly would not be compelling because the neighborhood could have been jury-rigged or p-hacked to include and exclude other flies even though flies may have been quite densely packed near the fly that was shot. With perturbation neighborhoods as we have defined them, based on balls of increasing radii, and where the geometry is used in the most natural way possible to determine these neighborhoods, this worry about jury-rigging or p-hacking is eliminated.

To sum up, by coordinating probability and geometry, perturbation neighborhoods expand the inferences we are entitled to draw about what is and is not within the reach of chance. With the fly-on-the-wall example, let's say the wall in question is the Wall of China, thousands of miles in length. Let's say it is carpeted with flies except for a circular area that's 20 feet in diameter. Outside the circular area, flies are so dense that a bullet cannot help but hit one of them. Inside the circular area, there's only one fly at the exact center. It is highly unlikely for that fly in the circular area to be hit by chance, and if it is hit, chance would be excluded. While this conclusion makes good intuitive sense, the formal apparatus of perturbation neighborhoods described here brings rigor to this exclusion of chance (and to the concomitant design inference).

B.4 The Uniformizability (or Normalizability) Problem

Perturbation neighborhoods, just now discussed in Appendix B.3, raise a technical problem that has been debated in intelligent-design circles and that we now address. The fly-on-the-wall example remains our go-to intuition pump for perturbation neighborhoods. As it is, all actual walls are finite and naturally lend themselves to uniform probabilities, which assign equal probability to geometrically equivalent pieces of the wall. For an actual wall Ω with a prespecification V and a neighborhood $B_s(c)$, the relevant probability for a chance-elimination argument is the uniform probability P for the entire wall. In that case, the probability $P(B_s(c))$ will be positive and the conditional probability for matching the prespecification V within $B_s(c)$ will be well-defined and given by $P(V^*|B_s(c))$. This conditional probability will then be a uniform probability on $B_s(c)$,[11] and it will be used to determine whether V^* is sufficiently improbable within $B_s(c)$ to warrant a chance-elimination argument.

But what if the wall is infinite in length? In that case, no uniform probability exists for the wall as a whole. A uniform probability P may then be defined for a large finite portion of the wall; and provided that $B_s(c)$ is included within such a portion, the probability $P(V^*|B_s(c))$ will be well-defined and constitute a uniform probability on the neighborhood $B_s(c)$ so long as the finite portion of the space on which P is defined includes $B_s(c)$. But P, in that case, will not be a uniform probability on the wall as a whole. Instead, it will only be a uniform probability for that portion of the wall where it assigns positive probability. The rest of the wall, indeed the infinite remaining portion of the wall, will then get assigned by P a probability of zero. That's the way it is with probability measures in general: on spaces infinite in extent, they concentrate most of the probability on finitely constrained subspaces.[12]

An "infinite wall" arises in cosmological fine-tuning arguments. Certain constants of nature required for the universe to be life permitting occupy a very narrow band in a configuration space of physical constants. The configuration space is infinite in extent. Yet any wide neighborhood around this narrow band is overwhelmingly sparse of constant values that lead to life-permitting universes. The probability of life-permitting constants within such a wide neighborhood is thus extremely small.

More formally, suppose we let V prespecify those constants that make for a life-permitting universe, let c denote an exemplary instance of such constants, and let $B_s(c)$ be the ball of radius s around c. In that case, $P(V^*|B_s(c))$ will be a very small probability provided that P is defined as a uniform probability on $B_s(c)$ or some larger finite neighborhood around $B_s(c)$. But what if the entire probability space Ω of physical constants is infinite in extent, with physical constants going from zero to positive infinity (as with the strength of gravity) or even from minus infinity to positive infinity (as with the curvature of the universe, which can be positive, zero, or negative)? In that case, conventional probability theory says that no uniform probability exists on Ω. Notwithstanding, as we'll see shortly, there's good reason to think that conventional probability theory breaks down here because of certain artificial limitations to it, and that a more general probability theory can set matters right.

Philosophers Timothy and Lydia McGrew, in collaboration with mathematician Eric Vestrup, have argued that because of this "normalizability problem" (the terminology is theirs and it denotes the inability to put uniform probabilities on spaces infinite in extent), small-probability chance-elimination arguments break down for cosmological fine tuning, are incoherent, and therefore must be rejected.[13] It matters not, according to their argument, that uniform probabilities apply and work locally. The absence of uniform probabilities for the space of physical constants as a whole allegedly undermines such probability-based cosmological fine-tuning arguments.

The term "normalizability," as used by the McGrews and Vestrup to denote the ability to put a uniform probability on a metric space, seems unfortunate in that the term applies in many different mathematical contexts (such as to a change of scale in which measured values are all made to reside within a "normalized range"). As it is, in the study of uniform probabilities, the term "uniformizable" already exists, predates the McGrew-Vestrup usage, and denotes the ability to put a uniform probability on a metric space.[14] So let's use it instead.

In my [WmD's] work on uniform probability from the late 1980s, I developed the concept of uniformizability for compact metric spaces. This focus on compact spaces was for convenience but not strictly necessary. In particular, Prohorov's theorem conveniently characterized convergence of probability measures on these spaces.[15] Yet there is a reasonably straightforward way to extend uniform probability measures from compact metric spaces (the sorts of spaces for which the McGrews and Vestrup agree that uniform probabilities exist) to noncompact ones that are a countable union of compact spaces (the spaces that arise in cosmological fine tuning and that the McGrews and Vestrup regard as non-uniformizable). Suppose, for simplicity, we focus on the real line (i.e., $\mathbb{R} = \Omega$). Let's take as our metric the ordinary absolute-value metric and suppose c is a particular physical constant that is life permitting (the multidimensional case adds no substantive technical challenges).

Now consider a sequence of intervals that increase to cover the entire real line (say $[-n,n]$ as n goes to infinity). Each of these intervals will have a uniform probability U_n that induces a probability on the entire real line. Specifically, each U_n will be concentrated on the interval $[-n,n]$, where it acts as a uniform probability, and it will assign 0 probability to the set-theoretic complement of these intervals, which is to say to $(-\infty,n) \cup (n,\infty)$. Think of U_n as defined on the entire real line $\mathbb{R}$, but as a uniform probability when restricted or conditionalized to $[-n,n]$.

From functional analysis, all these probability measures U_n reside in a unit ball of linear functionals that, according to the Banach-Alaoglu Theorem, is compact in the weak* ("weak star") topology.[16] This theorem applies to the case at hand because probability measures are always linear functionals residing in this unit ball. It follows that these probability measures U_n will have a limit point in the weak* topology, call it U. U likewise resides in the unit ball of linear functionals. Note that the U_n will not actually converge to U. Rather, U will be a non-uniquely determined limit point of the probability measures U_n. As such, it will be canonical only for those sets on which the value of U across all such weak limits is uniquely defined.[17]

One thing we do lose with U is countable additivity. As a probability measure, U will only be finitely additive. Countable additivity is mathematically convenient when you can get it. But given that we have no empirical access to infinity, finite additivity is all we can ever really confirm in applications of probability to the real world. Note that commonsense axiomatizations of probability, as given in Appendix A.1, are stated in terms of finite rather than countable additivity. Probability can be done without countable additivity. The mathematician Jimmie Savage, for instance, developed probability theory by only requiring finite additivity.[18] In consequence, the finite additivity of U should not count against its status as a uniform probability on the real line $\mathbb{R}$.

Even though U is not uniquely defined in terms of the U_n, it will be canonical on many sets that we are apt to consider. For all bounded sets, for instance, it will assign a probability of 0. In other words, for any real numbers $a < b$, $U((a,b)) = 0$. This makes sense in that U, as a limit point of the U_n, will tend to dissipate all probability to infinity. At the same time, to any half-space of the form $(-\infty,a)$ or (b,∞), U will assign a probability value of ½. Thus $U((-\infty,a)) = U((b,\infty)) = ½$. Likewise, for a set like $\ldots \cup (-4,-3) \cup (-2,-1) \cup (0,1) \cup (2,3) \cup (4,5) \cup \ldots$ (i.e., a set of alternating unit intervals), U will assign a value of ½. All such results hold for well-behaved subsets of $\mathbb{R}$ on which the limiting behavior of the U_n is clear and consistent. The entire sequence of U_ns, when applied to such sets, converges, eliminating the need to look to subsequences to get convergence.

Where U will behave anomalously is for sets A that are unions of intervals like $A_m = ((2m)!,(2m+1)!)$ for $m = 1, 2, 3$, etc. The exclamation mark (!) here denotes the factorial, so that $3! = 3×2×1$, $4! = 4×3×2×1$, etc. Factorials increase superexponentially. If we now let $A = A_1 \cup A_2 \cup A_3 \cup ...$, we find that limsup (limit superior) $U_n(A)$ as n goes to infinity is ½, but liminf (limit inferior) $U_n(A)$ as n goes to infinity is 0. That's because as the union of the A_ms is successively taken, what belongs to A swamps what doesn't belong to A, and then vice versa. The interval $((2m)!,(2m+1)!)$ swamps $(0,(2m)!)$ and the complementary interval $((2m+1)!,(2m+2)!)$, which is not included in A, then swamps $(0,(2m+1)!)$.

What this means is that even though $U(A)$ will be well defined, we won't be able to determine what it is. This may seem mysterious, but the Banach-Alaoglu theorem, which asserts the compactness of the unit ball for linear functionals in the weak* topology, depends on the Axiom of Choice, and this axiom often leads to counterintuitive results even though it plays an indispensable role in real and functional analysis.

We want to suggest, indeed urge, that the finitely additive uniform probability U resolves the uniformizability problem for the configuration spaces that arise in cosmological fine tuning. These spaces, even though not compact metric spaces, are an increasing countable union of compact metric subspaces, each of which has a uniform probability. All those uniform probabilities on these subspaces are consistent with each other in the sense that if one takes a uniform probability that's defined on a larger compact space and then conditions it on a smaller space contained within it, the conditional probability will equal the uniform probability on the smaller space.[19]

Consequently, the collection of uniform probabilities defined on ever-increasing compact subspaces will allow for the analysis of perturbation neighborhoods exactly as laid out in Appendix B.3.

In essence, we just take the fly on the wall, and if the wall is infinite, truncate the wall so that it's really big but finite, and then do the analysis on this finite "sub-wall."

Frankly, this move has always struck us as perfectly adequate for dealing probabilistically with fine-tuning arguments even in the absence of the finitely additive uniform probability U. That's because the uniform probabilities U_n that have U as a limit point, as they are dissipated out over ever-increasing swaths of the configuration space of physical constants, only make the probabilistic analysis of perturbation neighborhoods more compelling, lowering the probability. In the limit, the bands of physical constants compatible with a life-permitting universe come to have probability not just close to zero but exactly zero with respect to U.

Treating U as a finitely additive uniform probability on all of the real line $\mathbb{R}$ avoids probabilistic contradiction. But it does more than merely establish probabilistic consistency. It answers the uniformizability problem. U's assignment of zero probability to all finitely bounded sets makes good intuitive sense. Indeed, what else should a uniform probability on all of $\mathbb{R}$ assign to finitely bounded sets? A finite number, however, large, if divided by infinity is zero. Infinitesimals are not an option for conventional probability theory, which treats them as zero.[20] Because to any finitely bounded set, U assigns probability 0, for any ball of radius r centered at c, namely $B_s(c)$, $U(B_s(c)) = 0$. In consequence, it is not possible to condition U on $B_s(c)$ to recover the uniform probability on $B_s(c)$.

Although sets of probability zero, by representing events that apparently never happen, might seem trivial, in fact probabilists can find them interesting and worth analyzing. Often, this means finding related probability measures that assign non-zero probability to otherwise zero-probability events. With uniformizability, this means putting uniform probability measures on finitely bounded sets. These will be ordinary, countably additive uniform probability measures, whose limiting behavior is encapsulated in the non-countably additive uniform probability measure that assigns zero probability. This approach to uniformizability argues for the plausibility of using ordinary uniform probabilities to gauge the improbability of cosmological fine tuning.

B.5 Stopping and Waiting Times

Stopping times and waiting times are both widely used in probability theory. A stopping time denotes the first time that a stochastic process reaches an event that meets a given condition. A waiting time denotes the elapsed time between two successive events that meet a given condition. If we think of restarting a stochastic process right after an event that meets a certain condition has happened, then a waiting time can be a stopping time. As it is, the mathematical literature distinguishes between stopping and waiting times, whereas the evolutionary biological literature tends to conflate them.[21] In this appendix, we'll be mathematical purists and focus explicitly on stopping times, though bearing in mind that usage varies in other contexts.

Stopping times are correlated with probabilistic resources—longer stopping times imply greater probabilistic resources. As the name suggests, stopping times measure how long before something of interest happens, at which point the search for the thing of interest can be broken off, or stopped. The process by which something of interest happens, however, acts over a period of time, which typically can be broken into multiple joint opportunities to bring about the thing of interest. These multiple opportunities may thus be treated as probabilistic resources. Even so, stopping times add a complication to probabilistic resources that we now want to explore.

Stopping times measure the first time a stochastic process achieves a certain result. A stochastic process is an indexed collection of random variables.[22] Stopping times apply to stochastic processes that are indexed by nonnegative numbers. Stopping times can therefore be continuous, taking on nonnegative real values (i.e., from the interval $[0,\infty)$ to represent continuous time), or they can be discrete, taking on nonnegative integer values (i.e., from the set $\{0,1,2,...\}$ to represent discrete time). We'll limit ourselves here to discrete stopping times that arise from random variables X_1, X_2, X_3,... (See Appendix A.6.)[23] Note that this sequence can go on indefinitely, that it constitutes a discrete-time stochastic process, and that for convenience we omit X_0 (the stochastic process thus begins with X_1).

We assume that each random variable X_i in this stochastic process will take on values in a probability space Θ (capital Greek theta), which we think of as a search space. Given a rejection region R that is a subset of Θ and that is induced by a specification, our aim, then, is to determine how long it takes for the stochastic process to reach R. The stopping time for $X_1, X_2, X_3, \ldots$ to reach R will then be the first positive integer N such that $X_N \in R$ (implying that $X_i \notin R$ for $i < N$).

Note that N is itself a random variable, defined in terms of the random variables $X_1, X_2, X_3, \ldots$ as well as the rejection region R. As random variables, each of $X_1, X_2, X_3, \ldots$ will be defined on some other probability space, call it Ω. Thus, for any $\omega \in \Omega$ and any X_i in this sequence of random variables, $X_i(\omega)$ will take on a value in Θ. Technically, then, N will then map Ω to $\{1,2,3,\ldots\}$ (the indexing set for the random variables), which for any $\omega \in \Omega$ will then be defined as

$$N(\omega) =_{\text{def}} \min\{ \, n \text{ a positive integer} \mid X_n(\omega) \in R$$
$$\text{and } X_i(\omega) \notin R \text{ for } i < n \, \}.$$

In case this set over which a minimum is being taken is empty, meaning that no $X_i(\omega)$ for any i ever reaches R, then $N(\omega)$ is equated with ∞ (infinity).

Everything we've written about probabilistic resources carries over seamlessly to stochastic processes of random variables $X_1, X_2, X_3, \ldots$ that are independent and identically distributed. In that case, $P(X_i \in R)$ will equal some positive probability p, which will be the same for each i. Moreover, the stopping time probability $P(N = n)$, which is the probability that the first X_i to reach R is X_n, will then be $p \times (1-p)^{n-1}$. The stopping time probability in this case therefore follows a geometric distribution. The average stopping time,

$$\sum_{n=1}^{\infty} n \times P(N = n),$$

will then just be the mean of the geometric distribution, which is $1/p$.

All this suggests that if the maximal number of opportunities k for the X_i's to reach R is sharply less than $1/p$ ($k \ll 1/p$), then the probabilistic resources will be so low that we should think it highly unlikely that the stochastic process will reach R. More formally, we imagine that there's no practical way to run the stochastic process more than k times, so practically speaking we are dealing with finitely indexed stochastic process X_1, X_2, ..., X_k. In that case, the probability that even one among k of these X_i's will land in R is 1 minus the probability that none of them does, which is then just $1 - (1 - p)^k$, which for $k \ll 1/p$, as we are assuming, will be close to $k \times p$, which in turn will be close to zero. This is exactly what we've been saying throughout this chapter about probabilistic resources when they are insufficient to overturn a probability bound, except in the different idiom of stochastic processes and search.

Things are not always this simple, however. Stochastic processes of random variables X_1, X_2, X_3,... need not be independent or identically distributed. In such cases, the average stopping time can be misleading. Imagine random variables X_1, X_2, X_3,... where $P(X_1 \in R)$ is .999999 and $P(X_k \in R)$ is .000001 for some k, and where $P(X_2 \in R) = P(X_3 \in R) = \cdots = P(X_{k-1} \in R) = 0$.[24] Imagine further that k is much larger than 1,000,000—let's say 10^{100}. Finally, let's assume X_1 and X_k are mutually exclusive with respect to R, so that if one random variable is in R, the other is not, and vice versa. In that case, for the stopping time N induced by this stochastic process, $P(N = 1) = P(X_1 \in R)$ and $P(N = k) = P(X_k \in R)$, so the average stopping time for this process is

$$1 \times P(N = 1) + k \times P(N = k) = .999999 + 10^{94} \approx 10^{94}.$$

Thus, even though this stochastic process reaches the rejection region in just one step with large probability (i.e., .999999, implying that the probability of not reaching the rejection region is a mere 1 in a million), the process will nonetheless on average take impossibly long to stop, namely, 10^{94} steps on average. Averages like this are misleading. In this example, we would be wrong to think that because this average is so large, the stochastic process has a small probability of reaching the rejection region.[25] Instead, we should think the process will tend to reach the rejection region very quickly and with high probability.

Given a rejection region R within a space Θ, it is helpful to think of a stochastic process $X_1, X_2, X_3, \ldots$ on this space as a search for R. Searches in the computer science and engineering literature have four key components: (1) an initialization, which says where to start the search; (2) a feedback mechanism, which says for any query how well the search is doing (in evolutionary computing, this takes the form of a fitness function); (3) an update rule, which uses the feedback in the second component to determine where next to query the search space; and (4) a stop criterion, which says when to break off the search.

The stop criterion has two aspects. On the one hand, it can determine when the search has reached R, is therefore successful, and thus can be broken off. This criterion is therefore directly pertinent to stopping times. On the other hand, the stop criterion will also stipulate an upper bound k on the number of steps that the search can be continued, and beyond which the search runs out of steam because the resources for conducting any actual (as opposed to purely theoretical) search will always be limited. This upper bound will therefore always be less than the absolute probabilistic resources 10^{150}, the reason being that any query will require the activity of at least one proton acting at a rate not greater than the time it takes for light to cross the proton's diameter.

Given a stochastic process $X_1, X_2, X_3, \ldots$ active in a search space Θ with a rejection region R that's the object of the search, it now makes sense to think of probabilistic resources as taking two forms, one endogenous to the stochastic process, the other exogenous to it. Thus, given the stop criterion, we determine a number k such that $X_1, X_2, \ldots, X_k$ generously exceeds the longest we can reasonably expect to run the search to find R. This number k then counts the endogenous probabilistic resources associated with this stochastic process. But additionally, we can also determine a number m, which is chosen to generously exceed the most times we can expect to independently run the stochastic process. This number m then counts the exogenous probabilistic resources associated with this stochastic process.

So, endogenous probabilistic resources tell us how long we can run the stochastic process, and exogenous probabilistic resources tell us how many times we can run the process. As a simple example to illustrate these two types of probabilistic resources, imagine that you are on your laptop crunching some numbers to solve a problem. Finding the solution to the problem constitutes a successful search. But your computer is only so fast, and you only have so much time to solve the problem. The endogenous probabilistic resources k then capture this limitation on your machine. But yours is not the only laptop. How many other laptops are out there that might be able, independently, to try to solve the problem? The exogenous probabilistic resources m capture this limitation on the number of solution attempts across computer laptops.

We assume that endogenous probabilistic resources k stay fixed across different independent opportunities m (= exogenous probabilistic resources) to run the stochastic process in an effort to reach the rejection region R. In such circumstances, the total probabilistic resources are $k \times m$, a number that, given the constraints on search described previously, will never exceed the absolute probabilistic resources 10^{150}.

Once all these probabilistic resources are factored into the stochastic process, how probable is it that the process, when run up to k steps at a time and when each such run is re-run up to m times, will reach R, thereby successfully concluding the search? The most important probability we need a handle on under these circumstances is a probability we'll call q, defined as follows:

$$P(X_1 \in R \text{ or } X_2 \in R \text{ or ... or } X_k \in R) = q.$$

In general, this probability won't have a neat and clean form. But once we have it, we're in business. That's because we are assuming that the number of attempts to re-run this k-step stochastic process is m and that these attempts will be probabilistically independent. Under these conditions, the probability that at least one of these m attempts successfully reaches R will be $1 - (1 - q)^m$. If this last number is small, which is to say far less than $\frac{1}{2}$, and given that we've factored in all relevant probabilistic resources, then we should think that the search for R is misbegotten and destined to fail. On the other hand, to the degree that $1 - (1 - q)^m$ is large, we should think that the search for R is likely to be successful.

There are then two relevant stopping times for such a stochastic process. One is how quickly we can expect a single run of the stochastic process to successfully find R. For the stopping time N defined in terms of this stochastic process and rejection region, it then follows that

$$P(N \leq i) = P(X_1 \in R \text{ or } X_2 \in R \text{ or } ... \text{ or } X_i \in R)$$

for any i. This also means that $P(N \leq k) = q$.

The other stopping time will then be how long it takes for multiple runs (no more than m) of length k before the stochastic process finds the rejection region R. Because multiple runs are independent and identically distributed, this second stopping time will follow the geometric distribution, as described earlier in this appendix.

B.6 Probabilistic Hurdles to Irreducible Complexity in Biology

This and the next two appendices (B.6, B.7, and B.8) are drawn from Chapter 7 of The Design of Life *by William A. Dembski and Jonathan Wells (2008). The authors are grateful to Discovery Institute Press for permission to reprint this material here, which has been lightly edited for continuity with the rest of this book.*

In attempting to coordinate the successive evolutionary changes needed to bring about irreducibly complex biological systems and, in particular, molecular or biochemical machines, the Darwinian mechanism encounters a number of hurdles. These include the following:

(1) *Availability*. Are the parts needed to evolve an irreducibly complex biochemical system such as the bacterial flagellum even available?

(2) *Synchronization*. Are these parts available at the right time so that they can be incorporated when needed into the evolving structure?

(3) *Localization*. Even with parts that are available at the right time for inclusion in an evolving system, can the parts break free of the systems in which they are currently integrated and be made available at the "construction site" of the evolving system?

(4) *Interfering Cross-Reactions*. Given that the right parts can be brought together at the right time in the right place, how can the wrong parts that would otherwise gum up the works be excluded from the "construction site" of the evolving system?

(5) *Interface Compatibility*. Are the parts that are being recruited for inclusion in an evolving system mutually compatible in the sense of meshing or interfacing tightly so that, once suitably positioned, the parts work together to form a functioning system?

(6) *Order of Assembly*. Even with all and only the right parts reaching the right place at the right time, and even with full interface compatibility, will these parts be assembled in the right order to form a functioning system?

(7) *Configuration*. Even with all the right parts slated to be assembled in the right order, will they be arranged in the right way to form a functioning system?[26]

To see what's at stake in overcoming these hurdles, imagine you are a contractor who has been hired to build a house. If you are going to be successful at building the house, you will need to overcome each of these hurdles. First, you have to determine that all the items you need to build the house (e.g., bricks, wooden beams, electrical wires, glass panes, and pipes) exist and thus are *available* for your use. Second, you need to make sure that you can obtain all these items within a reasonable period of time. If, for instance, crucial items are back-ordered for years on end, then you won't be able to fulfill your contract by completing the house within the appointed time. Thus, the availability of these items needs to be properly *synchronized*. Third, you need to transport all the items to the construction site. In other words, all the items needed to build the house need to be brought to the *location* where the house will be built.

Fourth, you need to keep the construction site clear of items that would ruin the house or interfere with its construction. For instance, dumping radioactive waste or laying high-explosive mines on the construction site would effectively prevent a usable house from ever being built there. Less dramatically, imagine that a garbage-filled river periodically flooded, dumping excessive amounts of junk on the site (items irrelevant to the construction of the house, such as tin cans, broken toys, and discarded news-papers). In that case, it might become so difficult to sort through the clutter, and thus to find the items necessary to build the house, that the house itself might never get built. Then, too, there would be the river water itself, which might easily spoil many of the building materials. In this scenario, the water and river junk left behind at the construction site after each flood hinder the construction of a usable house and may thus be described as producing *interfering cross-reactions*.

Fifth, procuring the right sorts of materials required for houses in general is not enough. As a contractor, you also need to ensure that they are properly adapted to each other. Yes, you'll need nuts and bolts, pipes and fittings, electrical cables and conduits. But unless the nuts properly fit the bolts, unless the fittings are adapted to the pipes, and unless the electrical cables fit inside the conduits, you won't be able to construct a usable house. Yes, each part taken by itself can make for a perfectly good building material capable of working successfully in some house or other. But your concern here is not with some house or other but with the house you are actually building. Only if the parts at the construction site are adapted to each other such that they interface correctly will you be able to build a usable house. In short, as a contractor you need to ensure that the parts you bring to the construction site not only are of the type needed to build houses in general but also share *interface compatibility* so that they can work together effectively.

Sixth, even with all and only the right materials at the construction site, you need to make sure that you put the items together in the correct order. Anybody who has put together an item of furniture from its individual parts, reassembled a machine after taking it apart to fix it, or built a model airplane realizes that the parts cannot be put together in any old order. Instead, if the parts are put together in the wrong order, it's often necessary to reverse the steps in which they were put together, getting the parts back by themselves before they can then be put together in the right order. Thus, in building the house, you need first to lay the foundation. If you first erect the walls and then try to lay the foundation under the walls, your efforts to build the house will fail. The right materials require the right *order of assembly* to produce a usable house.

Seventh, and last, even if you are assembling the right building materials in the right order, the materials also need to be arranged appropriately. That's why, as a contractor, you hire masons, plumbers, and electricians. You hire these subcontractors not merely to assemble the right building materials in the right order but also to position them in the right way. For instance, it's all well and good to take bricks and assemble them in the order required to build a wall. But if the bricks are oriented at strange angles, or if the wall is built at a slant so that the slightest nudge will cause it to topple over, then no usable house will result even if the order of assembly is correct. In other words, it's not enough for the right items to be assembled in the right order; in addition, as they are being assembled, they need to be properly *configured*.

Now, as a building contractor, you find none of these hurdles insurmountable. That's because, as an intelligent agent, you see the big picture. You can look ahead to where you're going and what your final product will be. You can therefore coordinate all the tasks needed to overcome these hurdles. You have an architectural plan for the house. You know what materials are required to build the house. You know how to procure them. You know how to deliver them to the right location at the right time. You know how to secure the location from vandals, thieves, debris, weather, and anything else that would spoil your construction efforts. You know how to ensure that the building materials are properly adapted to each other so that they work together effectively once put together. You know the order of assembly for putting the building materials together. And, through the skilled laborers you hire (such as the subcontractors), you know how to arrange these materials in the right configuration. All this *know-how* results from intelligence and is the reason you can build a usable house.

But the Darwinian mechanism of random variation and natural selection has none of this know-how. All it knows is how to randomly modify biological structures and then to preserve those random modifications that happen to be useful at the moment (usefulness being measured in terms of survival and reproduction). The Darwinian mechanism is an instant gratification mechanism. If the Darwinian mechanism were a building contractor, it might put up a wall because of its immediate benefit in keeping out intruders from the construction site even though by building the wall now, no foundation could be laid later and, in consequence, no usable house could ever be built. That's how the Darwinian mechanism works, and that's why it is so limited. It is a trial-and-error tinkerer for which each act of mindless tinkering needs to maintain or enhance present advantage or select for a newly acquired advantage. It cannot make present sacrifices to achieve future as-yet unrealized benefits.

Imagine, therefore, what it would mean for the Darwinian mechanism to clear these seven hurdles in evolving a bacterial flagellum. (See Section 7.7.) We start with a bacterium that has no flagellum, no genes coding for proteins in the flagellum, and no genes homologous to genes coding for proteins in the flagellum. Such a bacterium is supposed to evolve, over time, into a bacterium with the full complement of genes needed to put together a fully functioning flagellum. Is the Darwinian mechanism adequate for coordinating all the biochemical events needed to clear these seven hurdles and thereby evolve the bacterial flagellum? To answer this question, let's run through these seven hurdles in turn, assessing the challenge each poses to the Darwinian evolution of the bacterial flagellum.

We start with availability. Can the Darwinian mechanism overcome the *availability hurdle*? To overcome this hurdle, the Darwinian mechanism needs to form novel proteins. The bacterial flagellum, if it evolved at all, evolved from a bacterium without any of the genes, exact or homologous, for the proteins constituting the flagellum. Now, the Darwinian mechanism may in some cases be able to modify existing proteins or recruit them wholesale for new uses. But, as a general-purpose mechanism for producing novel proteins, the evidence goes against the Darwinian mechanism. As we saw in Section 7.6, research into the folding characteristics of protein domains indicates that certain classes of proteins are highly unlikely to evolve by Darwinian processes. We have no compelling reason to think that the sort of proteins found in the bacterial flagellum are any different. Thus it appears unlikely the Darwinian mechanism could generate the novel proteins required in the evolution of the bacterial flagellum.

Consider next the *synchronization hurdle*. Darwinian evolution has a long time to work and may not be affected by short-term deadlines (though astrophysics imposes long-term deadlines, as with the Sun turning into a red giant in about five billion years, causing it to expand and burn up everything in its path, including Earth). So it may not be crucial when a specific protein or anatomical structure becomes available for evolution—unless it becomes available so prematurely that it decays before it can be put to use. But note that development is not so forgiving. As an organism develops from a fertilized egg to an adult, it needs specific building blocks at the correct times or it will die. Although evolution may be relatively insensitive to the synchronization hurdle, development is not.

The *localization hurdle*, compared to the synchronization hurdle, seems considerably more difficult for the Darwinian mechanism to clear. The problem here is that items originally assigned to certain systems need to be reassigned and recruited for use in a newly emerging system. This newly emerging system starts as an existing system that then gets modified with items previously incorporated into other systems. But how likely is it that these items break free and get positioned at the construction site of another system, thereby transforming it into a newly emerging system with a novel or enhanced function? Our best evidence suggests that this repositioning of items previously assigned to different systems is improbable and becomes increasingly improbable as more items need to be repositioned simultaneously at the same location. There are two reasons for this. First, the construction site for a given biochemical system tends to maintain its integrity, incorporating only proteins pertinent to the system and keeping out stray proteins that could be disruptive. Second, proteins don't break free of systems to which they are assigned as a matter of course; rather, a complex set of genetic changes is required, such as gene duplications, regulatory changes, and point mutations.

The *interfering cross-reaction hurdle* intensifies the challenge to the Darwinian mechanism posed by the previous hurdle. If the bacterial flagellum is indeed the result of Darwinian evolution, then evolutionary precursors to the flagellum must have existed along the way. These precursors would have been functional systems in their own right, and in their evolution to the flagellum would have needed to be modified by incorporating items previously assigned to other uses. These items would then need to have been positioned at the construction site of the given precursor. Now, as we just saw with the localization hurdle, there is no reason to think that this is likely. Indeed, foreign proteins floating around in places where they are not expected to be tend to get broken down and recycled (e.g., by protein scavengers known as proteasomes). But suppose the construction site becomes more open to novel proteins (thus lowering the localization hurdle and thereby raising the probability of overcoming it). In that case, by welcoming items that could help in the evolution of the bacterial flagellum, the construction site would also welcome items that could hinder its evolution. It follows that to the degree that the localization hurdle is easy to clear, to that degree the interfering cross-reaction hurdle is difficult to clear, and vice versa.[27]

The *interface-compatibility hurdle* raises yet another difficulty for the Darwinian mechanism. The problem is this: For the Darwinian mechanism to evolve a system, it must redeploy parts previously targeted for other systems. But that's not all. It also needs to ensure that those redeployed parts mesh or interface properly with the evolving system. If not, the evolving system will malfunction and thus no longer confer a selectable advantage. The products of Darwinian evolution are, after all, kludges. In other words, they are systems formed by sticking together items previously assigned to different uses. Now, if these items were built according to common standards or conventions, there might be reason to think that they could work together effectively. But natural selection, as an instant gratification mechanism, has no inherent capacity for standardizing the products of evolution. Yet without standardization, what evolution can manufacture and innovate becomes extremely limited.

Think of cars manufactured by different auto makers—say, a Chevrolet Impala and a Honda Accord. Although these cars will be quite similar and have subsystems and parts that perform identical functions in identical ways, the parts will be incompatible. You can't, for instance, swap a piston from one car for a piston in the other or, for that matter, swap bolts, nuts, and screws from the two vehicles. That's because these cars were designed independently according to different standards and conventions. Of course, at the Chevrolet plant that builds the Impala, there will be standardization ensuring that different parts of the Impala and other Chevrolet models have compatible interfaces. But across automobile manufacturers (e.g., Chevrolet and Honda), there will be no (or very little) standardization to which the construction of parts must adhere. In fact, common standards and conventions that facilitate the interface compatibility of distinct functional systems points not just to the design of the systems but also to a common design responsible for the standardization.

But the Darwinian mechanism is incapable of such common design. As an instant gratification mechanism, its only stake is in bringing about structures that constitute an immediate advantage to an evolving organism. It has no stake in ensuring that such structures also adhere to standards and conventions that will allow them to interface effectively with other structures down the line. Note that the universality of DNA and proteins is no answer to this standardization objection. DNA's very universality guarantees that different types of cells will lack any constraint on standardization. To make an analogy, a lathe can take a suitable piece of wood and rotate it, forming it into a leg of any length and diameter for a stool. But the lathe doesn't determine the length and diameter, and only the right length and diameter will work with a given stool. (Imagine one of the existing legs is broken and the lathe is being used to build a new leg.) The lathe is universal for woodworking in the same way that DNA is universal for protein building.

Let's now apply these considerations to the bacterial flagellum. Evolutionists sometimes argue that the bacterial flagellum evolved from a microsyringe known as the type III secretion system (T3SS for short—see Section 7.7 for the background on this evolutionary model). In this model, a pilus or hairlike structure gets redeployed and attached to the T3SS, eventually to become the whiplike tail that moves the bacterium through its watery environment. Yet before the pilus attached to the T3SS, these two systems had to evolve independently. Consequently, short of invoking sheer blind luck, there is no reason to think that these systems should work together— any more than there is to think that independently designed cars would have swappable parts.

This is a vulnerability of Darwinian theory, and it can be tested experimentally: take an arbitrary T3SS and pilus and determine the extent of the genetic modifications needed for the pilus to extrude through the T3SS's protein delivery system (which is how the pilus is supposed to interface with the T3SS). At present (and this holds true a decade and a half after the publication of *The Design of Life*), there are no sound theoretical or experimental reasons to think that the Darwinian mechanism can overcome the interface-compatibility hurdle.

Clearing the *order-of-assembly hurdle* is another implausible stretch for the Darwinian mechanism. The Darwinian mechanism is said to work by accretion and modification, adding novel parts to already functioning systems as well as modifying existing parts, and thereby forming new systems with enhanced or novel functions. But consider what happens when novel parts are first added to an already functioning system. In that case, the earlier system becomes a subsystem of a newly formed supersystem. Moreover, the order of assembly of the subsystem will, at least initially (before subsequent modifications), be the same as when the subsystem was a standalone system. There's a lot to juggle here, and the Darwinian mechanism gives no evidence of being up to the task.

In general, just because the parts of a subsystem can be put together in a given order doesn't mean those parts can be put together in the same order once embedded in a supersystem. In fact, in the evolution of systems such as the bacterial flagellum, we can expect the order of assembly of parts to undergo substantial permutations (certainly, this is the case with the model for the evolution of the bacterial flagellum from the T3SS). How, then, does the order of assembly undergo the right permutations? For most biological systems, the order of assembly is entrenched and does not permit substantial deviations. The burden of evidence is therefore on the Darwinist to show that for an evolving system, the Darwinian mechanism coordinates not only the emergence of the right parts but also their assembly in the right order. Darwinists have shown nothing of the sort.

Finally, we consider the *configuration hurdle*. In the design and construction of human artifacts, this hurdle is one of the more difficult to overcome. On the other hand, in the evolution of irreducibly complex biochemical systems such as the bacterial flagellum, this is one of the easier hurdles to overcome. Here's why. In the actual assembly of the flagellum and systems like it, the biochemical parts do not come together haphazardly. Rather, they self-assemble in the right configuration when random collisions allow specific, cooperative, local electrostatic interactions to lock the flagellum together, one piece at a time.

Thus, in the evolution of the bacterial flagellum, once the interface-compatibility and order-of-assembly hurdles are cleared, so is the configuration hurdle. There's a general principle here: for self-assembling structures, such as biological systems, configuration is a byproduct of other constraints (such as interface compatibility and order of assembly). But note, this is not to say that the configuration of these systems comes for free. Rather, it is to say that the cost of their configuration is included in other costs.

B.7 The Origination Inequality

The seven hurdles described just now in Appendix B.6 should not be understood as merely subjective or purely qualitative challenges to the Darwinian mechanism. It is possible to assess objectively and quantitatively the challenge these hurdles pose to the Darwinian mechanism. Associated with each hurdle is a probability:

p_{avail} The probability that the types of parts needed to evolve a given irreducibly complex biochemical system become available (the *availability probability*)

p_{synch} The probability that these parts become available at the right time so that they can be incorporated when needed into the evolving system (the *synchronization probability*)

p_{local} The probability that these parts, given their availability at the right time, can break free of the systems in which they are currently integrated and be localized at the appropriate site for assembly (the *localization probability*)

p_{i-c-r} The probability that other parts, which would produce interfering cross-reactions and thereby block the formation of the irreducibly complex system in question, get excluded from the site where the system will be assembled (the *interfering-cross-reaction probability*)

p_{i-f-c} The probability that the parts recruited for inclusion in an evolving system interface compatibly so that they can work together to form a functioning system (the *interface-compatibility probability*)

p_{o-o-a} The probability that even with the right parts reaching the right place at the right time, and even with full interface compatibility, they will be assembled in the right order to form a functioning system (the *order-of-assembly probability*)

p_{config} The probability that even with all the right parts being assembled in the right order, they will be arranged in the right way to form a functioning system (the *configuration probability*)

Note that each of these probabilities is conditional on the preceding ones. Thus, the synchronization probability assesses the probability of synchronization *on condition that* the needed parts are available. Thus, the order-of-assembly probability assesses the probability that assembly can be performed in the right order *on condition that* all the parts are available at the right time and at the right place without interfering cross-reactions and with full interface compatibility. Consequently, the probability of an irreducibly complex system arising by Darwinian means cannot exceed the following product (and note that because the probabilities are conditional on the preceding ones, this product slips in no unwarranted assumption of probabilistic independence):

$$p_{avail} \times p_{synch} \times p_{local} \times p_{i-c-r} \times p_{i-f-c} \times p_{o-o-a} \times p_{config}.$$

If we now define p_{origin} (the *origination probability*) as the probability of an irreducibly complex system originating by Darwinian means, then the following inequality holds:

$$p_{\text{origin}} \leq p_{\text{avail}} \times p_{\text{synch}} \times p_{\text{local}} \times p_{\text{i-c-r}} \times p_{\text{i-f-c}} \times p_{\text{o-o-a}} \times p_{\text{config}}.{}^{28}$$

This is the origination inequality.

Because probabilities are numbers between zero and one, this inequality tells us that if even one of the probabilities to the right of the inequality sign is small, then the origination probability must itself be small (indeed, no bigger than any of the probabilities on the right). It follows that we don't have to calculate all seven probabilities to the right of the inequality sign to ensure that p_{origin} is small. It also follows that none of these probabilities needs to be calculated exactly. It is enough to have reliable upper bounds on these probabilities. If any of these upper bounds is small, then so is the associated probability and so is the origination probability. And if the origination probability is small, then the irreducibly complex system in question is both highly improbable and specified (all these irreducibly complex systems are specified in virtue of their biological function). It follows that if the origination probability is small, then the system in question exhibits specified complexity; and since specified complexity is a reliable empirical marker of actual design, it follows that the system itself is designed.

The origination inequality demonstrates that intelligent design is a testable scientific theory. Consider, for instance, the interface-compatibility probability. Scientists can bring together existing biochemical systems (anything from individual proteins to complex biochemical machines) and determine experimentally the likelihood that, and the degree to which, their interfaces are compatible. So too, they can take apart existing biochemical systems, perturb them, and then try to put them back together again. To the degree that these systems tolerate perturbation, they are candidates for Darwinian evolution. Conversely, to the degree that these systems are sensitive to perturbation, they are inaccessible to Darwinian evolution. Experiments like this can be conducted on actual biochemical systems or using computer simulations that model biochemical systems.

For example, in their simulation of how gene duplications might facilitate protein evolution, Michael Behe and David Snoke have delineated a strategy for estimating interface-compatibility probabilities.[29] What they do is calculate the size of populations and number of generations for duplicated genes to evolve and produce novel proteins capable of interfacing (meshing) with an evolving irreducibly complex system. With very modest assumptions about the number of genetic changes that duplicated genes must undergo to produce novel interface-compatible proteins, they find that large populations (circa 10^{20}) and large numbers of generations (circa 10^8) are required before such a protein gets fixed in the population.[30]

Although Behe and Snoke did not convert waiting times to probabilities, they could have. A longer waiting time for fixation corresponds to a smaller probability of fixation in a given generation.[31] And since the waiting time for fixation that Behe and Snoke calculate is large, the probability of evolving a novel interface-compatible protein will be small. In light of the origination inequality, the work of Behe and Snoke therefore argues for the design of the bacterial flagellum.

Darwinists tacitly embrace the origination inequality whenever they invoke high probability events to support their theory. For instance, in arguing that antibiotic resistance in bacteria results from the Darwinian mechanism and not from intelligent design, Darwinists attempt to show that the probability of the genetic changes needed for antibiotic resistance is large (e.g., that the same bacteria exposed to the same antibiotic regime reliably experience the same resistance-conferring mutations).[32] But having embraced the origination inequality when it confirms their theory, Darwinists tend to shun it when it disconfirms their theory. Indeed, one will be hard-pressed to find a Darwinist who concedes that some low-probability transformation constitutes an obstacle to Darwinian processes. Obviously, there's a double-standard at play here.

Design theorists, by contrast, are more naturally inclined to approach the origination inequality with a fair and open mind. As design theorists, we are skeptical of Darwinian theory. Yet we readily concede that the Darwinian mechanism succeeds at certain modest evolutionary tasks, such as conferring antibiotic resistance (though see the endnote in the previous paragraph). Both the math, and the empirical evidence, confirm as much in certain cases. But we then apply the same empirical and probabilistic analysis and find that the Darwinian mechanism is incapable of substantially more demanding tasks, such as building novel molecular machines and systems.

With the probabilities that make up the origination inequality, there is no inherent obstacle to deriving reliable, experimentally confirmed estimates for them. Both Darwinists and design theorists have a stake in estimating these probabilities. The origination inequality does not stack the deck either for or against the intelligent design of irreducibly complex biochemical systems. So long as each of its probabilities is shown to be large or to lack reasonable estimates, such a system fails to exhibit specified complexity. In this case, researchers—Darwinists or otherwise—would be right to reject that such systems display marks of intelligent design.

But at the same time, should any of the probabilities in the origination inequality prove sufficiently small, then the system with those probabilities exhibits specified complexity. In that case, since specified complexity is a reliable empirical marker of intelligence, design theorists would be within their rights to insist that such systems do in fact display marks of intelligent design. Thus, we see that the origination inequality makes for a level playing field in deciding between Darwinism and intelligent design.

B.8 The Drake Equation

The origination inequality described in Appendix B.7 parallels the Drake equation, which comes up in the search for extraterrestrial intelligence (SETI). In 1960, astrophysicist Frank Drake organized the first SETI conference. At that conference, he introduced the now-famous Drake equation:

$$N = N^* \times f_p \times n_e \times f_l \times f_i \times f_c \times f_L.^{33}$$

Here is what the terms of this equation mean:

N The number of technologically advanced civilizations in the Milky Way Galaxy capable of communicating with Earth

N^* The number of stars in the Milky Way Galaxy

f_p The fraction of stars that have planetary systems

n_e The average number of planets per star capable of supporting life

f_l The fraction of life-supporting planets where life evolves and thus actually becomes present

f_i The fraction of planets with life where intelligent life evolves

f_c The fraction of planets with intelligent life where civilizations arise and develop advanced communications technology

f_L The fraction of a planetary lifetime during which communicating civilizations exist

The Drake equation gauges how likely SETI researchers are to find signs of intelligence from distant space: the bigger that N is, the more likely they are to succeed; the smaller that N is, the less likely they are to succeed.

Just as there are seven terms on the right of the origination inequality, so there are seven terms on the right of the Drake equation. Just as these terms in the origination inequality determine its applicability, so with the Drake equation. Moreover, in both the origination inequality and the Drake equation, the seven terms on the right occur in a particular order, reflecting the dependence of terms on previous terms. For instance, the fraction of planets on which intelligent life evolves depends on the fraction of planets on which life as such evolves.

Despite these and other similarities between the Drake equation and the origination inequality—not least that both are used for discovering signs of intelligence—there is a notable difference. For the Drake equation to convince us that the search for extraterrestrial intelligence is likely to succeed, *none of the terms* on the right side of that equation must get too small. Only then will SETI researchers stand a reasonable chance of discovering signs of extraterrestrial intelligence.

By contrast, with the origination inequality, to determine that an irreducibly complex biological system exhibits specified complexity, and therefore is the product of intelligent design, it is enough to show that *even one term* on the right side of this inequality is sufficiently small. That's because all the terms in the origination inequality are probabilities and therefore cannot exceed 1. Accordingly, a term can shrink the product, but never grow it. Thus, once a term has shrunk a probability in the origination inequality to something too small for chance, there's no digging out of that hole.

This difference greatly diminishes the applicability of the Drake equation compared to the origination inequality. The origination inequality requires only that a single term be securely estimated and found to be small. For the Drake equation, in order to securely infer a high probability of our discovering extraterrestrial life, all of the terms need to be known, rather than just one. Moreover, astrobiologists are far from meeting that challenge. Indeed, few of the terms of the equation can be securely estimated, making the Drake equation difficult, if not impossible, to apply. Michael Crichton emphasized this point sardonically in his widely publicized Caltech Michelin Lecture:

> The only way to work the [Drake] equation is to fill in with guesses. And guesses—just so we're clear—are merely expressions of prejudice. Nor can there be "informed guesses." If you need to state how many planets with life choose to communicate, there is simply no way to make an informed guess. It's simply prejudice. As a result, the Drake equation can have any value from "billions and billions" to zero. An expression that can mean anything means nothing... I take the hard view that science involves the creation of testable hypotheses. The Drake equation cannot be tested.[34]

Yet even if the terms of the Drake equation can be estimated (as opposed to merely guessed at), the equation itself tends more readily to point up the failure of the search for extraterrestrial intelligence than its success. For the Drake equation to undercut the likelihood of finding an extraterrestrial intelligence, it's enough that even one of the equation's terms be estimated and turn out to be small. On the other hand, for the Drake equation actually to confirm the likelihood of finding an extraterrestrial intelligence, *all* its terms have to be estimated, and all of them need to turn out to be large.

To recap, the origination inequality confirms an intelligence active in the formation of irreducibly complex biological structures provided that *even one* of its terms can be estimated and turns out to be small. That's because as soon as even one term on the right side of the origination inequality shows itself to be small, the product of terms on the right side (as a product of probabilities) must be at least as small. Consequently, the origination probability, which is bounded above by this product of probabilities, must also be at least that small. In this way, the origination inequality is better suited than the Drake equation for discovering signs of intelligence.

THE FIRST EDITION OF *THE DESIGN INFERENCE*

ALTHOUGH THIS SECOND EDITION OF *THE DESIGN INFERENCE* supersedes the first, various aspects of the first edition are worth revisiting here to help readers put this new edition into context. We start with the endorsements, which give some sense of the initial energy and enthusiasm that greeted the first edition (Appendices C.1 and C.2). The preface that follows gives as good a capsule summary of what this project is about as exists to date (Appendix C.3). The acknowledgments (Appendix C.4) remain heartfelt and, now twenty-five years later, underscore the transitoriness of life.

None of the material presented in this appendix is, strictly speaking, necessary to understand the second edition. Readers who are unfamiliar with the first edition but have a historical bent may find this appendix interesting. At the same time, readers of the first edition may find it useful to see how some things have changed from the first to the second edition. All the key ideas of the first edition remain, yet some unnecessary complications have fallen away, some more powerful technical machinery has been introduced, and some terminology has been revised. Such changes were briefly described in the introduction and are more thoroughly addressed in Appendix C.5.

C.1 Endorsements Appearing in the Hardback First Version

As the century and with it the millennium come to an end, questions long buried have disinterred themselves and come clattering back to intellectual life, dragging their winding sheets behind them. Just what, for example, is the origin of biological complexity and how is it to be explained? We have no more idea today than Darwin did in 1859, which is to say no idea whatsoever. William Dembski's book is not apt to be the last word on the inference to design, but it will surely be the first. It is a fine contribution to analysis, clear, sober, informed, mathematically sophisticated and modest. Those who agree with its point of view will read it with pleasure, and those who do not, will ignore it at their peril.

—David Berlinski, mathematician and philosopher,
author of *The Tour of the Calculus*

Dembski has written a sparklingly original book. Not since David Hume's *Dialogues Concerning Natural Religion* has someone taken such a close look at the design argument, but it is done now in a much broader post-Darwinian context. Now we proceed with modern characterizations of probability and complexity, and the results bear fundamentally on notions of randomness and on strategies for dealing with the explanation of radically improbable events. We almost forget that design arguments are implicit in criminal arguments "beyond a reasonable doubt," plagiarism, phylogenetic inference, cryptography, and a host of other modern contexts. Dembski's analysis of randomness is the most sophisticated to be found in the literature, and his discussions are an important contribution to the theory of explanation, and a timely discussion of a neglected and unanticipatedly important topic.

—William Wimsatt, philosopher of biology,
University of Chicago

In my view, Dembski has given us a brilliant study of the precise connections linking chance, probability, and design. A lucidly written work of striking insight and originality, *The Design Inference* provides significant progress concerning notoriously difficult questions. I expect this to be one of those rare books that genuinely transforms its subject.

—Jon P. Jarrett, philosopher of physics,
University of Illinois at Chicago

C.2 Subsequent Endorsements for the First Edition

Dembski's book is a serious and valuable attempt to evaluate the scientific status of the concept of design. The oft repeated observation that teleological questions lie beyond the remit of science has perhaps obscured an important distinction between the detection of whether there is purpose and the discovery of that purpose. Although the latter is beyond the remit of science, Dembski argues in connection with the former, that the fact of design may be scientifically detectable. His lucid and rigourous analysis of the design inference based on the concept of specified complexity deserves to be read by all who have an interest in this subject.

—John Lennox, philosopher of mathematics,
University of Oxford

The question of whether you can an infer a "designer" from the study of nature is anathema to many scientists. It seems to hearken back to a dark and superstitious era, when we invoked God as a convenient explanation for all we did not know. Without a systematic, quantitative means for evaluating the evidence, however, such assertions reflect more a philosophical prejudice than a statement of science. Through his work on establishing the complexity-specification criterion, Dr. Dembski is making a critical step in moving the discussion away from such subjectivity, and into the realm of rational analysis in the best tradition of scientific inquiry.

—Robert Kaita, Principal Research Physicist, Plasma
Physics Laboratory, Princeton University

Even the most vociferous critics of the venerable teleological argument for God's existence (the so-called argument from design), such as Hume, Kant, and Darwin, have testified to the intuitive force of that argument. But they have insisted that the impression of intelligent design in nature, however irresistible, must be ignored or suppressed because that impression cannot be given a rigorous formulation. Now a brilliant newcomer, trained in philosophy, mathematics, and science, claims to have met successfully the challenge of the critics. Drawing upon design-detection techniques in such diverse fields as forensic science, artificial intelligence, cryptography, SETI, intellectual property law, random number generation, and so forth, William Dembski argues that specified complexity is a sufficient condition of inferring design and that, moreover, such specified complexity is evident in nature. Never since the time of Paley has the teleological argument seemed so compelling. Both the proponents of natural theology as well as her critics will be forced to deal with the argument of this groundbreaking and important book.

—William Lane Craig, visiting scholar in philosophy,
Biola University

The publication of William A. Dembski's *The Design Inference* by Cambridge University Press is one of the most significant events in recent intellectual history. It signals a breakthrough: the acknowledgement on the part of the academic and scientific establishment, for the first time in over sixty years, of the fact that the concept of intelligent design may have a legitimate role to play in our scientific understanding of nature. Dembski's pioneering work on the philosophy and theory of information is something no serious student of the foundations and methodology of science can afford to ignore. This book will substantially alter the shape of contemporary discussions about the nature and presuppositions of science.

—Robert Koons, professor of philosophy,
University of Texas at Austin

When we wrote *The Mystery of Life's Origin: Reassessing Current Theories*, we pointed out the immense improbabilities of an accidental origin of life. The inference that this somehow implies an intelligent designer was reasonable, but too subjective. In *The Design Inference*, Dr. William Dembski has provided a clear conceptual basis for evaluating the significance of such improbable events as the origin of life. The importance of this extremely insightful contribution cannot be overstated. I think this new paradigm will become the norm by which improbable events can be evaluated in many different areas of scientific investigation where assessing the possibility of intelligent intervention is necessary.

—Walter Bradley, distinguished professor of engineering (emeritus), Baylor University, co-author of *The Mystery of Life's Origin*

Dembski has written the most important book on the topic of design to appear in living memory. One simply cannot understand current issues in science and religion without reading it. It is seminal. It also has the virtue of being what only the great books are: readable and clear. Anyone interested in science cannot afford to miss this book.

—John Mark Reynolds, president, The Saint Constantine School (formerly at Biola and head of the Torrey Honors Institute)

Professor Bill Dembski's thesis in *The Design Inference* is an important breakthrough in the philosophy of science. It deserves, and will readily command, a broader readership. I am recommending his book both to my professional colleagues and to friends and students who are interested in how probable reason operates in ordinary inferences and in our assumptions about the nature of science. Dembski's writing is precise, forceful, and appeals both to scholars and to reflective lay readers.

—John Angus Campbell, professor of communications (emeritus), University of Memphis

William Dembski's book *The Design Inference* shows, contrary to popular opinion, that design inferences are NOT unscientific, but are actually employed routinely in several scientific disciplines. Better yet, Dembski provides clear objective criteria for inferring design, and thus dispels the myth that design is inherently arbitrary and subjective. This groundbreaking book will go a long way toward liberating biology from the straitjacket of Darwinism and promoting future progress.

—Jonathan Wells, senior fellow, Discovery Institute

In *The Design Inference* William Dembski asks, how do we know that something has been purposely arranged? How do we know there is a mind behind a particular event? In answer Dembski constructs the formal logical arguments necessary for rigorous evaluation of design hypotheses. This book should be on the shelf of every person who has intuitively grasped that life and the universe are not the products of chance, but has thus far lacked the conceptual tools to rigorously defend that idea.

—Michael Behe, professor of biochemistry, Lehigh
University, author of *Darwin's Black Box*

Detecting design and distinguishing it from natural causation is one of the things science frequently does, but understanding of this subject has been impeded by ideological prejudice. William Dembski's superb analysis brings design theory into mainstream science. I predict that it will have an enormous influence over the science and philosophy of the 21st century. Dembski is one of the most important of the "design" theorists who are sparking a scientific revolution by legitimating the concept of intelligent design in science. At some point not far in the future, scientists will be saying "of course biological organisms are intelligently designed," and "of course neo-Darwinism was never more than a pseudoscientific philosophical ideology like Freudianism and Marxism." When that happens, William Dembski will deserve a lot of the credit.

—Phillip E. Johnson, late professor of law, University of
California at Berkeley, author of *Darwin on Trial*

From ancient times, every civilization has had the intuition that the world is designed, but now Dembski has offered rigorous criteria for distinguishing intelligent design from natural causes. This is a real intellectual breakthrough, and makes the detection of design a proper part of science itself.

—Chuck Colson, late chairman, Prison Fellowship Ministries

C.3 Preface to the First Edition

Highly improbable events don't happen by chance. Just about everything that happens is highly improbable. Both claims are correct as far as they go. The aim of this monograph is to show just how far they go. In *Personal Knowledge*, Michael Polanyi (1962, p. 33) considers stones placed in a garden. In one instance the stones spell "Welcome to Wales by British Railways," in the other they appear randomly strewn. In both instances the precise arrangement of stones is vastly improbable. Indeed, any given arrangement of stones is but one of an almost infinite number of possible arrangements. Nevertheless, arrangements of stones that spell coherent English sentences form but a minuscule proportion of the total possible arrangements of stones. The improbability of such arrangements is not properly referred to chance.

What is the difference between a randomly strewn arrangement and one that spells a coherent English sentence? Improbability by itself isn't decisive. In addition what's needed is conformity to a pattern. When the stones spell a coherent English sentence, they conform to a pattern. When they are randomly strewn, no pattern is evident. But herein lies a difficulty. Everything conforms to some pattern or other—even a random arrangement of stones. The crucial question, therefore, is whether an arrangement of stones conforms to the right sort of pattern to eliminate chance.

This monograph presents a full account of those patterns capable of successfully eliminating chance. Present statistical theory offers only a partial account of such patterns. To eliminate chance the statistician sets up a *rejection region* prior to an experiment. If the outcome of the experiment falls within the rejection region, chance is eliminated. Rejection regions are patterns given prior to an event. Although such patterns successfully eliminate chance, they are by no means the only ones. Detectives, for instance, routinely uncover patterns after the fact—patterns identified only after a crime has been committed, and which effectively preclude attributing the crime to chance. (See Dornstein [1996] and Evans [1996].)

Although improbability is not a sufficient condition for eliminating chance, it is a necessary condition. Four heads in a row with a fair coin is sufficiently probable as not to raise an eyebrow; four hundred heads in a row is a different story. But where is the cutoff? How small a probability is small enough to eliminate chance? The answer depends on the relevant number of opportunities for patterns and events to coincide—or what I call the relevant *probabilistic resources*. A toy universe with only 10 elementary particles has far fewer probabilistic resources than our own universe with 10^{80}. What is highly improbable and not properly attributed to chance within the toy universe may be quite probable and reasonably attributed to chance within our own universe.

Eliminating chance is closely connected with design and intelligent agency. To eliminate chance because a sufficiently improbable event conforms to the right sort of pattern is frequently the first step in identifying an intelligent agent. It makes sense, therefore, to define design as "patterned improbability," and the design inference as the logic by which "patterned improbability" is detected and demonstrated. So defined, the design inference stops short of delivering a causal story for how an intelligent agent acted. But by precluding chance and implicating intelligent agency, the design inference does the next best thing.

Who will want to read this monograph? Certainly anyone interested in the logic of probabilistic inferences. This includes logicians, epistemologists, philosophers of science, probabilists, statisticians, and computational complexity theorists. Nevertheless, a much broader audience has a vital stake in the results of this monograph. Indeed, anyone who employs small-probability chance-elimination arguments for a living will want to know the results of this monograph. The broader audience of this work therefore includes forensic scientists, SETI researchers, insurance fraud investigators, debunkers of psychic phenomena, origin-of-life researchers, intellectual property attorneys, investigators of data falsification, cryptographers, parapsychology researchers, and programmers of (pseudo-) random number generators.

Although this is a research monograph, my aim throughout has been to write an interesting book, and one that as little as possible duplicates existing texts in probability and statistics. Even the most technical portions of this monograph will be interlaced with examples accessible to non-technical readers. I therefore encourage non-technical readers to read this monograph from start to finish, skipping the technical portions. Readers with a background in probability theory, on the other hand, I encourage to read this monograph thoroughly from start to finish. Small-probability arguments are widely abused and misunderstood. In the analysis of small-probability arguments the devil is in the details. Because this monograph constitutes a sustained argument, a sustained reading would be optimal.

[Some specific reading suggestions for approaching various chapters and sections were presented next, along with one-sentence descriptions for each chapter. Because this second edition has been thoroughly rewritten, reorganized, and expanded, those suggestions are no longer helpful and have been removed.]

C.4 Acknowledgments for the First Edition

The beginnings of this monograph can be traced back to an interdisciplinary conference on randomness in the spring of 1988 at Ohio State University, at a time when I was just finishing my doctoral work in mathematics at the University of Chicago. Persi Diaconis and Harvey Friedman were the conveners of this conference. I shall always be grateful to Persi for bringing this conference to my attention and for urging me to attend it. Indeed, most of my subsequent thinking on the topics of randomness, probability, complexity, and design has germinated from the seeds planted at this conference.

The main mathematical tools employed in this monograph are probability theory and a generalization of computational complexity theory. Probability theory was the subject of my doctoral dissertation in mathematics at the University of Chicago (1988). This dissertation was completed under the direction of Patrick Billingsley and Leo Kadanoff. As for computational complexity theory, I was introduced to it during the academic year 1987–88, a year devoted to cryptography at the computer science department of the University of Chicago. Jeff Shallit, Adi Shamir, and Claus Schnorr were present that year and helped me gain my footing. Subsequently, I had the good fortune of receiving a post-doctoral fellowship in mathematics from the National Science Foundation (DMS-8807259). This grant allowed me to pursue complexity theory further at the computer science department of Princeton University (Fall 1990). My sponsor, Andrew Yao, made my stay at Princeton not only possible but also congenial.

The monograph itself is a revised version of my philosophy dissertation from the University of Illinois at Chicago (1996). This dissertation was completed under the direction of Charles Chastain and Dorothy Grover. I want to commend Charles and Dorothy as well as Michael Friedman and Walter Edelberg for the effort they invested in my philosophical development. This monograph would be much poorer without their insights and criticisms. I also wish to thank individual members of the philosophy departments at the University of Chicago and Northwestern University. On the south side, David Malament and Bill Wimsatt stand out. Their willingness to read my work, encourage me, and help me in such practical matters as getting a job went well beyond the call of duty. On the north side, I have profited enormously from seminars and discussions with Arthur Fine, David Hull, and Tom Ryckman. Tom Ryckman deserves special mention here, not only for his generous spirit, but also for his example of philosophical integrity.

Many people have shaped and clarified my thinking about design inferences over the past seven years. In this respect I want to single out my co-conspirators in design, Stephen Meyer and Paul

Nelson. As the co-authors of a planned volume for which the present monograph will serve as the theoretical underpinnings, they have exercised an unmatched influence in shaping my thoughts about design inferences. Their continual prodding and testing of my ideas have proved a constant source of refreshment to my own research. I cannot imagine the present monograph without them. This monograph, as well as the joint work with Meyer and Nelson, has been supported by a grant from the Pascal Centre at Redeemer College, Ancaster, Ontario, Canada and by a grant from the Center for the Renewal of Science and Culture at the Discovery Institute, Seattle.

Others who have contributed significantly to this monograph include Diogenes Allen, Douglas Axe, Stephen Barr, Michael Behe, Walter Bradley, Jon Buell, John Angus Campbell, Lynda Cockroft, Robin Collins, Pat Detwiler, Frank Döring, Herman Eckelmann, Fieldstead & Co., Hugh Gauch, Bruce Gordon, Laurens Gunnarsen, Charlie Huenemann, Stanley Jaki, Jon Jarrett, Richard Jeffrey, Phillip Johnson, Bob Kaita, Dean Kenyon, Saul Kripke, Robert Mann, John Warwick Montgomery, J. P. Moreland, Robert Newman, James Parker, III, Alvin Plantinga, Philip Quinn, Walter J. ReMine, Hugh Ross, Siegfried Scherer, Brian Skyrms, Paul Teller, Charlie Thaxton, Jitse van der Meer, J. Wentzel van Huyssteen, Howard van Till, Jonathan Wells, C. Davis Weyerhaeuser, John Wiester, A. E. Wilder-Smith, Leland Wilkinson, Mark Wilson, Kurt Wise, and Tom Woodward.

Finally, I wish to commend three members of my family: my parents, William J. and Ursula Dembski, to whom this monograph is dedicated, and without whose patience, kindness, and support this work would never have gotten off the ground; and my wife, Jana, who entered my life only after this work was largely complete, but whose presence through the final revisions continually renews my spirit.

C.5 Transitioning from the First to the Second Edition

Readers of the first edition of *The Design Inference* will be struck by how little of its actual text was retained in the second edition. Second editions of books often lightly edit existing text and then perhaps add a few extra sections or an occasional new chapter. By contrast, this second edition incorporates very little actual text from the first edition, rewrites and updates only select portions of that earlier content, and then adds a considerable amount of new material, especially on the nuts and bolts of specified complexity and on its application to evolutionary biology. The first edition was just under 90,000 words in length. The second exceeds 180,000 words in length.

There are good reasons for such a big difference—both in sheer length and in depth of exposition—between the first and second edition. Twenty-five years is a long time, and much has happened in the interim. Ideas become clearer over time. Notations become less clumsy. Unnecessary complications fall away. New, more compelling applications are found. The fact is that the design inferential apparatus developed in the first edition now allows for a much cleaner treatment and broader use. The first edition was a revised version of a doctoral dissertation in the philosophy of science and the foundations of probability. Philosophers tend to be sticklers. In consequence, I [WmD] ended up dotting a lot of i's and crossing a lot of t's that weren't necessary, even for a technical reader.

Expositions and examples in the first edition tended to be longer than needed. For instance, to illustrate probabilistic resources, a key concept in the book, I [WmD] imagined a "probabilistic reform" of the criminal justice system. Accordingly, prisoners would be given a fair coin and then had to toss so-and-so many heads in a row to, on average, get out of prison after a certain amount of time. I calculated, for instance, that with 8 hours a day, 6 days a week, at a rate of one toss every 10 seconds, 23 heads in a row would, on average, get a prisoner out of prison in 10 years. I also noted that, because of chance fluctuations, this number could vary quite a bit for individual prisoners, some getting out of prison much sooner than 10 years, others much later. Laying this all out and doing the elementary probability calculations took three pages in the first edition (pp. 175–177). Here you have it in half a paragraph. The second edition offers better, more concise expositions and examples than the first edition.

Some of the technical work in the first edition could be dropped with no loss to the book's main argument. For instance, Chapter 4 of the first edition developed a formal theory of complexity. It arose out of some work I had done on conditional logic. This formalism characterized probability measures and computational complexity measures. Chapter 4 provided a pleasing unifying framework for the measures needed to draw design inferences. It also helped to get the first edition published in that it led the acclaimed Notre Dame philosopher Philip Quinn to write a glowing referee report for Cambridge University Press as the book was being vetted for publication. (I was a postdoctoral fellow in the philosophy of religion at the University of Notre Dame in the 1990s, and during my time there, Phil confided that he had been one of the anonymous reviewers of the book and that he very much liked the chapter on complexity.) But this chapter, while tangentially relevant, is not needed to explicate the design inference or the subsequent ideas built on top of it. And so, that chapter has been dropped in the second edition.

Although this second edition condenses and streamlines all the essential aspects of the first edition, it is now twice as long as the first edition. How did this happen? This new edition contains many more examples to illustrate the different facets of the design inference. It also requires additional space to motivate and justify the design inferential logic. It is more rigorous than the old edition, and rigor comes at a cost of additional exposition. The design inference, as presented in the first edition, elicited many objections. The important ones needed to be addressed in this second edition. That took up space. And then there were the outright additions: the detailed treatment of specified complexity (Chapter 6), applications of the design inference to evolutionary biology (Chapter 7), and special topics related to the design inference (Appendix B). So, while the second edition reads more briskly than the first, there's a lot more to read.

Between the first and second edition, there were two main conceptual changes, one having to do with specification, the other with probabilistic resources. In both cases, these concepts became much clearer and easier to work with in the second edition. Specification in the first edition was defined in terms of *detachability*. The idea with detachability was that an inquirer seeking to draw a design inference had to have background knowledge independent of an observed event and yet sufficient to reconstruct a pattern describing the event. The motivating idea here was that a specification should not be a pattern that is simply read off an event. Rather, the pattern should have an objective, independent status, as characteristic of the patterns required for design inferences.

Detachability in the first edition was unpacked in terms of a conditional independence condition and a tractability condition. The idea with conditional independence was that an inquirer had information (background knowledge) that could be known independently of the event in question, which is to say, without affecting or altering its probability. And yet that information could then be used in a tractability condition to show that the pattern describing the event is identifiable with little difficulty, difficulty being measured by a complexity measure.

It was philosopher Rob Koons who pointed out to me that conditional independence did no real work here, and that all the real work instead was handled by tractability. In subsequent years, and in this second edition, the tractability condition came to mean that specifications are patterns with short description lengths. Patterns thus became tractable to the degree that they could be simply described. This all made good sense, matched up nicely with well-established design inferences, and made the concept of specification much more practically useful. Terminologically, this meant that the terms "detachability," "conditional independence," and "tractability" disappeared from the second edition, superseded by "minimum description length."

The other major conceptual change between the first and second edition concerns probabilistic resources. In the first edition, probabilistic resources took the form of specificational and replicational resources. The guiding metaphor for understanding these types of probabilistic resources was that of an archer shooting arrows at targets. The more arrows in the archer's quiver (replicational resources) and the more targets to shoot at (specificational resources), the more opportunities for some arrow to succeed at hitting some target by chance. Replicational resources were also the basis of what were called "saturated probabilities," which incorporated such probabilistic resources into events.

All these ideas are still there implicitly in the second edition, but they are now arranged differently and more coherently. Replicational resources in the first edition now coincide with probabilistic resources as such. Specificational resources in the first edition are now built into the specification-induced rejection regions needed to eliminate chance and draw design inferences. And finally, the saturated probabilities needed to determine whether an event has sufficiently small probability for a design inference now become the probabilities of these rejection regions.

The Explanatory Filter, a flowchart representation of the design inference, was perhaps the most memorable aspect of the first edition of this book. The Filter attracted a lot of attention, mainly because it took a complicated design inferential apparatus and made it simple, reconstructing how to infer design via just three decision nodes. Of course, in the background of the Filter is the entire design inferential apparatus, which is needed to justify it. The Explanatory Filter appears in this second edition, but it looks different from the first edition. In the first edition, the decision nodes focused on contingency, small probability, and specification. In this edition, there are now only two decision nodes, one focusing on specification, the other on small probability, with the need for a third node disappearing because we take chance to include necessity (as when probabilities collapse to zero and one). The two filters are equivalent, but the Filter in the second edition is conceptually sounder than the first, precluding some of the misconceptions that the first seemed to invite.

The second edition of *The Design Inference* is in every way conceptually cleaner, more compelling, and more unified than the first. Where the first edition struggled to hammer out a technical apparatus and invent a suitable vocabulary for making design inferences scientifically clear and rigorous, the second edition finds itself articulating a mature theory. I wish that, like Athena bursting full blown from the head of Zeus, this second edition could have been available from the start. If so, it might have avoided much needless controversy. But ideas are path dependent. They improve over trajectories that exhibit sudden advances but also painstaking incremental change. The path from the first to the second edition of this book is no exception.

ENDNOTES

Foreword

1. Dawkins, *The Blind Watchmaker*, 1.
2. Dawkins, *The Blind Watchmaker*, 21.
3. Dawkins, *The Blind Watchmaker*, 6.
4. Behe, *Darwin's Black Box*.
5. Darwin, *On the Origin of Species*, ch. 6.
6. Byers, "Life as 'Self-Motion.'"
7. Lents, *Human Errors*.
8. Pearl and Mackenzie, *The Book of Why*.
9. Bergsneider et al., "What We Don't (but Should) Know."
10. Greitz, "Radiological Assessment."
11. Egnor et al., "The Cerebral Windkessel and its Relevance."
12. Egnor, "The Cerebral Windkessel as a Dynamic Pulsation Absorber."
13. Egnor, "The Cerebral Windkessel as a Dynamic Pulsation Absorber."
14. Egnor et al., "A Quantitative Model of the Cerebral Windkessel."
15. *Expelled: No Intelligence Allowed*, a 2008 documentary narrated by Ben Stein about the academic pressures design scientists have faced. See https://www.imdb.com/title/tt1091617.

Introduction

1. Ian Barbour, a Templeton Prize winner and critic of intelligent design, in a Freudian slip even referred to it as "the divine inference"!

2. Dembski, *The Design Inference,* 228. Emphasis in the original.

3. Michael Behe recounts Dawkins' charge of laziness along with some of the backstory in a discussion with Michael Duduit at Union University back in 1998: https://www.uu.edu/unionite/spring98/darwin.htm (last accessed February 28, 2022).

4. Dawkins, *The Blind Watchmaker,* 9.

5. Dawkins, *The Blind Watchmaker,* 49.

6. Dawkins, *The Blind Watchmaker,* 141.

7. Dawkins, *The Blind Watchmaker,* 5–6.

8. For the original announcement of the Templeton book award competition, see https://web.archive.org/web/20080206101208/http://lists.ucla.edu/pipermail/religionlaw/1998-November/013503.html. For the announcement of winners as it appeared on the Templeton website, see https://web.archive.org/web/19991127214936/http://www.templeton.org/pcrs_winners.asp. The Philadelphia Center for Religion and Science administered the competition: https://web.archive.org/web/19991011211435/http://www.pc4rs.org/. Last accessed March 9, 2022.

9. Even though university presses are organized as non-profit corporations, they still like it when their books earn money!

Chapter 1

1. Cicero, *De Natura Deorum*, 213. Strictly speaking, Cicero places these words into the mouth of a Stoic philosopher named Balbus. Balbus here is debating an Epicurean philosopher and an Academic skeptic. Just how convincing Cicero found Balbus' argument from small probability is unclear. That he thought it had some force is clear. Cicero found merit in all three philosophical approaches taken in the dialogue.

2. Laplace, *Essai Philosophique,* 1307.

3. Reid, *Lectures on Natural Theology*, 46, emphasis added. Reid prefaces this quote with "Cicero, in his tract *De Natura Deorum*, speaks thus..." Reid thus seems to suggest to his readers that he is offering a direct quote from Cicero. Yet a careful examination of *De Natura Deorum* reveals that no portion of the quote appears anywhere in that work. Reid's

lectures on natural theology were transcribed by a student, so whether the words ascribed to Cicero were meant to be an exact quote or merely "in the spirit of Cicero" got lost in the transcription. Note that the Coan Venus (or "Coan Aphrodite") was a sculpture by the famous Athenian sculptor Praxiteles (c. 395–c. 330 BC).

4. De Moivre, *The Doctrine of Chances*, v. Note that piquet is played with 32 cards, so De Moivre's calculation is correct, namely, the probability comes to 1 in 32! (i.e., 32 factorial, or $32 \times 31 \times \ldots \times 2 \times 1$), or roughly 1 in 2.631×10^{35}, as De Moivre calculates. For an ordinary deck of 52 cards, the probability would come to 1 in 52!, or roughly 1 in 8.066×10^{67}.

5. Fisher, *Experiments in Plant Hybridisation*, 53.

6. Dawkins, *The Blind Watchmaker*, 139–141. Emphasis in the original.

7. This distinction addresses the worry that it's always possible to find any pattern one likes in a data set so long as one looks hard enough. Although there may be no limit to the patterns one can invent and afterwards impose on data, there are strict limits to the patterns with the right probabilistic and descriptive properties for eliminating chance, as will become evident later in this book. Distinguishing patterns by their ability to underwrite particular forms of inference is not new. Nelson Goodman (*Fact, Forecast, and Fiction*), for instance, distinguishes the patterns that lead to successful inductive inferences from those that do not, referring to the former as projectable predicates (cf. "all emeralds are green") and to the latter as non-projectable predicates (cf. "all emeralds are grue" where grue means green before the year 2000, blue thereafter).

8. See https://www.roulette17.com/stories/record-reds-blacks-in-a-row (last accessed March 28, 2022). This record number of 32 reds in a row is widely reported on the internet, but details about the actual casino and witnesses is scant. Even so, the number, though extreme, doesn't seem completely implausible if we factor in all the casinos around the world that have roulette tables.

9. See the chapter on recurrent events and renewal theory (ch. 8), and specifically the section on "success runs" (sec. 8.7), in Feller, *An Introduction to Probability Theory*, 322–326. The key formula here is 7.7 (324), which gives the mean recurrence time of runs of a given length.

10. See https://www.worldcasinodirectory.com/countries (last accessed February 27, 2023).

11. See https://www.gamblingsites.com/blog/17-facts-about-roulette-that-will-surprise-your-friends-12024 (last accessed March 29, 2022).

12. See https://www.bestuscasinos.org/blog/how-many-casinos-are-in-the-world-over-9284-by-the-way (last accessed March 2, 2023).

13. Gardner, "Arthur Koestler," 87–89.

14. The psychologist Carl Jung regarded striking (or improbable) coin-cidences as carriers of meaning not reducible to ordinary physical causality. He used the term *synchronicity* to describe such coincidences, seeing in them a meaningful connection between coinciding external events. Such occurrences were for Jung inexplicable from a standard cause-and-effect perspective, yet they could be deeply meaningful to the person experiencing them. Here is an example of synchronicity as recounted by Jung (*Synchronicity*, 10):

    > I noted the following on April 1, 1949: Today is Friday. We have fish for lunch. Somebody happens to mention the custom of making an "April fish" of someone. That same morning I made a note of an inscription which read: "Est homo totus medius *piscis* ab imo." In the afternoon a former patient of mine, whom I had not seen for months, showed me some extremely impressive pictures of fish which she had painted in the meantime. In the evening I was shown a piece of embroidery with fish-like sea-monsters in it. On the morning of April 2 another patient, whom I had not seen for many years, told me a dream in which she stood on the shore of a lake and saw a large fish that swam straight towards her and landed at her feet. I was at this time engaged on a study of the fish symbol in history. Only one of the persons mentioned here knew anything about it.

    Do instances of synchronicity like this arise from a designing intel-ligence and lead to a successful design inference (as laid out in Chapters 5 and 6)? It depends on the specified patterns identified and the calculated probabilities of the events matching them. Under the right circumstances, synchronicity could lead to a design inference. But note, for the examples that Jung considers, it's often quite difficult to form reasonable estimates of the underlying probabilities. Indeed, what is the probability of all this fish imagery coinciding the way it does?

15. What is known as the birthday paradox captures the underlying proba-bility theory here. See Feller, *An Introduction to Probability Theory*, 33. The move from 10^{34} to the probability of 1 in 10^{34} is an application of our "reciprocal rule," which we develop and justify in the chapter on probabilistic resources, and especially in Section 4.6.

16. There are ways to rig tossing a fair coin to achieve a preferred face with a probability greater than 50 percent. Depending on the approach, one can even move the probability up to 100 percent. Because coin tossing

is a matter of classical physics and thus ultimately deterministic given initial conditions, coin tossing machines exist that always give the same result depending on the side of the coin that was face up (Diaconis et al., "Dynamical Bias in the Coin Toss," 211–212).

It's also possible for people without using machines to skew the probabilities toward a greater preponderance of heads over tails (or vice versa). Magicians can develop a skill at coin tossing to the point of getting 70 percent heads (or tails). See Nicholas J. Johnson's piece titled "How to Rig a Coin Toss" at https://www.conman.com.au/post/7-ways-to-rig-a-coin-toss (last accessed February 9, 2023).

Diaconis et al., "Dynamical Bias in the Coin Toss," 211, have even shown that "for natural flips, the chance of coming up as started is about .51." In other words, if heads was facing up right before the coin was put in motion, natural flips tend to produce 51 percent heads (similarly for tails). But this approach requires coins to be caught by the hand, rather than allowing them to bounce on the table. For the purposes of this book, we'll make the traditional assumption that coins vigorously tossed by humans will have the propensity to come up heads or tails with equal probability. This assumption in no way undermines our work on the design inference in this book.

17. Borel, *Probabilities and Life*, 1. Borel's statement of this law reads simply as "Phenomena with very small probabilities do not occur." In context, however, he clearly was referring to events or phenomena that had been specified in advance. Indeed, in proposing this law, he focused on small-probability events characterized by known scientific theories, so these theories were providing prespecifications.

18. Knobloch, "Emile Borel as a Probabilist," 228.

19. Borel, *Probabilities and Life*, 28.

20. See Freedman et al., *Statistics*, 533–534.

21. Borel, *Probabilities and Life*, 26–28.

22. For the dialogue transcript of the film, see http://www.script-o-rama.com/movie_scripts/r/rosencrantz-and-guildenstern-are-dead-script.html (last accessed April 25, 2022). The film was based on Tom Stoppard's play by the same name, which was first performed in Edinburgh in 1966 and then in London and on Broadway in 1967. Stoppard wrote the screenplay for the film, and also directed it. In the play, Rosencrantz and Guildenstern likewise flip a long sequence of heads in a row, but it maxes out at 92 rather than, as in the film, at 157.

23. Keynes, *A Tract on Monetary Reform*, 80. Italics in the original.

24. Paul Nelson was the first in the intelligent design movement to articulate the ruin of practical reason that results from explaining away specified events of small probability by attributing them to chance willy-nilly.

25. Howson, *Objecting to God,* 84. His reference is to Dembski's 2004 book *The Design Revolution.*

26. Dave Farina, "Elucidating the Agenda of James Tour: A Defense of Abiogenesis," YouTube, https://www.youtube.com/watch?v=SixyZ7DkSjA starting at the 31-minute mark (last accessed April 13, 2023).

27. BBC Horizon, "A War on Science," https://topdocumentaryfilms.com /war-science starting at the 37-minute mark (last accessed April 21, 2023). *Horizon* is the UK's equivalent to PBS's *Nova.*

28. These remarks by Miller are no longer available at PandasThumb.org, but a record of them can be found at https://uncommondescent.com /evolution/ken-miller-blame-the-bbcs-bad-editing (last accessed April 25, 2023).

29. For the frequency distributions of Latin letters in different languages, see http://wiki.stat.ucla.edu/socr/index.php/SOCR_LetterFrequencyData (last accessed February 13, 2023).

30. See Wright, *Gadsby.*

31. For the first chapter of *Gadsby*, see http://spinelessbooks.com/gadsby /01.html (last accessed February 13, 2023). Gadsby in its entirety is available online on this website.

32. Quoted from http://spinelessbooks.com/gadsby/index.html (last accessed February 13, 2023).

33. Even so, the ability to see that probabilities can characterize side effects of design does not require being an actuary. Some simple programming skills suffice. For instance, to generate sequences of alphabetic letters that appear completely random (i.e., the letters appear drawn from a probability distribution that makes them equiprobable and probabilistically independent), take distinct English texts and "convolve" them. In other words, align the texts (ignoring spaces and punctuation) and then, letter by aligned letter, add them together cyclically, so that A+B=C, B+B=D, A+B+C=F, A+Z=A, B+Y=A, D+E+Y=H, and so on. Mathematically, this treats A = 1, B = 2, ..., Z = 26, and then adds these numbers for corresponding letters via modular arithmetic with modulo 26— see Ireland and Rosen, *A Classical Introduction to Modern Number Theory,* ch. 3.

   Convolving two independent English texts in this way produces a result that looks fairly random (a fact that underlies the running key, Vigenère, and one-time pad cryptosystems—compare Appendix B.1), though it may not pass some randomness checkers. But convolving increasingly many texts makes the outcome indistinguishable from chance for any randomness checker. The point to appreciate is that such convolutions of texts are entirely intentional: the texts themselves as well as the convolution of the texts are all intended. For a full arsenal of randomness checkers to test such convolutions for randomness, see the NIST technical report by Rukhin et al., *A Statistical Test Suite for Random and Pseudorandom Number Generators*.

34. See the Wikipedia article titled "Spelling of Shakespeare's Name" at https://en.wikipedia.org/wiki/Spelling_of_Shakespeare%27s_name (last accessed February 13, 2023). The figure of more than 80 spellings of Shakespeare is widely touted, but we have yet to see a complete listing of the diverse spellings. There were certainly a lot of them, and the Wikipedia article lists a goodly number, including six different spellings that Shakespeare used when signing his name: Willm Shakp, William Shaksper, Wm Shakspe, William Shakspere, Willm Shakspere, and (closest to the current spealling) William Shakspeare. The Elizabethans were far less obsessed with orthography than we are.

35. Dembski, "Randomness by Design" and Dembski, *Being as Communion*, ch. 16.

36. Statistician and ID critic David Bartholomew viewed chance as a force independent of intelligence. In his metaphysics of chance, he therefore failed to appreciate that expected chance behavior has no rational basis for constraining actual chance behavior unless chance itself is constrained by more than the laws of probability (such as the laws of large numbers, abbreviated LLN). One such constraint is, of course, intelligence. For a review of Bartholomew's *God, Chance and Purpose*, see Dembski, "God's Use of Chance."

37. See Hacking, *Logic of Statistical Inference*, 89.

38. Some statisticians shy away from the language of "elimination," preferring instead the language of "disconfirmation." Thus, it's not that a hypothesis ever gets definitely eliminated. Rather it gets sufficiently disconfirmed so that another hypothesis will seem better confirmed and thus more plausible. In practice, however, this nuance makes little difference. Disconfirm a hypothesis strongly enough, and it is for all intents and purposes eliminated.

39. Freedman et al., *Statistics*, ch. 28.

40. False negatives are commonplace in the practical business of inferring design. Suppose a husband and wife are walking on a path beside a high cliff. As the wife steps on a patch of ice, the husband pushes her, causing her to fall to her death. A police officer might see the patch of ice where

she fell and conclude accidental death. A false negative here does not mean a positive inference could not be drawn in the future if the officer obtains more information. For example, a witness might come forward with information that the husband took out a $10,000,000 life insurance policy the previous day and that the husband's diary entry from the day before that fantasizes about what he will do with "the 10 million bucks once she's gone."

Chapter 2

1. In fact, simply establishing priority, even without any official filing, is enough to establish copyright. For instance, on a blog, it's enough to establish copyright by posting original content that is time-stamped. If other websites then copy that material subsequently, they will be in violation of copyright. Interestingly, search engines provide an extra measure of copyright protection here by devaluing duplicate content, treating the content as it appeared with the original website as canonical, and then treating the websites that repeat that content as less authoritative. More on this in our main text.

2. Wikipedia has an entire entry on this topic: https://en.wikipedia.org /wiki/Fictitious_entry.

3. See https://www.worldwidewords.org/weirdwords/ww-nih1.htm (last accessed February 10, 2023): "If you have a copy of the *New Oxford American Dictionary*, please disregard the entry for *esquivalience*, which is supposedly the wilful avoidance of one's official responsibilities or the shirking of duties. It has been discovered to be a fake entry."

4. Edited by Harris and Levey, *The New Columbia Encyclopedia*, 1850, has the following entry:

    > Mountweazel, Lillian Virginia, 1942-1973, American photographer, b. Bangs, Ohio. Turning from fountain design to photography in 1963, Mountweazel produced her celebrated portraits of the South Sierra Miwok in 1964. She was awarded government grants to make a series of photo-essays of unusual subject matter, including New York City buses, the cemeteries of Paris and rural American mailboxes. The last group was exhibited extensively abroad and published as *Flags Up!* (1972). Mountweazel died at 31 in an explosion while on assignment for *Combustibles* magazine.

Quoted from https://grammarpartyblog.com/2012/01/30/the-incredible-story-of-lillian-virginia-mountweazel-and-dictionary-tomfoolery. The word *mountweazel* has become a synonym for fictitious entries in general: "A mountweazel is a bit of fake information deliberately added to a reference work like a map or dictionary to root out anyone who's illegally copying them." Quoted from https://emmawilkin.com/words-of-the-week-2/2020/2/26/mountweazel. Henry Alford made this definition official in his August 2005 piece for *The New Yorker* titled "Not a Word": https://www.newyorker.com/magazine/2005/08/29/not-a-word (all links in this note last accessed February 10, 2023).

5. Monmonier, *How to Lie with Maps*, ch. 4.

6. Not admitting to deliberately inserting such errors, Chambers in his book of mathematical tables nonetheless notes that errors in the tables "that are known to exist form an uncomfortable trap for any would-be plagiarist." Comrie, *Chambers' Shorter Six-Figure Mathematical Tables*, vi.

7. McKeon, *The Basic Works of Aristotle*, 1106.

8. See https://pixabay.com/blog/posts/how-to-clear-a-youtube-content-id-claim-with-a-pix-190 (last accessed December 2, 2022).

9. See http://copy-cats.work/misc2/hasselblad-1600-f-germany-salyus-s-ussr (last accessed August 5, 2022). See also https://gizmodo.com/incredible-soviet-rip-offs-of-western-technologies-973280252 (last accessed August 5, 2022).

10. See Walter J. Boyne's 2009 article by that title for *Air Force Magazine*: https://www.airforcemag.com/article/0609bomber (last accessed August 5, 2022). For the obsessive lengths to which the Soviets went in copying the B-29, see https://www.amusingplanet.com/2020/07/the-soviet-bomber-that-was-reverse.html (last accessed May 1, 2023).

11. *The Times—Princeton-Metro*, NJ, May 23, 1994, A1.

12. *New York Times*, August 23, 1994, A10.

13. See https://dnatesting.com/what-do-my-results-mean (last accessed April 12, 2022).

14. Lehman *The Perfect Murder,* 20. Emphasis added.

15. See Freedman et al., *Statistics*, 463, and Fisher, *Experiments in Plant Hybridisation*, 53. A careful reevaluation of Mendel's data by Edwards ("Are Mendel's Results Really Too Close?") still concluded that "the segregations are in general closer to Mendel's expectations than chance would dictate."

16. The Slutsky case was discussed in the PBS *NOVA* program that aired October 25, 1988, titled "Do Scientists Cheat?" A fuller account appeared in Janny Scott, "At UC San Diego: Unraveling a Research Fraud Case."

17. All quotes in this paragraph are from Kenneth Chang, "Similar Graphs Raised Suspicions on Bell Labs Research," *New York Times*, May 23, 2002, https://www.nytimes.com/2002/05/23/us/similar-graphs-raised-suspicions-on-bell-labs-research.html (last accessed April 13, 2022).

18. For a full-length treatment of the Schön case, see Reich, *Plastic Fantastic*.

19. See Reinsel, "Parapsychology: An Empirical Science," 194.

20. Campbell, *Madoff Talks,* 2.

21. For the full story of Charles Ponzi and his scheme, see Zuckoff, *Ponzi's Scheme*.

22. Perhaps the most engaging account of Madoff's fraud is that of Markopolos. Markopolos came to the Madoff case not as a newcomer but as a Cassandra who for years had been warning people, and especially the SEC, that Madoff was running a Ponzi scheme, though hardly anyone would listen (which is why his book is titled *No One Would Listen*). Henriques (*The Wizard of Lies*) and Campbell (*Madoff Talks*) are also worth reading. Henriques' book was turned into a 2017 HBO film starring Robert De Niro and Michelle Pfeiffer. Netflix, in 2023, released the four-episode documentary *Madoff: The Monster of Wall Street*, which is comprehensive and outstanding.

23. See Tversky and Kahnemann, for whom "losses and disadvantages have greater impact on preferences than gains and advantages" ("Loss Aversion in Riskless Choice," 1039). The field of behavioral economics has celebrated their work. See, for instance, Lewis, *The Undoing Project*.

24. See Arvedlund, "Don't Ask, Don't Tell" and Ocrant, "Madoff Tops Charts." Both articles came out in May of that year.

25. Henriques, *The Wizard of Lies,* ch. 7. This seems the best account of Markopolos' first encounter with Madoff's monthly-return numbers. In Markopolos (*No One Would Listen*, Appendix B), he considers even more extreme returns, namely, 7 months of losses over the course of 174 months (14½ years). Plugging in these numbers in the same binomial probability calculation as given at the end of this section yields an improbability even more ludicrous than what we got there, which came to roughly 1 in 40 billion. If we take the S&P 500 baseline as 30 percent of monthly returns being negative and then plug in the numbers in this more extreme case, the probability of getting 7 or fewer months of losses over 174 months total is roughly 1 in 400 quintillion. That's 1 in 400,000,000,000,000,000,000.

26. Markopolos' testimony before Congress is riveting and can be viewed here: https://www.youtube.com/watch?v=uw_Tgu0txS0. As he puts it, "I gift wrapped and delivered the largest Ponzi scheme in history to the SEC."

27. See Chaitin, "On the Length of Programs," but also Kolmogorov, "Three Approaches" and Solomonoff, "A Formal Theory of Inductive Inference, Part I," and "A Formal Theory of Inductive Inference, Part II."

28. For the connection between entropy in statistical mechanics and entropy in information theory, see Yockey, *Information Theory and Molecular Biology*, 66–67 (but note the errors in formulas 2.27 and 2.28). See also Zurek, *Complexity, Entropy and the Physics of Information*. For its connection to biology, especially to polynucleotides and polypeptides, see Yockey, *Information Theory, Evolution, and the Origin of Life*, chs. 4 and 11.

29. For further discussion of randomness see Dembski, "Randomness by Design" and *The Design Inference*, as well as Section 5.10.

30. Singh, *The Code Book*, 317.

31. Budiansky, *Code Warriors*, 212–213.

32. See Barron, *Breaking the Ring*.

33. Seifer, *Wizard: The Life and Times of Nikola Tesla*, 157 and 220–223.

34. Cocconi and Morrison, "Searching for Interstellar Communications." See also Fred Kaplan's "An Alien Concept" in *Nature* celebrating the 50th anniversary of scientific SETI.

35. Claims like this are represented in the Drake equation, which attempts to get at the probability that an extraterrestrial intelligence could make itself known to the inhabitants of Earth short of directly visiting Earth. For the Drake equation and its relation to an origination inequality that gets at the probability of complex biological systems forming on Earth, see Appendix B.8.

36. See Seth Shostak, "SETI and Intelligent Design," posted December 1, 2005, https://www.space.com/1826-seti-intelligent-design.html (last accessed December 2, 2022).

37. See the Wikipedia article on Arthur C. Clarke's *Space Odyssey* monolith at https://en.wikipedia.org/wiki/Monolith_(Space_Odyssey) (last accessed December 2, 2022).

38. The Wikipedia article on the Arecibo message can be found at https://en.wikipedia.org/wiki/Arecibo_message (last accessed December 2, 2022).

39. The actual question, as Fermi stated it, was "Where is everybody?" Fermi blurted out this question over lunch in 1950 at Los Alamos to Emil

Konopinski, Edward Teller, and Herbert York. See Eric M. Jones, "Where Is Everybody: An Account of Fermi's Question," Los Alamo National Laboratory, technical report, March 1985, https://sgp.fas .org/othergov/doe/lanl/la-10311-ms.pdf (last accessed January 4, 2023). Fermi's question is now commonly also called the Fermi paradox.

40. Crick and Orgel, "Directed Panspermia," 341. Panspermia of the undirected variety goes back to the ancient Greek philosopher Anaxagoras and before the twentieth century was mooted by scientists such as Lord Kelvin and Herman von Helmholtz—see Kolb and Clark, *Astrobiology for a General Reader*, ch. 10. Other scientists contemporary with Crick and Orgel who have taken panspermia seriously, albeit of the undirected variety, are Fred Hoyle and Chandra Wickramasinghe—see, for instance, Wickramasinghe et al., *Our Cosmic Ancestry in the Stars*.

41. See Venter, *Life at the Speed of Light*, especially the chapters titled "Digitizing Life," "First Synthetic Genome," and "Life by Design." Venter tends to take undue credit for the designs in his synthetic life forms. It would be more accurate to say that his designs rework existing designs. His work is that of an editor rather than an author. Yet editors still leave behind evidence that can form the basis of a design inference.

42. *Expelled: No Intelligence Allowed*, a 2008 documentary featuring Ben Stein. See https://www.imdb.com/title/tt1091617/ (last accessed January 4, 2023).

43. This excursus is adapted from Chapter 32 of *The Design Revolution* by William A. Dembski.

44. Hume, *Dialogues Concerning Natural Religion*.

45. Russell, *The Problems of Philosophy,* 98.

46. See Hume, *An Enquiry Concerning Human Understanding*.

47. Pennock, "The Wizards of ID: Reply to Dembski," 654.

48. Wilkins and Elsberry, "The Advantages of Theft over Toil," 718.

49. Sober, "Testability," 73, n. 20 (italics in original).

50. Reid, *Lectures on Natural Theology*, 50. Punctuation, capitalization, and spelling have been adjusted here to conform to current American usage.

51. Reid, *Lectures on Natural Theology*, 45.

52. Plantinga, *Warrant and Proper Function*.

Chapter 3

1. In fact, Stanley Milgram ("The Small World Problem") found that people in the U.S. are connected on average with three friendship links. He called this the "small world problem."

2. In the theory of computation, complexity is measured both spatially (in terms of storage capacity needed to solve a problem) and temporally (in terms of number of computational steps—which is how computer science measures time—needed to solve a problem). Both space complexity and time complexity play significant roles in computer science. Insofar as computer science studies the efficient performance of algorithms, most of the emphasis is on time complexity—see, for instance, Wegener (*Complexity Theory*) and Arora and Barak (*Computational Complexity*). The complexity measures that measure minimum description length and that receive most of the attention in this book are space complexity measures.

3. Grünwald and Roos, "Minimum Description Length Revisited."

4. Seibt, *Algorithmic Information Theory*.

5. This result follows from the failure of the halting problem in computer science to have an algorithmic solution. As Cover and Thomas (*Elements of Information Theory*, 483) put it: "One of the consequences of the nonexistence of an algorithm for the halting problem is the noncomputability of Kolmogorov complexity. The only way to find the shortest program in general is to try all short programs and see which of them can do the job. However, at any time some of the short programs may not have halted and there is no effective (finite mechanical) way to tell whether or not they will halt and what they will print out. Hence, there is no effective way to find the shortest program to print a given string." Note that Kolmogorov complexity is the minimum description length measure of algorithmic information theory or AIT.

6. Shermer, *Why Darwin Matters,* 38–39.

7. Peterson, *The Jungles of Randomness,* 5.

8. A bemused Amos Tversky made this point about the psychology of coin tossing at a conference on randomness that I [WmD] attended in April of 1988. Two months later, I got to see Tversky again at my graduation ceremony at the University of Chicago, when I received my doctorate in mathematics. Tversky, along with John Maynard Smith, received honorary doctorates at that ceremony. Tversky was a towering figure in psychology and economics, and he would have received the Nobel Prize with his collaborator Daniel Kahneman had he lived long enough, but sadly Tversky died of cancer in 1996.

9. Hardy and Wright, *An Introduction to the Theory of Numbers*, 128.

10. This remark was made at the Interdisciplinary Conference on Randomness convened at Ohio State University, April 11–16, 1988. Persi Diaconis and Harvey Friedman organized it at a time when "chaos theory" was all the rage. No proceedings were ever published, but the conference remains significant for assembling philosophers, mathematicians, psychologists, computer scientists, physicists, and statisticians in an attempt to understand randomness. I was simply an attendee. The conference, however, did inspire my paper "Randomness by Design," published with *Nous*. Science journalist James Gleick covered the conference for the *New York Times*.

11. It was at a 2001 summer symposium at Calvin College (now Calvin University) that Rob Koons pressed me [WmD] on the need to strip specification of superfluous elements and focus entirely on the complexity of generating the patterns that would count as specifications. In the end, the most convenient way to cash out such complexity was in terms of description length.

12. See https://www.esbnyc.com/about/facts-figures (last accessed May 3, 2023).

13. An attorney friend I [WmD] know advises that when one is on the witness stand or being deposed, one should respond to questions about one's unpopular beliefs not by rehearsing those beliefs in detail but by invoking some well-known figure and saying that one's beliefs coincide with those of that figure. This is a legal version of prespecification.

14. Smith, *The Theory of Evolution*, 265–266.

15. See https://en.wikipedia.org/wiki/Motor_vehicle (last accessed August 3, 2022).

16. Let's avoid ascribing this succession of Chevy Malibus to chance if the cars are within a mile of a Chevy dealership or service center, in which case they might naturally tend to congregate around these locations. Given this proviso, here's one way to approach the probability of these cars occurring in succession: Consider an urn model in which distinct cars are selected at random in proportion to their prevalence. If we now arrange cars in the order selected from the urn (with or without replacement) and if we note, as we have, that there are many different makes and models of cars, no one of which predominates, then an elementary combinatorial calculation suggests that getting ten new Chevy Malibus in a row is a specified event of small probability, and should not be ascribed to chance. The details are easily worked out.

17. With continuous probability distributions in which the probability of outcomes is zero, we look to where probability density functions get small rather than to where probabilities themselves vanish.

18. By probability measures and probability distributions we mean essentially the same thing, though the emphasis is a bit different. A probability measure is a countably additive, nonnegative real-valued function on certain subsets of Ω, namely, the measurable or Borel subsets of Ω—see Bauer, *Probability Theory and Elements of Measure Theory*. With a probability measure, the focus is thus on the mathematical object that assigns probability to subsets, or events, of Ω. By contrast, with a probability distribution, the focus is on how probability is smeared, or "distributed," across the possibility space Ω, concentrating probability in some regions, withdrawing it from others. Note that because the hypothesis H is the only one being considered, we could simply have defined the probability measure/distribution in question to be P rather than $P(\cdot|H)$, thus suppressing the reference to H.

The notation $P(\cdot|H)$ is in deference to Bayesian probability reasoning, which prefers this notation. Nonetheless, this notation can be problematic, such as when trying to define conditional probability for $P(\cdot|H)$ (see Appendix sections A.2–A.4). Suppose E and F are events. The conditional probability of E given F will then be $P(E\cap F|H)/P(F|H)$, but it really doesn't make sense to write $P(E|F|H)$ (a double conditioning, first on F and then on H), nor is it particularly clear to write $P(E|F\&H)$ or $P(E|F\cap H)$. Thus, in these situations, if it needs to be made clear that the probability measure P depends on, or is defined in terms of, a chance hypothesis H, it's better simply to write $P_H(\cdot)$, in which case the conditional probability of E given F is then just $P_H(E|F)$. And of course, given that H is the only hypothesis under consideration, there's always the option to write the probability measure/distribution as simply $P(\cdot)$, making the dependence on H tacit rather than explicit.

19. Statistically speaking, expectation refers to a mean or average, which for symmetric distributions like the normal distribution coincide with the mode, which identifies where the probability is most concentrated. Yet our analysis of events that are "too probable" lends itself more naturally to working with the mode (rather than the mean or even a median, if the latter exists). This will become clear momentarily as we use probability density functions to determine rejection regions.

20. Without this restriction, for any positive function g on Ω (given only some very broad regularity conditions on g, such as that g is integrable on Ω with respect to the measure μ), $v = (f/g)\cdot d\mu$ defines a measure on Ω which then allows g to be treated as a density such that $g\cdot dv$ also represents the probability distribution $P(\cdot|H)$. This then essentially

allows any positive function on g to serve as a density function representing $P(\cdot|H)$, thereby invalidating any inference based on where g is small and where it is large since where g is small and large could, by this device, be anywhere in Ω. For the modes and tails of this section to arise from specifications depends therefore on a canonical density function f being assigned to Ω, which is typically the case.

21. See, for instance, Bauer (*Probability Theory*, sec. 2.9), titled "Measures and Densities," which focuses on the Radon-Nikodym theorem. This theorem is typically what justifies such densities, and it is covered in any graduate text on probability theory or measure theory. See Cohn, *Measure Theory*, 123.

Chapter 4

1. All these probability calculations are easily confirmed, and depend only on undergraduate probability, analysis, and combinatorics. For non-mathematicians without ready access to equation solvers such as Stephen Wolfram's Mathematica, online calculators work well for such problems. See, for instance, https://keisan.casio.com/calculator (last accessed February 15, 2023).

2. Dam and Lin, *Cryptography's Role*, 380, n. 17.

3. To see that this is indeed the probability of an exhaustive search being successful in M queries when searching for a single item among K items (K assumed to be bigger than M), consider that the probability of failure on the first step is $(K-1)/K$, probability of failure on the second step given failure on the first step is $(K-2)/(K-1)$, ..., failure on the M-th step is $(K-M)/(K-M+1)$. Multiplying all these together yields the probability of failure to find the item in M steps and equals $(K-M)/K$ (because all the denominators but the first cancel). In consequence, the probability of at least one success in M steps is $1 - ((K-M)/K) = M/K$.

4. All the computer programs assumed by the NRC for the purpose of breaking cryptographic keys were running on conventional rather than quantum computers. Quantum computation changes some of these numbers, but, as we'll show later in this chapter, not substantially. Moreover, it does not change the underlying logic, whether for computational resources or probabilistic resources. In any case, quantum computation, in contrast to conventional computation, remains a distant promise rather than a fulfilled technological achievement.

5. See https://en.wikipedia.org/wiki/TOP500 (last accessed July 4, 2022).

6. See Marks et al., *Introduction of Evolutionary Informatics,* 47–48. Granted, Shor's ("Polynomial-Time Algorithms") algorithm, if it could be implemented on a full-scale quantum computer, could potentially break certain prime-number based cryptosystems with very large key-spaces. But the algorithm's success would not result from it doing an exhaustive search but from it taking advantage of certain number-theoretic constraints on the cryptographic keys. Shor's algorithm bypasses rather than engages in exhaustive search. The 3×10^{94} limitation on exhaustive search by Dam and Lin (*Cryptography's Role,* 380, n. 17) therefore still applies and is compelling.

7. But note, with 300 million Visa cards in circulation, the probability of by chance guessing *some* card's numerical data would be higher, though for practical purposes still quite low, and beyond the reach of any probabilistic resources available in practice.

8. An order of magnitude is a power of ten, whether positive or negative. Thus, with positive orders of magnitude, one order of magnitude is 10^1 = 10, two orders of magnitude is 10^2 = 100, and 20 orders of magnitude is 10^{20} = 100,000,000,000,000,000,000, or 100 quintillion. Thus, we say that 10,000 is three orders of magnitude bigger than 100, because 10,000 = 10^5, 100 = 10^2, and 10^2 must be multiplied by 10^3 to get up to 10^5. Orders of magnitude, however, apply not just to the large but also to the small—in other words, not just to positive powers of ten but also to negative powers of ten. Thus, conversely, $1/10,000 = 10^{-5}$ is three orders of magnitude smaller than $1/100 = 10^{-2}$ because $1/100$ needs to be multiplied by 10^{-3} to get down to 10^{-5}.

9. Practical impossibility arguments based on exhausting the probabilistic resources of the universe by looking to its total number of elementary particles always seem to stay in fashion. Stephen Wolfram, commenting on the current rage over ChatGPT and the ability of LLMs (large language models) like it to generate coherent text, offers such an argument. These LLMs depend on assigning probabilities to sequences of words. But the problem is that humans have simply not written enough words to calculate such probabilities empirically for any but the shortest sequences. Thus, the probabilities must be inferred and interpolated. Here is Wolfram's (*What Is ChatGPT Doing,* 19) appeal to probabilistic resources in the form of the universe's particles (2-grams are two-word combinations, 3-grams are three-word combinations, etc.):

    > In a crawl of the web there might be a few hundred billion words; in books that have been digitized there might be another hundred billion words. But with 40,000 common words, even the number of possible 2-grams is already 1.6 billion—and the number of possible 3-grams is 60 trillion. So there's no way we

can estimate the probabilities even for all of these from text that's out there. And by the time we get to "essay fragments" of 20 words, the number of possibilities is larger than the number of particles in the universe, so in a sense they could never all be written down.

10. Lloyd, "Ultimate Physical Limits to Computation," 1048.

11. See https://www.ncaa.com/news/basketball-men/bracketiq/2022-03-10/perfect-ncaa-bracket-absurd-odds-march-madness-dream (last accessed July 21, 2022). Note that Warren Buffett's Berkshire Hathaway and Dan Gilbert's Quicken Loans have teamed up to guarantee a $1 billion prize to anyone who fills out a perfect NCAA bracket—see https://www.scurichinsurance.com/now-thats-march-madness-buffett-gilbert-offer-1-billion-for-perfect-bracket (last accessed July 28, 2022). Granted, because college basketball teams differ in strength, as reflected in their seeding, it's possible to improve on the 1 in 10^{19} uniform probability implicit here by selecting stronger over weaker teams. But because seedings often do not reflect actual team strength and because upsets invariably happen, any such improvement in probability is minimal. In any case, Buffett and Gilbert's money is secure.

12. See https://en.wikipedia.org/wiki/Mega_Millions (last accessed July 21, 2022).

13. See the brief article "How Much Data Can a QR Code Store?" at http://qrcode.meetheed.com/question7.php (last accessed August 23, 2022).

14. "Eukaryotic proteins have an average size of 472 aa [= amino acids], whereas bacterial (320 aa) and archaeal (283 aa) proteins are significantly smaller (33-40% on average). Average protein sizes in different phylogenetic groups were: Alveolata (628 aa), Amoebozoa (533 aa), Fornicata (543 aa), Placozoa (453 aa), Eumetazoa (486 aa), Fungi (487 aa), Stramenopila (486 aa), Viridiplantae (392 aa)." Quoted from Tiessen et al., "Mathematical Modeling and Comparison," 1.

15. Roger Penrose, *The Emperor's New Mind,* 344. An indication of the enormity of these numbers is the following remark that Penrose makes about this $10^{(10^{123})}$ number: "Some perceptive readers may feel that I should have used the figure $e^{(10^{123})}$, but for numbers of this size, the e [= 2.718...] and the 10 are essentially interchangeable!"

16. Haug, Marks, and Dembski, "Exponential Contingency Explosion."

17. For $n = 1,000,000$ and $k = 500,000$, the number of ways of individuating k items among n items is expressed as "n choose k" and denoted by $C(n,k) = \binom{n}{k} = n!/k!(n-k)!$, which, by Stirling's formula, comes to

approximately $10^{(3 \times 10^5)}$. See Feller, *An Introduction to Probability Theory*, ch. 2.

18. Landman was responding to Dembski, "Specification: The Pattern That Signifies Intelligence." Landman published his critique of specified complexity, in which he failed to distinguish between mere possibilities and realistic possibilities as they relate to probabilistic resources, in a 2008 typescript that appeared at the following location on the web: http://www.riverrock.org/~howard/Dembski.pdf (last accessed December 5, 2022). Landman also included in his critique a mistaken dimensional analysis, in which he claimed that the units that appeared in the definition of specified complexity didn't check out. Yet he failed to consider what happens to units when one multiplies bit-operations by bits. Bit operations are the number of operations per bit, and as such the bit units cancel, leaving a dimensionless quantity, as required in Dembski ("Specification: The Pattern That Signifies Intelligence").

19. As estimated by the World Economic Forum. See https://www.weforum .org/agenda/2021/08/total-biomass-weight-species-earth (last accessed July 28, 2022).

20. "Bacteria are among the fastest reproducing organisms in the world, doubling every 4 to 20 minutes." Reported by the Pacific Northwest National Laboratory: https://www.pnnl.gov/science/highlights/highlight .asp?id=879 (last accessed July 28, 2022).

21. Behe (*The Edge of Evolution*, 143) estimates 10^{40} as an upper bound on "bacterial cells in the history of life on earth." Since bacterial cells vastly outnumber other organisms (each human, for instance, forms a single organism consisting of trillions of cells, and yet houses more bacterial cells than human cells), this estimate should also be reasonably close to the total number of organisms that have ever existed on planet Earth. In any case, 10^{48} is overkill, significantly overestimating the total number of organisms that have ever existed on planet Earth.

22. It might be simpler, given that universal probability bounds depend on absolute probabilistic resources and that local probability bounds depend on relative probabilistic resources, to refer to "absolute probability bounds" and "relative probability bounds," but the usage of universal and local probability bounds has come to acquire currency, and with universal probability bounds, the usage is now longstanding, going back to Emile Borel.

Chapter 5

1. John F. Sullivan, "Court in New Jersey Upholds Equal Odds for All," *New York Times*, July 23, 1985, B1. See https://www.nytimes.com/1985/07/23/nyregion/court-in-jersey-upholds-equal-odds-for-all.html (last accessed March 11, 2022). For a statistics textbook that examined the Caputo case, see Moore and McCabe, *Introduction to the Practice of Statistics*, 376–377. They concluded, "There is almost no chance that the Democrats would have won a truly random drawing so consistently." Moore and McCabe considered the Caputo case in the 1st (1989) and 2nd editions (1993) of their introductory statistics textbook, but dropped it in the 3rd (1999) and subsequent editions (there is currently a 2021 10th edition).

2. Sullivan, "Court in New Jersey Upholds," *New York Times*, July 23, 1985, B1.

3. Let p denote 1 in 50 billion and n denote 8,825,544. The order of ballot lines from one election to the next is assumed to be probabilistically independent. The probability of not getting at least 40 out of 41 Ds first should then be $(1-p)^n$. We want the probability of the complementary event, which is $1-(1-p)^n$. This probability then comes to roughly .000177, or about 1 in 5,650.

4. Legal scholars have debated the proper application of probabilistic reasoning to legal problems. Larry Tribe ("Trial by Mathematics"), for instance, has regarded the application of Bayes' theorem within the context of a trial as fundamentally unsound. Michael Finkelstein (*Quantitative Methods in Law*, 288 ff.) has taken the opposite view. Statistics has over time gained increasing authority in the law. As Finkelstein (*Basic Concepts of Probability*, vii) points out:

> When as a practicing lawyer I published my first article on statistical evidence in 1966, the editors of the *Harvard Law Review* told me that a mathematical equation had never before appeared in the review. This hardly seems possible—but if they meant a serious mathematical equation, perhaps they were right. Today all that has changed in legal academia. Whole journals are devoted to scientific methods in law or empirical studies of legal institutions. Much of this work involves statistics.

There appears no getting rid of the design inference within the law. Cases of bid-rigging (Finkelstein and Levin, *Statistics for Lawyers,* 51, 490–491), price-fixing (Finkelstein and Levenbach, "Regression Estimates of Damages," 79–106), and collusion often cannot be detected except with a design inference.

5. Swinburne, *The Existence of God*, 156–157. For the anthropic principle, the *locus classicus* is Barrow and Tipler, *The Anthropic Cosmological Principle*.

6. In a sense, the mad kidnapper has made things too easy for S to apply the GCEA. What if the kidnapper had said that only one particular set of ten card-draws would save S's life? Even if the kidnapper had said nothing about which card draws would save S's life, by examining the card-shuffling machine, S could learn that only one particular set of card-draws from the card-shuffling machine would save S's life, namely, the ten aces of hearts. Originally, the kidnapper explicitly prespecified the event E. But the machine itself implicitly prespecifies E. In either case, if E happens, the GCEA's logic will defeat ascribing E's occurrence to chance.

7. Howson and Urbach, *Scientific Reasoning,* 178.

8. Howson and Urbach, *Scientific Reasoning,* 178–180.

9. The letter "B" here is in reference to Borel (as in Emile Borel, a founder of modern probability theory). The sets to which probabilities are assigned are typically called "Borel sets." See, for instance, Cohn, *Measure Theory,* 4.

10. Monod, *Chance and Necessity*.

11. Dawkins, *The Blind Watchmaker,* 49.

12. Can the activity of a designing intelligence be characterized by a chance distribution? We saw in Section 1.4 that aspects of human behavior can be captured stochastically. But we also have no reason to think that every aspect of human intelligence can be captured by probability distributions that adequately characterize all our behaviors. That is certainly the dream of materialistic cognitive neuroscience or computational intelligence, but no chance hypothesis H has been articulated in sufficient detail to make this dream seem plausible.

13. This is standard symbolic or predicate logic in which for a predicate q, $\sim(\forall X)q(X)$ is logically equivalent to $(\exists X)\sim q(X)$. Basically, negation slips past quantifiers, changing universal to existential quantifiers, and vice versa.

14. This is essentially the same as the deductive form of the design inference that appeared in the first edition of this book—see Dembski (*The Design Inference*, 222), although there *nec* wasn't assimilated into *ch* as we do here in this second edition.

15. Needless to say, SP2 has a double reference: first to the "sp" at the start of the word "specification" and second to the abbreviation "SP" for "small probability."

16. Tracing the etymology of *intelligent* back still further, the *l-i-g* that appears in it derives from the Indo-European root *l-e-g*. This root appears in the Greek verb *lego*, which by the time of the New Testament typically meant "to speak." Yet its primitive Indo-European meaning was to lay, from which it came to mean to pick up and put together. And from there it came to mean to choose and arrange words, and therefore to speak. The root *l-e-g* has several variants. It appears as *l-o-g* in *logos*. It appears as *l-e-c* in *intellect* and *select*. And, as noted, it appears as *l-i-g* in *intelligent*.

17. Douglas Robertson ("Algorithmic Information Theory"), for instance, regards the defining feature of intelligence to be the "creation of new information." Dembski, *Being as Communion*, develops this idea at length.

18. Wittgenstein, *Culture and Value*, 1e.

19. See Mazur, *Learning and Behavior*, and Schwartz et al., *Psychology of Learning and Behavior*.

20. See Polkinghorne, *Belief in God in an Age of Science*, ch. 3.

21. Smith, "Irreducible Wholeness and Dembski's Theorem."

22. Behe, *Darwin's Black Box*.

23. Haught, *God After Darwin*, 71.

24. See Mackie, *The Cement of the Universe: A Study of Causation*. Here Mackie (especially beginning on 62) analyzes the requirements for something to qualify as a cause, laying out his famous INUS condition (the acronym stands for "an Insufficient but Non-redundant part of an Unnecessary but Sufficient condition"). The details here need not detain us. Suffice it to say that his analysis focuses on the logic of causation and, though drawing inspiration from David Hume, is theologically neutral. Mackie was, however, an ardent atheist who disputed with Alvin Plantinga over the problem of evil, Mackie arguing that an omnicompetent deity has no excuse for allowing evil, Plantinga arguing that free will could justify evil. Compare Mackie, "Evil and Omnipotence," with Plantinga. *God, Freedom, and Evil*.

25. Dembski, *Being as Communion*.

26. See Wilkins and Elsberry, "The Advantages of Theft over Toil."

27. Does any combination lock really allow a billion attempts to open it? The lock mechanism would surely wear out well before all those attempts could be completed.

28. Ruse, *Can a Darwinian Be a Christian?*, 121.

29. Sober, "Intelligent Design and Probability Reasoning."

30. Note that because we are dealing here with logical operators as they work together with probability, we'll be using in this section the logical operators ~, &, and ∨ (respectively *not, and,* and *or*) in place of the corresponding set-theoretic operators of complementation, intersection, and union (see Appendix A.1).

31. For the full details, see Adams, *The Logic of Conditionals,* ch. 1.

32. Sober, "Intelligent Design and Probability Reasoning," 69, 72.

33. Sober, "Intelligent Design and Probability Reasoning," 69.

34. Ultimately, the reason for Sober's mistake in identifying the wrong event here is that he has no concept of specification. It is important in probabilistic reasoning to determine whether an observed event corresponds to a specification and whether a small probability is assigned not just to that observed event but to the event identified with the specification. In our notation, it's not the probability of the event E that matters (E in this example being the exact sequence of 3,800 roulette wheel spins of which 100 are double zeros) but the probability of a rejection region induced by a specification V for which $V^* \supset E$. As it is, V^* in this example coincides with the event described by Sober when he wrote "suppose we spin the wheel 3,800 times and obtain a sequence of outcomes in which there are 100 double zeros." In fact, the pattern V here is a prespecification and the corresponding event V^* is the mode of a probability density function—in this case for a binomial distribution (see Section 3.6). The relevant rejection region R induced by this prespecification is therefore exactly V^*. Sober tacitly identified V^*, whose probability is roughly .040, but then substituted for it an event of much smaller probability, an exact sequence of 100 double zeros among 3,800 roulette wheel spins, which has a small but irrelevant probability of 1.475×10^{-201}.

35. Sober, "Intelligent Design and Probability Reasoning," 70. Sober credits Richard Royall with inspiring this example.

36. Thus Sober ("Intelligent Design and Probability Reasoning," 70) will write, "to say whether an observation is evidence for or against a hypothesis, we have to know what the other hypotheses are that we should consider."

Chapter 6

1. Orgel, *The Origins of Life,* 189.
2. Davies, *The Fifth Miracle,* 112.

3. Crick, "On Protein Synthesis," 144.

4. Crick, "On Protein Synthesis," 143.

5. Dawkins, *The Blind Watchmaker*, 9, 15.

6. See Bradley and Thaxton, "Information and the Origin of Life," 206–208.

7. Bradley and Thaxton, "Information and the Origin of Life," 208.

8. See Dembski, "Intelligent Design as a Theory of Information."

9. See Ewert, Dembski, and Marks, "Algorithmic Specified Complexity" and "Algorithmic Specified Complexity in the Game of Life."

10. Interestingly, specified complexity could in principle take on negative values if the description length exceeds the complexity as measured by small probability. In practice, however, events of complexity n bits can be described in n bits because sequences of bits can be described in terms of themselves. In the balancing of description length and complexity in specified complexity, there are often bookkeeping details to be kept track of that, depending on the computational environment and idiosyncrasies of the descriptive language, could for some events allow description length to exceed complexity as measured in bits. See Section 7.6 for an example of negative specified complexity.

11. Billingsley, *Convergence of Probability Measures,* ch. 1.

12. An introduction to Algorithmic Information Theory, which is also called Kolmogorov complexity, can be found in Cover and Thomas, *Elements of Information Theory*, ch. 14. For a comprehensive account of this theory, see Li and Vitányi, *An Introduction to Kolmogorov Complexity and Its Applications*.

13. All the details about these assumptions are readily found in Cover and Thomas, *Elements of Information Theory*.

14. Prefix-free languages are also called *prefix codes* or *instantaneous codes*. See Cover and Thomas, *Elements of Information Theory,* 106.

15. For Turing completeness as it applies to Algorithmic Information Theory or, equivalently, to Kolmogorov Complexity, see Cover and Thomas, *Elements of Information Theory,* secs. 14.1 and 14.2.

16. See Cover and Thomas, *Elements of Information Theory,* 467, Theorem 14.2.1, which characterizes the universality of Kolmogorov complexity.

17. Except for the largest unabridged dictionaries, 200,000 words is adequate for any dictionary of the English language. Moreover, any specialty words not in such a dictionary can be redescribed with those 200,000 words. We are here, as is our habit, trying to be over-generous so that no

one can charge us with trying to stack the numbers in our favor. Note that Stephen Wolfram (*What Is ChatGPT Doing*, 18), in his discussion of ChatGPT, regards 40,000 words as adequate for understanding how this artificial-intelligence model assigns probabilities to sequences of words: "There are about 40,000 reasonably commonly used words in English. And by looking at a large corpus of English text (say a few million books, with altogether a few hundred billion words), we can get an estimate of how common each word is. And using this we can start generating 'sentences', in which each word is independently picked at random, with the same probability that it appears in the corpus."

18. Cover and Thomas, *Elements of Information Theory,* 109.

19. Outcomes, as defined in Appendix A.1, are the bare individual elements making up a probability space, and so they will all be mutually exclusive and exhaustive. In consequence, the sum of their probabilities will be one. On the other hand, events, made up of outcomes, can overlap, in which case they will fail to be mutually exclusive and thus will not in general have their sums bounded above by one. See Appendix A.1.

20. See Cover and Thomas, *Elements of Information Theory,* sec. 5.9.

21. This result applies to outcomes and not to events. Mathematically, an event is a set of outcomes (see Appendix A.1). We cannot easily apply a code to events consisting of possible outcomes.

22. This dataset is available online from https://www.kaggle.com/datasets /rtatman/english-word-frequency (last accessed February 14, 2023).

23. This was the point behind the concept of *detachability* in the first edition of this book (Dembski, *The Design Inference,* 5.3).

24. Note that we are intentionally minimizing complexity. We could have included question marks, semicolons, colons, numerals, etc., but that would have caused the number of characters to exceed 29, thus increasing complexity and decreasing probability, as evident in the subsequent calculations.

25. To leave virtually no English word unaccounted for, we are going with the 200,000 or so English words that constitute the standard Unix file titled simply `words`. This is a comprehensive newline-delimited list of words found in English dictionaries. It is very large as dictionaries go, containing just about anything that could be construed as a word, including acronyms. Many spell-check programs are based on it. See https://en.wikipedia.org/wiki/Words_(Unix) (last accessed June 23, 2023). Even so, most spell-check programs employ far fewer words than this. Wikipedia's "spell checker" entry explains the rationale for going with fewer words:

It might seem logical that where spell-checking dictionaries are concerned, "the bigger, the better," so that correct words are not marked as incorrect. In practice, however, an optimal size for English appears to be around 90,000 entries. If there are more than this, incorrectly spelled words may be skipped because they are mistaken for others. For example, a linguist might determine on the basis of corpus linguistics that the word baht is more frequently a misspelling of bath or bat than a reference to the Thai currency. Hence, it would typically be more useful if a few people who write about Thai currency were slightly inconvenienced than if the spelling errors of the many more people who discuss baths were overlooked.

See https://en.wikipedia.org/wiki/Spell_checker (last accessed June 22, 2023). When it comes to specified complexity in texts, it makes sense to go with a larger dictionary since measures of specified complexity in texts must not unduly limit the chance formation of valid English words and sentences. The more words, the more we give chance a chance.

26. See Kennedy and Churchill, *The Voynich Manuscript*. See also the declassified book on the Voynich manuscript from the National Security Agency by Mary D'Imperio, *The Voynich Manuscript: An Elegant Enigma*.

27. For Zipf's law, see Debowski, *Information Theory Meets Power Laws*, sec. 1.3.

28. Timm and Schinner ("A Possible Generating Algorithm") consider this possibility.

29. This occurrence has been dubbed "Smiley Face Hill." For a recent image of it, see https://www.oregonlive.com/pacific-northwest-news/2020/12/who-planted-a-giant-smiley-face-of-trees.html (last accessed April 14, 2023).

30. These forest swastikas have merited their own Wikipedia entry: https://en.wikipedia.org/wiki/Forest_swastika (last accessed April 14, 2023).

31. For images and discussion of the "Face on Mars," see the Wikipedia entry on Cydonia: https://en.wikipedia.org/wiki/Cydonia_(Mars) (last accessed April 14, 2023).

32. For the Old Man of the Mountain, also known as the Great Stone Face or the Profile, see https://en.wikipedia.org/wiki/Old_Man_of_the_Mountain. For the Badlands Guardian, see https://en.wikipedia.org/wiki/Badlands_Guardian. For the Face on Moon South Pole, see https://en.wikipedia.org/wiki/Face_on_Moon_South_Pole. Each of these Wikipedia entries

includes images of these formations. All entries were last accessed April 14, 2023.

33. For Mount Rushmore, see https://en.wikipedia.org/wiki/Mount_Rushmore. For examples of carved Buddhas, see https://en.wikipedia.org/wiki /Leshan_Giant_Buddha and https://en.wikipedia.org/wiki/Buddhas_of _Bamiyan. For Decebalus, see https://en.wikipedia.org/wiki/Rock _sculpture_of_Decebalus. All last accessed April 14, 2023.

34. See https://blogs.scientificamerican.com/observations/6-strange-facts-about-the-interstellar-visitor-oumuamua (last accessed April 15, 2023).

35. The 1980 Pennsylvania Lottery Scandal merits its own Wikipedia page: https://en.wikipedia.org/wiki/1980_Pennsylvania_Lottery_scandal (last accessed April 15, 2023).

36. For a more in-depth account of the Ontario Lottery Retail Scandal, as it has been called, as well as for a conventional statistical analysis of the probabilistic problems raised by the scandal, see Rosenthal, "Statistics and the Ontario Lottery Retail Scandal," available online at http://probability.ca/jeff/ftpdir/chancelotpub.pdf (last accessed April 15, 2023).

37. There's much on the web about whether pi and *e* are encoded in the first verse of Old Testament and the first verse of the New Testament respectively. We leave it to readers to hunt down such references. Some commentators, like the authors, are deeply skeptical of finding pi in Genesis 1:1 and *e* in John 1:1. Others think that there is something to this coincidence. Bible codes were the rage in the late 1990s. One of the authors of the present book [WmD] wrote a critique of such attempts to find hidden messages in the Bible. See https://www.firstthings.com /article/1998/08/004-the-bible-by-numbers.

Chapter 7

1. See Behe, *Darwin's Black Box* and *The Edge of Evolution*. For the record, neither author of this book holds to common descent or universal common ancestry. In our view, the evidence from paleontology and molecular biology suggests that fundamental divisions exist among different types of organisms, especially at the phylum level for animals. We follow Stephen Meyer (*Signature in the Cell* and *Darwin's Doubt*) here as well as Ewert ("The Dependency Graph of Life"). That said, from the vantage of intelligent design, the degree of continuity or discontinuity in ancestral relations among organisms is a minor issue compared to the Darwinian claim that organisms develop novel structures and functions

via an unguided process absent all real design. Behe and the authors of this book agree that biology is chock full of design, that overwhelming evidence supports this claim, and that the design inference is crucial for adjudicating such evidence.

2. Ruse, *Monad to Man*.

3. Mathematician Jason Rosenhouse is an exception. According to him, the patterns evident in biological systems are no different from the patterns we invent/impose on cloud formations. Thus, for Rosenhouse, to see the bacterial flagellum as a bidirectional motor-driven propeller is to see a pattern that in no essential way differs from seeing clouds shaped like a dragon: "How can we be confident that in using function as a specification we are not doing the equivalent of looking at a fluffy, cumulus cloud and seeing a dragon?... Saying of a flagellum that it resembles an outboard motor is comparable to saying of a cloud that it resembles a dragon." For Rosenhouse, all such biochemical machines, despite exhibiting clear engineering patterns, require no fundamentally different type of explanation from clouds that resemble dragons. Natural selection produces the one, natural weather the other.

Clearly, there is a fundamental difference in the types of patterns here, namely, between clouds on which we impose patterns versus functional biological systems that exhibit clear engineering patterns. Indeed, the only plausible reason for conflating these types of patterns is to avoid the force of specification in the design inference. Rosenhouse's position is very much the minority view, even among scientists who think the design inference is inapplicable to biological systems. The majority see the problem of design in biology as simply a failure of improbability, a point on which Rosenhouse will agree with the majority when convenient, though he will also claim that there's no way to reasonably calculate or even estimate "the probability of evolving a particular complex system." Rosenhouse therefore sees the design inference as failing in biology with respect to both specification and improbability.

Thus, for Rosenhouse, if the probabilities can be calculated or estimated for such evolving systems, they will be large rather than small; and if they can't be calculated or estimated, well then, too bad for the design inference, which depends on making explicit such probabilities. This note responds mainly to Jason Rosenhouse, "A Reply to Dembski's Review of My Book," The Panda's Thumb (July 3, 2022): http://pandasthumb.org/archives/2022/07/Dembski-response.html (last accessed February 16, 2023). But for context, see also Rosenhouse, *The Failures of Mathematical Anti-Evolutionism*.

4. We are talking here about biological systems that are specified—after all, the focus of this book is on design inferences where improbability and specification together triangulate on design. Unspecified aspects of life can be radically contingent and highly improbable—yet without calling for a design inference. Stephen Jay Gould's *Wonderful Life* metaphor of "replaying the tape of life," in which the history of biology might be fundamentally altered if different stochastic elements played different roles in the course of natural history could lead to such improbability, but the resulting differences in evolutionary outputs across different possible evolutionary histories would be unspecified. To the degree that they are specified, such as in convergent evolution, they would undercut Gould's radical contingency thesis. Simon Conway Morris (*Life's Solution*) has taken this approach in direct opposition to Gould, and while not embracing intelligent design, Conway Morris is finding teleology in biology.

5. Dawkins, *The Blind Watchmaker*, 49.

6. But note, just because individual steps in a progression are reasonably probable does not mean that the entire progression is reasonably probable. Depending on the joint probabilities of the steps in the progression, the entire progression may or may not be reasonably probable. In the case of Dawkins' WEASEL example, the progression to the target phrase METHINKS IT IS LIKE A WEASEL occurs with high probability given 40 or so steps. The reason for this high probability of the entire progression, however, is that cumulative selection, programmed intentionally to achieve the target phrase, guarantees moving in steps that increasingly approach the target.

    But imagine tossing a die repeatedly. Rolling a 1 through 5 gets you to roll again. So, the probability of rolling again at each step is reasonably probable, namely 5/6. But to progress through a thousand rolls of the die, each time getting a 1 through 5, has probability $(5/6)^{1000}$, which is less than 1 in 10^{79}. In this case, nothing like cumulative selection is helping the process along. And even with cumulative selection available, its ability to render a progression probable depends on what is being selected for. The target phrase METHINKS IT IS LIKE A WEASEL will not occur if selection is instead selecting for a different target phrase.

7. Dawkins, *The Blind Watchmaker*, 47–48.

8. In his book *Climbing Mount Improbable*, Dawkins changes the metaphor to climbing a mountain, but without strengthening his case for the power of cumulative/natural selection. On one side of the mountain is a sheer face. Climbing it there would mean jumping to the top from the bottom, which is highly improbable and corresponds to single-step selection. But on the other side of the mountain is a gradual ascent that allows climbing

it baby step by baby step. Each baby step is highly probable and getting to the very top via these baby steps is also highly probable. Getting up the mountain with high probability by means of a gradual ascent corresponds to cumulative selection.

But that's the point: cumulative selection can only get you up the mountain if it has such a gradual winding path up to its peak. And just as there's no requirement for mountains to guarantee such a gradual ascent, there's nothing about biological systems to guarantee they can and must evolve gradually. In either case, to answer whether a gradual ascent actually obtains requires empirical evidence and conceptual analysis. Thus, Dawkins' metaphor of climbing a mountain, while rhetorically useful in making Darwinian cumulative selection seem plausible and powerful, assumes the very thing that needs to be demonstrated. Are there such gradual step-by-step pathways, each one beneficial or at least neutral, all the way from one biological form to a fundamentally different one? Dawkins and other mainstream evolutionists routinely retreat to imaginative illustrations and vague just-so stories instead of providing detailed, explicit, step-by-step pathways.

9. Rosenhouse, *The Failures of Mathematical Anti-Evolutionism,* 194.

10. From a companion piece to Rosenhouse (*The Failures of Mathematical Anti-Evolutionism*) published May/June 2022 at the *Skeptical Inquirer,* https://skepticalinquirer.org/2022/05/the-failures-of-mathematical-anti-evolutionism (last accessed February 20, 2023).

11. Eigen and Schuster, *The Hypercycle,* 15–28.

12. Berlinski, *Black Mischief: Language, Life, Logic, Luck,* 345–350.

13. Küppers, *Information and the Origin of Life,* ch. 8.

14. Satinover, *The Quantum Brain,* 89–92.

15. Yarus, *Life from an RNA World,* 64–68.

16. Dobzhansky, "Discussion of G. Schramm's paper," 310.

17. Such scenarios, in which self-organizational processes pass the baton to Darwinian processes, have been touted for decades. See, for instance, Deamer (*Origin of Life,* sec. 2), and a generation earlier Kauffman (*The Origins of Order*, ch. 7).

18. NASA's official definition of life states, "Life is a self-sustaining chemical system capable of Darwinian evolution." See https://astrobiology .nasa.gov/research/life-detection/about (last accessed June 24, 2023). Origin-of-life researchers Cooper, Walker, and Cronin ("A Universal Chemical Constructor," 101) define life as "a complex chemical system with memory."

19. Eric Anderson ("A Factory That Builds Factories That Build Factories That...," 85–86) has argued effectively that a robust form of self-replication, in which an environment is not artificially stacked to help replication, cannot be simple and cannot be handled by any "miracle molecule." As he concludes, "Self-replication, contrary to the materialist abiogenesis story, is not the beginning feature, a rudimentary trait that a single molecule could handle. Rather it is a culminating trait, one of the most dazzlingly hightech traits in the biosphere. The accumulated evidence, taken together, strongly suggests that self-replication lies at the end of a very complicated, deeply integrated, highly sophisticated, thoughtfully planned, carefully controlled engineering process."

20. Such other naturalistic mechanisms would include symbiogenesis, gene transfer, genetic drift, the action of regulatory genes in development, and self-organizational processes. Yet none of these comes close to natural selection in assuming the role of a probability amplifier to invalidate design inferences. If there is a designer substitute in biology, it is natural selection. Other naturalistic mechanisms don't even rank. This means that if natural selection fails to rescue the chance hypothesis in evolutionary biology, then chance is indeed in trouble. Our argument is that it does indeed fail in this regard.

21. Dennett, *Darwin's Dangerous Idea*, 21.

22. "The God Delusion Debate" between Richard Dawkins and John Lennox, organized by Larry Taunton and held in 2007, available online at https://www.youtube.com/watch?v=zF5bPI92-5o (last accessed February 21, 2023).

23. Kingsolver, *Small Wonder*, 96. In what sense is something that explains everything a good explanation of anything?

24. Earman, *Bayes or Bust?*, 165.

25. Earman, *Bayes or Bust?*, 165.

26. As evident in Dembski and Ruse, *Debating Design: From Darwin to DNA*, 328–329, and Dembski, *The Design Revolution*, chs. 30 and 31.

27. Harold, *The Way of the Cell*, 205.

28. Medawar, *Advice to a Young Scientist*, 39. Emphasis in the original. Medawar adds, "A scientist who habitually deceives himself is well on the way toward deceiving others." To this, physicist and Nobel laureate Richard Feynman (*Surely You're Joking*, 343) would add, "The first principle is that you must not fool yourself—and you are the easiest person to fool." Of course, what's good for the goose is good for the gander—design theorists need to hold themselves to the same high

standard of not deceiving themselves. Thankfully, Darwinists are ever ready to oblige in trying to keep design theorists honest.

29. Mill, *A System of Logic.*

30. Mill, *A System of Logic*, 256.

31. Miller, *Only a Theory,* 77.

32. Darwin, *On the Origin of Species,* 82.

33. Rosenhouse, *The Failures of Mathematical Anti-Evolutionism,* 215.

34. Goodwin, *How the Leopard Changed Its Spots,* 35–36.

35. See, for instance, Altshuler and Linden, "Design of Wire Antennas," for the construction of high-performing crooked-wire antennas by means of genetic algorithms.

36. Lenski et al., "Mutator Genomes Decay," E9026. Those who think Lenski's experiment supports a robust view of Darwinian evolution will likely cite the development in this experiment of a Cit+ phenotype. Thus, to introduce an article examining this feature of Lenski's experiment, Van Hofwegen et al. ("Rapid Evolution of Citrate Utilization," 1022) write: "The isolation of aerobic citrate-utilizing *Escherichia coli* (Cit+) in long-term evolution experiments (LTEE) has been termed a rare, innovative, presumptive speciation event." But they then refute this claim by showing that "direct selection would rapidly yield the same class of *E. coli* Cit+ mutants" in about "100 generations." Lenski's evolution of Cit+ is therefore unimpressive. It suggests that Cit+ was readily evolvable and that it involved little specified complexity that wasn't already there.

37. Kauffman, *Investigations,* 19. Kauffman here was discussing the significance of the no free lunch theorems for biology. Dembski (*No Free Lunch*) has addressed this topic in greater detail. Our point in the present book, however, is to focus on the design inference, leaving the no free lunch theorems and their logical extension, conservation of information, for a sequel to this book. We cite Kauffman here to underscore Mill's method of difference, showing how it refutes a simplistic view of natural selection, in which natural selection is treated as a panacea that makes biological evolution work and that disqualifies design.

38. Kauffman, *Investigations,* 18.

39. Matthew Arnold seems to have appreciated such limitations on Darwinism. Soon after the publication of Darwin's *Origin* in 1859, Arnold remarked to a biology professor, "I cannot understand why you scientific people make such a fuss about Darwin. Why, it's all in Lucretius" (Judd, *The Coming of Evolution,* 13). The Roman poet Lucretius was a follower

of the Greek atomistic philosopher Epicurus. Epicurus had accounted for the origin of biological forms in terms of pure chance assemblages of atoms that occasionally produced stable configurations, which could then count as alive. Lucretius, in Book 5 of his *De Rerum Natura*, went further, anticipating natural selection by seeing most organic forms as perishing but a few as enduring thanks to chance endowing them with advantages conducive to survival, such as strength and cunning.

Neither Epicurus nor Lucretius had in the subsequent centuries been able to dampen support for design in biology. Until Darwin. And yet, Alfred Russel Wallace, Darwin's nineteenth-century co-discoverer of natural selection, seems likewise to have understood such limitations on Darwinism. Later in life, Wallace was explicit about seeing natural selection as inadequate to account for the full range of structures and functions of living forms. Wallace embraced common descent and agreed that natural selection played an important role in evolution. Yet ultimately, he saw natural selection as operating "within a larger teleological... framework." See Flannery, *Intelligent Evolution,* 26.

40. Darwin, *On the Origin of Species*, 189.

41. Rosenhouse, *The Failures of Mathematical Anti-Evolutionism,* 178.

42. Rosenhouse, *The Failures of Mathematical Anti-Evolutionism,* 159–160.

43. The boundaries are thus treated as reflecting barriers. Probabilists use reflecting barriers to modulate stochastic processes. Essentially, in this example, we are doing what would be a random walk on the hypercube, except that the walk is constrained by selection, which assigns higher fitness the higher up we move in the path (from all 0s to all 100s). As Jones and Smith (*Stochastic Processes,* 49) put it, "If the walk is bounded, then the ends of the walk are known as barriers, and they may have various properties... [The barrier could] be absorbing, which implies that the walk must end once a barrier is reached since there is no escape. On the other hand, the barrier could be reflecting, in which case the walk returns to its previous state." Such a reflecting barrier operates here when evolution on the hypercube attempts to venture out beyond the hypercube into the hyperspace of all 100-tuples of integers.

44. Jones and Smith, *Stochastic Processes,* 13–14.

45. Dennett, *Darwin's Dangerous Idea.*

46. In this section, we focus on dysteleology in the sense of poor or sub-standard design. Dysteleology also includes evil or malevolent design, as when a parasite benefits itself at the expense of a host organism. Darwin saw such design as perverse and as disconfirming divine design. As Darwin (*The Correspondence of Charles Darwin*, vol. 8, 224) put it

in a letter to Asa Gray: "I cannot persuade myself that a beneficent and omnipotent God would have designedly created the Ichneumonidae [parasitic wasps] with the express intention of their feeding within the living bodies of Caterpillars." Our focus in this book is on design as such, not on its goodness or wickedness. Malevolent design may be actual design. Substandard design may be actual design, but it can raise doubts about whether we're dealing with design at all. Hence the focus on it in this section. Malevolent design falls under the more general theological problem of natural evil. For a sustained treatment of that problem from a Judeo-Christian perspective, see Dembski, *The End of Christianity: Finding a Good God in an Evil World.*

47. Dawkins, *The Blind Watchmaker,* 1. We might say that biology is unique among the sciences in putting forward a developed theory purporting to explain the appearance of design, an appearance that is merely an appearance. Darwinian evolution is a theory whose main claim to fame is explaining things that appeared to be designed. For this reason, Darwinian evolution is sometimes called a designer mimic. It allegedly can produce objects for which humans would intuitively infer design. Historically, before the theory of Darwinian evolution gained traction, most humans did in fact infer design to explain the apparent design in biological systems. The controversy is over whether they were correct to draw such a design inference or if they were guilty of committing a false positive.

48. See, respectively, https://en.wikipedia.org/wiki/Rofecoxib, https://en.wikipedia.org/wiki/Firestone_Tire_and_Rubber_Company, and https://en.wikipedia.org/wiki/Windows_Vista (last accessed July 3, 2023).

49. Petroski, *Invention by Design,* 30.

50. If you're a sausage maker, for instance, you want a product that tastes good and also is cheap to make. But good taste correlates with quality of meat, and good quality meat is more expensive than poor quality meat. So improving taste will compete with keeping down costs. You therefore cannot produce the best sausage in the world at virtually no cost. Taste and cost are thus tradeoffs in a constrained optimization.

51. Doudna and Sternberg, *A Crack in Creation*, ch. 1. In the same vein, tech titan Bill Gates (*The Road Ahead,* 228) comments on the sophistication of DNA: "Biological information is the most important information we can discover, because over the next several decades it will revolutionize medicine. Human DNA is like a computer program but far, far more advanced than any software ever created."

52. Paul Nurse, "The Five Core Principles of Life," uploaded June 2023, https://www.youtube.com/watch?v=5EwVBC3VsRA (last accessed July 5, 2023).

53. Ricardo and Szostak, "The Origin of Life on Earth."

54. Shapiro, *Evolution: A View from the 21st Century,* 143.

55. Quoted from https://www.miragenews.com/how-nature-inspires-technology-the-field-of-1023128. See also Benyus, *Biomimicry,* and her YouTube video on biomimicry: https://www.youtube.com/watch?v=sf4oW8OtaPY (last accessed July 4, 2023).

56. For a critique of Benedictis' views as well as links to her TEDxMIT presentation, see Emily Reeves, "Verdicts of 'Poor Design' in Biology Don't Have a Good Track Record," https://evolutionnews.org/2021/05/verdicts-of-poor-design-in-biology-dont-have-a-good-track-record (last accessed February 25, 2023).

57. See Franze et al., "Müller Cells Are Living Optical Fibers."

58. This and the next paragraph are largely drawn from Dembski, *The Design Revolution,* ch. 6.

59. Lents' (*Human Errors,* ch. 1) reference to "too many bones" calls to mind the scene in the film *Amadeus* where the Austrian emperor criticizes Mozart for writing "too many notes." After listening to a performance of Mozart's opera *The Abduction from the Seraglio,* Emperor Joseph II confronts the composer: "Your work is ingenious. It's quality work. And there are simply too many notes, that's all. Just cut a few and it will be perfect." Frustrated with the emperor's simplistic view of his music, Mozart asks, "Which few did you have in mind, Majesty?" The scene highlights the incomprehension of Mozart's genius by some of his contemporaries. It suggests a similar incomprehension on the part of dysteleologists. For the scene, see https://www.youtube.com/watch?v=H6_eqxh-Qok (last accessed July 8, 2023).

60. Burgess, "Why the Ankle-Foot Complex Is a Masterpiece," 7. The quote by Lents is from Lents, *Human Errors,* 29.

61. Berg and Anderson, "Bacteria Swim."

62. Reported in conversation with microbiologist Scott Minnich.

63. Aizawa, "What Is Essential for Flagellar Assembly?," 91.

64. Alberts, "The Cell as a Collection of Protein Machines," 291.

65. Wilkins, "A Special Issue on Molecular Machines," 1146.

66. Xue et al., "A Delicate Nanoscale Motor," 1 and 6. Even though this article refers to the bacterial flagellum as having evolved, Pallen and

Matzke, in attempting to sketch its evolution, admitted, "The flagellar research community has scarcely begun to consider how these systems have evolved" ("From *The Origin of Species* to the Origin of Bacterial Flagella," 788). This assessment still holds.

67. Beeby et al., "Propulsive Nanomachines," 290, and Rosenhouse, *The Failures of Mathematical Anti-Evolutionism,* 151–152.

68. Hume, *Dialogues Concerning Natural Religion.* Such quotes from Hume's *Dialogues* are easily searched online.

69. Rosenhouse, *The Failures of Mathematical Anti-Evolutionism,* 273.

70. Dembski, "Science and Design."

71. See Sternberg, "On the Roles of Repetitive DNA Elements," and Wells, "Using Intelligent Design Theory to Guide Scientific Research."

72. All quotes from Collins, *The Language of God,* 136–137.

73. Thus in 1976, Dawkins (*The Selfish Gene,* 44–45): "The true 'purpose' of DNA is to survive, no more and no less. The simplest way to explain the surplus DNA is to suppose that it is a parasite, or at best a harmless but useless passenger, hitching a ride in the survival machines created by the other DNA." And again, in 1986, Dawkins (*The Blind Watchmaker,* 116): "Amazingly, only about 1 per cent of the genetic information in, for example, human cells, seems to be actually used." And yet again, in 2003, Dawkins (*A Devil's Chaplain,* 99): "Creationists might spend some earnest time speculating on why the Creator should bother to litter genomes with untranslated pseudogenes and junk tandem repeat DNA."

74. Dawkins (*The Greatest Show on Earth,* 333).

75. Richard Dawkins, "Jonathan Sacks and Richard Dawkins at BBC RE: Think Festival 12 September 2012": http://www.youtube.com/watch?v=roFdPHdhgKQ [12:57–13:11; 13:18–14:10].

76. ENCODE Project Consortium, "An Integrated Encyclopedia."

77. See https://www.encodeproject.org/help/project-overview (last accessed April 3, 2023).

78. ENCODE Project Consortium, "An Integrated Encyclopedia."

79. In an interview with Ed Yong for *Discover Magazine* two months after the *Nature* article: https://www.discovermagazine.com/the-sciences/encode-the-rough-guide-to-the-human-genome (last accessed April 3, 2023).

80. Pennisi, "ENCODE Project Writes Eulogy for Junk DNA."

81. See, for instance, Graur et al., "On the Immortality of Television Sets." Lents, *Human Errors,* ch. 3, follows Graur, dismissing ENCODE as a

massive exercise in "flawed reasoning." Casey Luskin details moves by Darwinists to circumvent the ENCODE results in his extensive typescript "The ENCODE Embroilment: Research on 'Junk DNA' Verifies Key Predictions of Intelligent Design," available at https://www.discovery.org/m/securepdfs/2021/10/Luskin-ENCODEandJunkDNA-101621.pdf (last accessed April 4, 2023).

82. The watchword in such situations seems to be that natural selection must always be given the preeminence and that it never pays to underestimate it. The impulse to ascribe to natural selection the attributes of deity, such as omnipotence and omniscience, seems present as well.

83. Theories of paradigmatic rank, such as Darwinism, have tremendous staying power even when they are tested and found wanting. Adherents are loath to abandon them. "Once it has achieved the status of paradigm," writes Thomas Kuhn (77), "a scientific theory is declared invalid only if an alternate candidate is available to take its place." To this, Kuhn has Max Planck (*Scientific Autobiography,* 33–34) add, "A new scientific truth does not triumph by convincing its opponents and making them see the light, but rather because its opponents eventually die, and a new generation grows up that is familiar with it." (Quoted also in Kuhn, 151) An alternate candidate to Darwinism is now in place in the form of intelligent design. But that doesn't mean Darwinism is going to fade gently into the night. That's why Kuhn's book was titled *The Structure of Scientific Revolutions.* To change a paradigm in science is to commit a revolutionary act, and revolutions are never placid affairs.

84. For overlapping genes in viruses and bacteria, see Wright et al., "Overlapping Genes." For overlapping genes in eukaryotes, see Nakayama et al., "Overlapping of Genes in the Human Genome."

85. Wright et al. ("Overlapping Genes") is interesting in providing the usual story but also in focusing on "engineered genomes" and "synthetic genomics," both of which are exercises in intelligent design. Designers are known to bring about informationally dense structures. Natural selection, by contrast, evinces no such capacity.

86. Keefe and Szostak, "Functional Proteins."

87. Yockey, "On the Information Content of Cytochrome *c.*"

88. Axe, "Estimating the Prevalence of Protein Sequences."

89. Miller, *Finding Darwin's God,* 145–147.

90. Behe, "Answering Scientific Criticisms of Intelligent Design," 141.

91. See https://en.wikipedia.org/wiki/Solving_chess (last accessed July 10, 2023).

92. We used this file in Section 6.8. See https://en.wikipedia.org/wiki /Words_(Unix) (last accessed June 23, 2023).

93. Meyer (*Darwin's Doubt*) carefully documents the absence of transitional forms to connect the phyla of the Cambrian.

94. Dembski, *The Design Inference,* sec. 2.3.

95. That happened very quickly after the publication of the first edition. See, for instance, the *BioScience* review of the first edition by Pigliucci, "Chance, Necessity, and the War Against Science."

96. We hesitate to mention relevant researchers and publications here because of the reprisals they may face. A case in point that we ourselves witnessed concerned the publication of Marks et al., *Biological Information: New Perspectives*, a book to which we were contributors and which ended up being published by the Singapore publisher World Scientific in 2013. That book, an ID-friendly proceedings of a 2011 Cornell conference on biological information, had been contractually agreed upon, completely typeset, and slated for release in a matter of weeks (Amazon listed the book for pre-order, ready to be shipped on March 31, 2012) when the original publisher, Springer Verlag, pulled it because ID opponents had learned of it and tried, with temporary success, to kill it. We were fortunate subsequently to get the anthology published with World Scientific, although Springer's about-face delayed publication by a year. I [WmD] have a screenshot of the Amazon listing in my files. Here are Springer's ISBNs for the book: ISBN-10: 3642284531; ISBN-13: 978-3642284533.

97. Jason Rosenhouse has tried to mount a principled objection to connecting irreducible complexity with small probabilities and therewith design inferences. Rosenhouse claims that irreducible and specified complexity are both bogus notions that ID proponents mistakenly use to limit Darwinian natural selection. In any synergy between these two forms of complexity, he regards specified complexity as the more bogus in that he sees it as saddling irreducible complexity with unnecessary probability calculations. Given Michael Behe's claim that no sound empirical evidence supports the gradual evolution of irreducibly complex biological systems, it would follow that irreducible complexity all by itself is sufficient to render such systems improbable. What need then for an improbability calculation in line with specified complexity and such as is required for a design inference?

But Rosenhouse, in raising this objection, confuses two forms of probability. There's the historical probability of failure in evolving irreducibly complex systems, and then there's the analytic determination of small probability and its associated specified complexity in evolving these systems. While this historical record of failure is enough to regard the evolution of these systems as improbable in an informal sense, specified complexity goes further by analyzing the particular probabilistic hurdles that explain this record of failure. For Rosenhouse's charge that irreducible complexity renders specified complexity superfluous and misleading, see his article "A Reply to Dembski's Review of My Book," July 3, 2022, http://pandasthumb.org/archives/2022/07/Dembski-response.html (last accessed April 7, 2023).

98. Pallen and Matzke list 38 flagellar proteins for *Salmonella typhimurium* ("From *The Origin of Species* to the Origin of Bacterial Flagella," 787).

99. "Of the forty or so proteins in the standard flagellum of *S. typhimurium* strain LT2 or *E. coli* K-12, only about half seem to be universally necessary." Pallen and Matzke, "From *The Origin of Species* to the Origin of Bacterial Flagella," 785.

100. University of Delaware biologist John McDonald provided a memorable example of this backward-evolution approach. Using it, he attempted to justify the evolution of irreducibly complex systems by considering simplifications of the standard mousetrap that preserve some level of function, even if greatly reduced. For instance, the hammer that crushes the mouse might be eliminated if the spring could be elongated and a portion of it poised precariously on the edge of the base, where it might snap and, perhaps not very effectively, attempt to kill a mouse. For McDonald's "full evolution" of the standard mousetrap from much simpler beginnings as well as for a critical reply to it, see Behe, *A Mousetrap for Darwin,* ch. 14. McDonald has continued to add to his example over time. Its current version is available online at https://udel.edu/~mcdonald/mousetrap.html (last accessed April 6, 2023).

101. But note, these new parts would have to be borrowed and retooled from some other sources. This is co-option. Gauger and Axe ("The Evolutionary Accessibility of New Enzyme Functions"), Axe and Gauger ("Model and Laboratory Demonstrations"), and Reeves et al. ("Enzyme Families–Shared Evolutionary History or Shared Design?") tested the feasibility of modifying a protein to perform a new function. Their research suggests that acquiring function in this way is beyond the reach of natural selection because too many changes are needed before some new function can be achieved—akin to the hypercube example in Section 7.4, where it gets stuck if five or more simultaneous changes are needed to find the nearest functional cell. So there is some substantial ID research to test this add-and-modify model.

102. Even Pallen and Matzke ("From *The Origin of Species* to the Origin of Bacterial Flagella") admit that the "ur-flagellum" required at least four

subsystems: 1) a secretion system, 2) a filament, 3) a motor, and 4) a chemotaxis apparatus.

103. Pallen and Matzke ("From *The Origin of Species* to the Origin of Bacterial Flagella," 788). Compare Gauger and Axe, "The Evolutionary Accessibility of New Enzyme Functions" as well as Axe and Gauger, "Model and Laboratory Demonstrations."

104. It needs to be stressed again (can it be stressed enough?) that intelligent design is not incompatible with evolution as such but only with unguided forms of it, which unduly restrict the types of complexity and information that may evolve. This is a point that Darwinists continually sidestep, whether deliberately or through a habit of slipshod thinking. Thus Pallen and Matzke ("From *The Origin of Species* to the Origin of Bacterial Flagella," 784), whose work we are citing extensively in this section, write:

> By even the most conservative estimate, there must therefore be thousands of different bacterial flagellar systems, perhaps even millions. Therefore, there is no point discussing the creation or ID of "the" bacterial flagellum. Instead, one is faced with two options: either there were thousands or even millions of individual creation events, which strains Occam's razor to [the] breaking point, or one has to accept that all the highly diverse contemporary flagellar systems have evolved from a common ancestor.

Pallen and Matzke clearly favor the latter option. But the latter option is also perfectly compatible with ID. Intelligent evolution (a form of ID) could evolve an "ur-flagellum," to use their terminology, and from there evolve millions of flagellar variants. The point at issue is whether Darwinian evolution can do the same. We argue no. Pallen and Matzke conveniently conflate evolution in a broad sense, which could be compatible with intelligent design, and evolution in a narrow sense, which is thoroughly Darwinian.

105. The problem of explaining the evolution of the bacterial flagellum from the T3SS is compounded because it makes better sense, from the vantage of natural history, to think the T3SS evolved (or devolved) from the bacterial flagellum rather than the other way around. Note that the T3SS provides no motility function—all it does is secrete proteins from a basal body. Bacterial flagella have presumably been around ever since bacteria needed to navigate their watery environments, which would be for billions of years. Type III secretion systems are essentially poison delivery systems for bacteria that infect animals (such as *Yersinia pestis*, the bubonic plague bacterium), and thus would not have been present until animals

came on the scene, which is considerably more recent than billions of years. Consistent with these observations, Minnich and Meyer ("Genetic Analysis," 222) conclude: "Phylogenetic analyses of the gene sequences show that the flagella proteins arose first and those of the TSS [T3SS] came later. In other words, if anything, the TSS (less complex) evolved from the flagellum (more complex)."

This direction of evolution from flagellum to T3SS makes it easier to overcome evolutionary obstacles: it is easier, evolutionarily, to simplify or discard structures and thereby alter functions than it is to build up increasingly complicated structures displaying novel functions by borrowing from existing materials, as would be required if the (simpler) T3SS evolved into the (more complicated) bacterial flagellum. A motorcycle, for instance, could shed its wheels and frame to leave only an engine, which left to run by itself could serve as a heater. Cave fish that lose sight and thereby suffer no loss of fitness because they exist away from sunlight and thus come merely to have eye nubs illustrate the same point. The shedding and concomitant repurposing of components by taking evolution in the direction of simplicity is no big challenge for Darwinian evolution. The big challenge is how to get eyes and flagella in the first place.

106. Pallen and Matzke ("From *The Origin of Species* to the Origin of Bacterial Flagella," 788).

107. The scenario sketched here is the most gradual way that a modify-and-exapt approach to evolution might work. Of course, it would in principle be possible for the modify-and-exapt approach to take two larger functioning systems and then to kludge them so that the kludged system is structurally and functionally different from the two systems that were kludged. But this is asking a lot. One might just as well ask that a motor and a bicycle created for different uses be kludged to form a functioning motorcycle. The problem with achieving such a result is that the motor and the bicycle would, earlier in their history, have been targeted for different uses and thus wouldn't be adapted to work with each other. A functioning motorcycle kludged in this way would thus be highly unlikely to arise because it would be highly unlikely for the motor and the bicycle to share, as we would say in Appendix B.6, *interface compatibility*. Indeed, any success in forming a functioning motorcycle from such a kludge would be due to purposive design.

108. "The Biggest Mystery in the History of the Universe | Joana Xavier on the Origin of Life," Evolution 2.0 Podcast with Perry Marshall, September 24, 2022, https://www.youtube.com/watch?v=0Xb33ZflqpI (last accessed January 4, 2023). Interestingly, philosopher Willard Quine, near the end of his life, wrote an article on naturalism in which he saw

naturalism as well able to accommodate entities such as intelligent designers. As Quine ("Naturalism; Or, Living Within One's Means," 252) wrote, "If I saw indirect explanatory benefit in positing sensibilia, possibilia, spirits, a Creator, I would joyfully accord them scientific status too, on a par with such avowedly scientific posits as quarks and black holes... The most we can reasonably seek in support of an inventory and description of reality is testability of its observable consequences in the time-honored hypothetico-deductive way." The only issue that naturalism should raise for intelligent design is testability of its observable consequences. The rest is ideological window-dressing.

109. See respectively Snoke ("Systems Biology as a Research Program") for systems biology, Savransky (*Engineering of Creativity*) for TRIZ, and Laufmann and Glicksman (*Your Designed Body*) for the merger between biology and engineering.

Epilogue

1. Medawar (*The Limits of Science*, 78–82) used the term *conservation of information* in the 1980s to describe computational systems that are only able to produce logical consequences from a given set of axioms or starting points. Such systems therefore create no novel information since everything in the consequences is already implicit in the starting points. Medawar's use of the term is the first that we know of, and he even referred to the *Law of Conservation of Information* to capture this inability of computation to generate novel information. But Medawar was focused on purely deterministic systems.

    For a broader range of stochastic systems, the closest precursor to our use of conservation of information occurred in Cullen Schaffer's "A Conservation Law for Generalization Performance." According to his *Law of Conservation of Generalization Performance*, a learner capable of "better-than-chance performance" must be using external information or otherwise "is like a perpetual motion machine." Conservation of information was in the air in the late 1990s, especially in connection with the then recently proven no free lunch theorems by David Wolpert and Bill Macready ("No Free Lunch Theorems"). But the focus of these theorems was not so much on information that improves particular searches (which is the focus of conservation of information) as on the inability of particular searches to do better than blind or purely random searches when search performance was averaged across search problems.

    In essence, no free lunch says that there's no universally optimal search algorithm but that any search will be good for some things, less good for others. Conservation of information, by contrast, says that what makes a particular search good is its inclusion of specific information adapted to the particular search. Moreover, this information will need to satisfy strict accounting rules so that the improvement of a search must be paid for with information extracted elsewhere, with the result that no net gain of information ultimately occurs. This is a powerful idea whose research potential has yet to be tapped.

2. See Dembski, "Intelligent Design as a Theory of Information" and Dembski, *No Free Lunch*, chs. 4 and 5.

3. For search as a general framework for understanding evolution as well as pushback against that view, see my [WmD's] 2014 University of Chicago lecture on conservation of information sponsored by Leo Kadanoff: https://www.youtube.com/watch?v=MN74Vn-R5fg (last accessed March 1, 2023). See also Dembski, *Being as Communion,* ch. 17.

4. Quoted from "Darwinism under the Microscope," PBS television interview of William Dembski and Eugenie Scott by Peter Robinson for the program *Uncommon Knowledge*, filmed December 7, 2001, on the Stanford University campus, with video available online at https://www.hoover.org/research/darwin-under-microscope-questioning-darwinism (last accessed February 27, 2023).

5. Thomas Schneider, "Evolution of Biological Information," 2794.

6. Ewert, Dembski, and Marks, "Climbing the Steiner Tree," 10. Available online at https://evoinfo.org/papers/steiner.pdf (last visited February 21, 2022).

7. Ewert, Dembski, and Marks, "Climbing the Steiner Tree."

8. In his expert witness report for the *Kitzmiller v. Dover* trial, Robert Pennock wrote, "My colleagues and I have demonstrated experimentally that a Darwinian mechanism can discover irreducibly complex systems." To support this claim, Pennock then cited Lenski et al., "The Evolutionary Origin of Complex Features," a *Nature* article on which he was a co-author. Likewise, Kenneth Miller, in his expert witness report for that trial, referenced this article to argue that evolutionary mechanisms can generate biological information. See https://billdembski.com/wp-content/uploads/2019/05/2005.09.Expert_Rebuttal_Dembski.pdf, p. 18. The article by Lenski and Pennock described the evolutionary computing program Avida.

9. Dawkins, *The Blind Watchmaker,* 316.

10. See https://www3.beacon-center.org/welcome/beacon-mission (last visited February 21, 2022). BEACON was originally housed at http://www.beacon.msu.edu when it started in February of 2010. It continues to this

day, but its website's traffic and domain authority are low. It goes without saying that our work to disprove the power of evolutionary algorithms to create search-specific information was accomplished with far less than $25 million!

11. Consider a 2020 manifesto titled "The Surprising Creativity of Digital Evolution" by Lehman et al. With 53 authors (including ID opponents Christoph Adami, Robert Pennock, and Richard Lenski), this article advocates for the power of Darwinian evolutionary processes to create novel biological information. Or consider Takagi et al.'s "The Coevolution of Cellularity and Metabolism Following the Origin of Life." From the title it sounds as though this is a straight-up biology article. But in fact, it describes "a series of artificial life simulations." Real biology is absent from this article.

      Conservation of information shows that all such efforts to use artificial life, digital evolution, evolutionary computing, evolutionary algorithms, etc. to explain the emergence of novel information, biological or otherwise, constitute a hopeless enterprise. But as Jonathan Wells (*Icons of Evolution* and *Zombie Science: More Icons of Evolution*) has underscored, even in science some disreputable ideas are too pleasing and comforting to be disowned. And some, like zombies, don't even let death stand in the way of their comeback.

12. See https://biologos.org (last accessed June 29, 2023).

13. See the Evolutionary Informatics Lab publications page: https://evoinfo .org/publications.html (last accessed June 23, 2023). Three of the seminal publications on conservation of information are Dembski and Marks, "Conservation of Information in Search: Measuring the Cost of Success" and "The Search for a Search: Measuring the Information Cost of Higher-Level Search," and Dembski, Ewert, and Marks, "A General Theory of Information Cost Incurred by Successful Search."

Appendix A

1. The asterisk, in mapping descriptions to events, thus makes the logic of descriptions preserve the logic of events, and therefore constitutes, as mathematicians would say, a homomorphism.

2. See Suppes, *Probabilistic Metaphysics,* as well as Eells, *Probabilistic Causality.*

3. It's possible to generalize further the range of entities to which probabilities can be applied so as to include propositions, which are abstract entities that convey meaning and make truth claims. Typically, proposi-

tions are represented linguistically via statements that express the propositions. We expand on this generalization in Section 6.7, where we need it for a fuller understanding of specified complexity. In general, however, it's enough for our purposes to focus on probabilities of events in that design inferences are drawn for events. Propositions can subsume events by affirming that an event has happened.

4. For a brief summary of Fisher's views on tests of significance and null hypotheses, see Fisher, *The Design of Experiments,* 13–17.

5. See Howson and Urbach, *Scientific Reasoning,* 178. For a similar criticism, see Hacking, *Logic of Statistical Inference,* 81–83.

6. Royall, *Statistical Evidence: A Likelihood Paradigm,* 61–62.

7. See just about any book on probability theory, such as Ross, *Introduction to Probability Models,* secs. 2.7–2.9.

8. For waiting times, see Ross, *Introduction to Probability Models,* sec. 5.3.

9. Uniform probabilities also apply to certain infinite metric spaces. A metric can induce a uniform probability on a space by assigning identical probability to geometrically equivalent subsets. For a general account of uniform probabilities, see Dembski, "Uniform Probability."

10. This section addresses the charge by Jason Rosenhouse (*The Failures of Mathematical Anti-Evolutionism,* 126) and others that the design inference defaults to uniform probabilities and thus cannot make do with non-uniform probabilities. That is not the case: a valid design inference must come to terms with the actual probabilities in play for any phenomenon under investigation. All the intelligent design literature admits this point. This section elaborates on that point by noting that uniform probabilities invariably arise as a baseline against which any further probabilistic analyses then need to be made.

11. See https://ncatlab.org/nlab/show/Polish+space (last accessed May 14, 2022). For the full theory, in terms of Polish spaces, see Cohn, *Measure Theory,* ch. 8.

12. Technically speaking, binary numbers with at least one 0 and with all 1s after a certain point need to be ignored in the one-to-one correspondence. That's because a number like 01111 (all 1s thereafter) is, as infinite binary numbers go, the same number as 10000 (all 0s thereafter). Both numbers in this case map to the real number ½. As it turns out, binary numbers ending in all 1s form a countable set and thus have probability zero in the correspondence between infinite binary numbers and the unit interval (i.e., the real numbers between zero and one). Such binary numbers that endlessly and without interruption repeat 1 can therefore

be excluded, ensuring that the correspondence is indeed one to one and that the spaces are indeed Borel isomorphic.

13. T. E. S. Raghavan shared this insight in teaching a statistics class that one of the authors (WmD) took with him at the University of Illinois at Chicago in the first half of 1982.

14. Stalnaker, *Inquiry*, 85.

15. Dretske, *Knowledge and the Flow of Information*, 12.

16. Shannon and Weaver, *The Mathematical Theory of Communication*, 8.

17. Shannon and Weaver, *The Mathematical Theory of Communication*, 34. Note that for copyright reasons, I [WmD] had this diagram redone. It matches point for point Shannon's original diagram.

18. Websites giving poker odds are widespread. See, for instance, https://mathworld.wolfram.com/Poker.html (last accessed January 28, 2023).

19. See, for instance, Cover and Thomas, *Elements of Information Theory*, 14.

20. AIT, or algorithmic information theory, is also known as Kolmogorov complexity. See Cover and Thomas, *Elements of Information Theory*, ch. 14. We address Kolmogorov complexity in Section 6.2.

Appendix B

1. Actually, artificial intelligence (AI) is starting to offer some insight into the problem of generating amino acid sequences that might form functional proteins. AI-based natural language models seem to have made some progress toward "generat[ing] artificial enzymes from scratch." See https://scitechdaily.com/limitless-possibilities-ai-technology-generates-original-proteins-from-scratch (last accessed March 6, 2023). For the *Nature Biotechnology* article on which this report was based, see Madani et al., "Large Language Models." A different approach to generating functional proteins from sequence data looks to multiple sequence alignments (MSAs) across different proteins. See Townsley et al., "PSICalc: A Novel Approach." But predicting which will map to functional proteins and which won't simply by inspecting the structure of DNA sequences does not at this time have a general solution.

2. That said, synonymous codons, which map to the same amino acid, need not produce the same protein since synonymous codons do not guarantee that the resulting amino acid sequences will fold the same way. One

factor in this is what are called translation efficiencies. See Plotkin and Kudla, "Synonymous But Not the Same."

3. Meyer (*Return of the God Hypothesis*, ch. 18) draws attention to boundary-valued fine tuning in the Wheeler-Dewitt equation.

4. Kaplansky, *Set Theory and Metric Spaces,* 16.

5. On the other hand, it might also be argued that the Darwinian selection-mutation mechanism does not make for a simply described evolutionary process, being informed by a highly contingent environment that admits no clear and simple description.

6. See Leslie, "Evil and Omnipotence," 17–18, for his fly-on-the-wall example.

7. For a topological treatment of metric spaces, see Munkres, *Topology,* 119–126.

8. Technically, this is the open ball of radius r around c. The closed ball of radius r around c is defined as $\{ x \in \Omega \mid d(c,x) \le r \}$.

9. Richard Hamming introduced this metric in "Error Detecting and Error Correcting Codes."

10. Axe, "Estimating the Prevalence of Protein Sequences," is a case in point.

11. One of the key results of Dembski, "Uniform Probability," is that for metric spaces that allow uniform probabilities, conditioning those probabilities on open subsets of the metric spaces in turn yields uniform probabilities on those open subsets. In other words, conditional uniform probabilities tend to remain uniform probabilities.

12. This is a property that probabilists refer to as *tightness*. See Billingsley, *Convergence of Probability Measures,* 9: "A probability measure P ... is *tight* if for each positive ε there exists a compact set K such that $P(K) > 1 - \varepsilon$." Compact sets are the most finitely constrained subspaces of any topological space. Tightness guarantees that most of the probability is finitely constrained but cannot guarantee that all of it is—asymptotically vanishing tendrils can, even with tightness, reach out infinitely in extent.

13. McGrew, McGrew, and Vestrup, "Probabilities and the Fine-Tuning Argument: A Skeptical View."

14. Dembski, "Uniform Probability."

15. Billingsley, *Convergence of Probability Measures,* 37.

16. Rudin, *Functional Analysis,* 68–69.

17. We owe this insight to an email exchange with Alex Pruss.

18. Dubins and Savage, *Inequalities of Stochastic Processes.*

19. This was one of the key results about the consistency of uniform probabilities in Dembski, "Uniform Probability."

20. Nonstandard approaches to probability theory exist, but they don't change the conclusions of this appendix. For a nonstandard approach to probability theory that uses ultrafilters, see Albeverio et al., *Nonstandard Methods in Stochastic Analysis and Mathematical Physics.*

21. For the mathematical literature that distinguishes the two, see Ross, *Introduction to Probability Models*, secs. 2.2 and 3.3. For the evolutionary biology literature that tends to equate waiting times with stopping times, see Gavrilets, "Waiting Time to Parapatric Speciation," and Hössjer et al., "On the Waiting Time Until Coordinated Mutations Get Fixed in Regulatory Sequences." Gavrilets' waiting time focuses on when a speciation event first happens. Hössjer et al.'s waiting time focuses on when a set of coordinated set of mutations first gets fixed in a population. For early work that lays the mathematical foundations for waiting times as applied to evolutionary biology, see Ulam, "How to Formulate Mathematically Problems of Rate of Evolution?"

22. For a general definition and treatment of stochastic processes, see Bhattacharya and Waymire, *Stochastic Processes*, ch. 1.

23. Continuous stochastic processes are invariably a (weak) limit of discrete stochastic processes, so there is no loss of generality in this assumption. See Prohorov's and Donsker's theorems in Billingsley (*Convergence of Probability Measures,* 37 and 68–73) as well as Skorokhod representation in Ethier and Kurtz (*Markov Processes,* 102).

24. How might a stochastic process like this arise? Imagine R represents finding an Easter egg. X_1 denotes the guess by someone who knows with .999999 probability where the Easter egg is. X_2, X_3,... X_{k-1} represent people who are misled about where the Easter egg is and thus are inveterately wrong in guessing where it is. And then X_k represents a person who with probability .000001 knows where the Easter egg is if X_1 is mistaken about where it is. Granted, this stochastic process is contrived, but it is a perfectly legitimate stochastic process, requiring only a few probabilistic details to be filled in.

25. A mathematician colleague on the faculty of SUNY Stony Brook notes that when multi-billionaire Jim Simons, founder of the quant trading firm Renaissance Technologies, was on faculty with its math department, the average annual income of math professors there was in the many millions of dollars. This colleague quickly added that the median salary was substantially lower. Averages can be misleading.

26. See Dembski, *No Free Lunch,* sec. 5.10, and Menuge, *Agents Under Fire,* ch. 4.

27. There are systems that automatically let in what's needed and keep out what's not needed. But can such systems avoid displaying specified complexity and implicating intelligent design? Take the lipid bilayer membrane around the cell wall. This membrane is itself a complex specified engineering marvel whose origin is in need of explanation. Moreover, it is fitted for the very specific needs of cells and is wholly unsuitable for the job of welcoming in the right kind of proteins needed for evolving a bacterial flagellum and keeping out the wrong ones. A radically different system would be required for such a task. And there's no reason to think that such a system could evolve without input from an actual intelligence and without displaying specified complexity.

28. Note that this inequality need not be a strict equality because it can be refined with additional terms. For instance, consider the *retention probability* p_{reten}, the probability that items available at the right time and in the right place stay at the right place long enough (i.e., are retained) for the bacterial flagellum (or whatever irreducibly complex system is in question) to be properly constructed. The retention probability is therefore conditional on the availability, synchronization, and localization probabilities and could be inserted as a factor after these terms in the origination inequality.

    Or consider the *proportionality probability* p_{propor}, the probability that items available at the right time, in the right place, and for long enough occur in the right proportion for the bacterial flagellum to be properly constructed. The protein that goes into the flagellum's whip-like tale requires tens of thousands of subunits; proteins for other parts of the flagellum require only a few hundred subunits. Without the right proportion of suitable parts (subunits), no functioning flagellum can be built. The proportionality probability is conditional on availability, synchronization, localization, and retention probabilities and could be inserted as a factor after these terms in the origination inequality.

29. Behe and Snoke, "Simulating Evolution by Gene Duplication."

30. Behe and Snoke, "Simulating Evolution by Gene Duplication," 11.

31. For reasoning probabilistically with waiting times, see Feller, *An Introduction to Probability Theory and Its Applications*, secs. 2.7, 11.3, and 17.2. See also Appendix B.5.

32. In fact, a strict Darwinian explanation for antibiotic resistance seems to be more the exception than the rule. In times of environmental stress, bacteria go into a programmed defense that constitutes a targeted search for gene combinations that will enable at least a few of the bacteria's descendants to survive (the genetic changes here are therefore not random

mutations as understood within neo-Darwinism). To see this, consider the following abstract from an article in *Cell*:

> According to classical evolutionary theory, phenotypic varia-
> tion originates from random mutations that are independent of
> selective pressure. However, recent findings suggest that organ-
> isms have evolved mechanisms to influence the timing or
> genomic location of heritable variability. Hypervariable contin-
> gency loci and epigenetic switches increase the variability
> of specific phenotypes; error-prone DNA replicases produce
> bursts of variability in times of stress. Interestingly, these mech-
> anisms seem to tune the variability of a given phenotype to
> match the variability of the acting selective pressure. Although
> these observations do not undermine Darwin's theory, they
> suggest that selection and variability are less independent than
> once thought.

Quoted from Rando and Verstrepen, "Timescales of Genetic and Epigenetic Inheritance."

33. See Sagan, *Cosmos,* 299.

34. Michael Crichton, "Aliens Cause Global Warming," Caltech Michelin Lecture, January 17, 2003, available online at https://stephenschneider. stanford.edu/Publications/PDF_Papers/Crichton2003.pdf (last accessed April 8, 2023).

BIBLIOGRAPHY

Adams, Ernest W., *The Logic of Conditionals: An Application of Probability to Deductive Logic* (Dordrecht: Reidel, 1975).

Aizawa, Shin-Ichi, "What Is Essential for Flagellar Assembly?" in Ken F. Jarrell, ed., *Pili and Flagella: Current Research and Future Trends* (Poole, UK: Caister Academic Press, 2009), 91–98.

Alberts, Bruce, "The Cell as a Collection of Protein Machines: Preparing the Next Generation of Molecular Biologists," *Cell* 92 (February 8, 1998): 291.

Albeverio, Sergio, Jens Erik Fenstad, Raphael Hoegh-Krohn, and Tom Lindstrom, *Nonstandard Methods in Stochastic Analysis and Mathematical Physics* (Orlando, FL: Academic Press, 1986).

Altshuler, Edward E., and Derek S. Linden, "Design of Wire Antennas Using Genetic Algorithms," in Y. Rahmat-Samii and E. Michielssen, eds., *Electromagnetic Optimization by Genetic Algorithms* (New York: Wiley, 1999), 211–248.

Anderson, Eric H., "A Factory That Builds Factories That Build Factories That...," in Thomas Y. Lo, Paul K Chien, Eric H. Anderson, Robert A. Alston, and Robert P. Waltzer, *Evolution and Intelligent Design in a Nutshell* (Seattle: Discovery Institute Press, 2020), 65–86.

Arora, Sanjeev, and Barak, Boaz, *Computational Complexity: A Modern Approach* (Cambridge: Cambridge University Press, 2009).

Arvedlund, Erin C., "Don't Ask, Don't Tell: Bernie Madoff Is so Secretive, He Even Asks His Investors to Keep Mum," *Barron's* (May 7, 2001), available online at https://www.sec.gov/news/studies/2009/oig-509/exhibit-0156.pdf (last accessed January 14, 2023).

Axe, Douglas D., "Estimating the Prevalence of Protein Sequences Adopting Functional Enzyme Folds," *Journal of Molecular Biology* 341 (2004): 1295–1315.

Axe, Douglas D., and Ann K. Gauger, "Model and Laboratory Demonstrations That Evolutionary Optimization Works Well Only If Preceded by Invention—Selection Itself Is Not Inventive," *Bio-Complexity* 2 (2015): 1–13.

Barron, John, *Breaking the Ring: The Rise and Fall of the Walker Family Spy Network* (New York: Avon, 1988).

Barrow, John, and Frank Tipler, *The Anthropic Cosmological Principle* (Oxford: Oxford University Press, 1986).

Bartholomew, David J., *God, Chance and Purpose: Can God Have It Both Ways?* (Cambridge: Cambridge University Press, 2008).

Bauer, Heinz, *Probability Theory and Elements of Measure Theory* (London: Academic Press, 1981).

Beeby, Morgan et al., "Propulsive Nanomachines: The Convergent Evolution of Archaella, Flagella, and Cilia," *FEMS Microbiology Reviews* 44, no. 3 (2020): 253–304.

Behe, Michael J., "Answering Scientific Criticisms of Intelligent Design," in M. J. Behe, W. A. Dembski, and S. C. Meyer, eds., *Science and Evidence for Design in the Universe* (San Francisco: Ignatius Press, 2000), 133–149.

Behe, Michael J., *Darwin's Black Box: The Biochemical Challenge to Evolution*, 2nd ed. (New York: Free Press, 2006). First edition published 1996.

Behe, Michael J., *The Edge of Evolution: The Search for the Limits of Darwinism* (New York: Free Press, 2007).

Behe, Michael J., *A Mousetrap for Darwin* (Seattle: Discovery Institute Press, 2020).

Behe, Michael J., and David W. Snoke, "Simulating Evolution by Gene Duplication of Protein Features that Require Multiple Amino Acid Residues," *Protein Science* 13 (2004): 1–14.

Benyus, Janine M., *Biomimicry: Innovation Inspired by Nature* (New York: HarperCollins, 1997).

Berg, Howard C., and Robert A. Anderson, "Bacteria Swim by Rotating their Flagellar Filaments," *Nature* 245 (1973): 380–382.

Bergman, George M., *An Invitation of General Algebra and Universal Constructions* (New York: Springer, 2015).

Bergsneider, Marvin et al., "What We Don't (but Should) Know about Hydrocephalus," *Journal of Neurosurgery: Pediatrics* 104, no. 3 (2006): 157–159.

Berlinski, David, *Black Mischief: Language, Life, Logic, Luck*, 2nd ed. (Cambridge, MA: Harcourt Brace Jovanovich, 1988).

Bhattacharya, Rabi N., and Edward C. Waymire, *Stochastic Processes* (Philadelphia: SIAM, 2009).

Billingsley, Patrick, *Convergence of Probability Measures*, 1st ed. (New York: Wiley, 1968). Note that a second edition appeared in 1999, but it is riddled with typos and introduces a revamped notation that is confusing. Stick with the first edition.

Borel, Emile, *Probabilities and Life*, trans. M. Baudin [1943] (reprinted New York: Dover, 1962).

Bradley, Walter L., and Charles B. Thaxton, "Information and the Origin of Life," in J. P. Moreland, ed., *The Creation Hypothesis: Scientific Evidence for an Intelligent Designer* (Downers Grove, IL: InterVarsity, 1994), 173–210.

Bronshtein, I. N., K. A. Semendyayev, Gerhard Musiol, Heiner Mühlig, *Handbook of Mathematics*, 6th ed. (Berlin: Springer, 2015).

Budiansky, Stephen, *Code Warriors: NSA's Codebreakers and the Secret Intelligence War Against the Soviet Union* (New York: Random House, 2016).

Burgess, Stuart, "Why the Ankle-Foot Complex Is a Masterpiece of Engineering and a Rebuttal of 'Bad Design' Arguments," *Bio-Complexity* 3 (2022): 1–10.

Byers, Sarah, "Life as 'Self-Motion': Descartes and 'the Aristotelians' on the Soul as the Life of the Body," *The Review of Metaphysics* 59, no. 3 (2006): 723–755.

Campbell, Jim, *Madoff Talks: Uncovering the Untold Story Behind the Most Notorious Ponzi Scheme in History* (New York: McGraw Hill, 2021).

Chaitin, Gregory J., "On the Length of Programs for Computing Finite Binary Sequences," *Journal of the Association for Computing Machinery* 13 (1966): 547–569.

Cicero, Marcus Tullius, *De Natura Deorum* [45 BC] trans. H. Rackham (Cambridge, MA: Harvard University Press, 1933).

Cocconi, Giuseppe, and Philip Morrison, "Searching for Interstellar Communications," *Nature* 184, no. 4690 (1959): 844–846.

Cohn, Donald L., *Measure Theory*, 2nd ed. (Boston: Birkhäuser, 2013).

Collins, Francis S., *The Language of God: A Scientist Presents Evidence for Belief* (New York: Free Press, 2006).

Comrie, Leslie John, *Chambers' Shorter Six-Figure Mathematical Tables* (Edinburgh: W. & R. Chambers, 1964).

Conway Morris, Simon, *Life's Solution: Inevitable Humans in a Lonely Universe* (Cambridge: Cambridge University Press, 2003).

Cooper, Geoffrey J. T., Sara I. Walker, and Leroy Cronin, "A Universal Chemical Constructor to Explore the Nature and Origin of Life," in Stoyan K. Smoukov, Joseph Seckbach, and Richard Gordon, eds., *Conflicting Models for the Origin of Life* (Beverly, MA: Scrivener Publishing, 2023), 101–130.

Cover, Thomas M., and Joy A. Thomas, *Elements of Information Theory*, 2nd ed. (New York: Wiley, 2006).

Crick, Francis, "On Protein Synthesis," *Symposium for the Society of Experimental Biology* 12 (1958): 138–163.

Crick, Francis, "The Origin of the Genetic Code," *Journal of Molecular Biology* 38 (1968): 367–379.

Crick, Francis, *Life Itself: Its Origin and Nature* (New York: Simon & Schuster, 1981).

Crick, Francis, and Leslie Orgel, "Directed Panspermia," *Icarus* 19 (1973): 341–46.

Dam, Kenneth W., and Herbert S. Lin, eds., *Cryptography's Role in Securing the Information Society* (Washington, DC: National Academy Press, 1996).

Darwin, Charles, *On the Origin of Species*, facsimile 1st ed. [1859] (reprinted Cambridge, MA: Harvard University Press, 1964).

Darwin, Charles, *The Correspondence of Charles Darwin*, volume 8 [1860] (Cambridge: Cambridge University Press, 1993).

Davies, Paul, *The Fifth Miracle* (New York: Simon & Schuster, 1999).

Dawkins, Richard, *The Selfish Gene* (Oxford: Oxford University Press, 1976).

Dawkins, Richard, *The Blind Watchmaker* (New York: Norton, 1986).

Dawkins, Richard, *Climbing Mount Improbable* (New York: Norton, 1996).

Dawkins, Richard, *A Devil's Chaplain: Reflections on Hope, Lies, Science, and Love* (New York: Houghton Mifflin, 2003).

Dawkins, Richard, *The Greatest Show on Earth: The Evidence for Evolution* (New York: Free Press, 2009).

Deamer, David W., *Origin of Life: What Everyone Needs to Know* (New York: Oxford University Press, 2020).

Debowski, Lukasz, *Information Theory Meets Power Laws: Stochastic Processes and Language Models* (Hoboken, NJ: Wiley, 2021).

Dembski, William A., "Uniform Probability," *Journal of Theoretical Probability* 3, no. 4 (1990): 611–626.

Dembski, William A., "Randomness by Design," *Nous* 25, no. 1 (1991): 75–106.

Dembski, William A., "On the Very Possibility of Intelligent Design," in J. P. Moreland, ed., *The Creation Hypothesis: Scientific Evidence for an Intelligent Designer* (Downers Grove, IL: InterVarsity Press, 1994), 113–138.

Dembski, William A., "Intelligent Design as a Theory of Information," *Perspectives on Science and Christian Faith* 49, no. 3 (1997): 180–191.

Dembski, William A., *The Design Inference: Eliminating Chance Through Small Probabilities*, 1st ed. (Cambridge: Cambridge University Press, 1998).

Dembski, William A., "Science and Design," *First Things* 86 (October 1998): 21–27.

Dembski, William A., *No Free Lunch: Why Specified Complexity Cannot Be Purchased without Intelligence* (Lanham, MD: Rowman and Littlefield, 2002).

Dembski, William A., *The Design Revolution: Answering the Toughest Questions About Intelligent Design* (Downers Grove, IL: InterVarsity, 2004).

Dembski, William A., "Specification: The Pattern That Signifies Intelligence," *Philosophia Christi* 7, no. 2 (2005): 299–343.

Dembski, William A., "God's Use of Chance," *Perspectives on Science and Christian Faith* 60, no. 4 (2008): 248–250.

Dembski, William A., *The End of Christianity: Finding a Good God in an Evil World* (Nashville, TN: Broadman and Holman, 2009).

Dembski, William A., *Being as Communion: A Metaphysics of Information* (Surrey, England: Ashgate, 2014).

Dembski, William A., and Michael Ruse, eds., *Debating Design: From Darwin to DNA* (Cambridge: Cambridge University Press, 2004).

Dembski, William A., and Jonathan Wells, *The Design of Life: Discovering Signs of Intelligence in Biological Systems* (Dallas: Foundation for Thought and Ethics/Discovery Institute Press, 2008).

Dembski, William A., and Robert J. Marks II, "Conservation of Information in Search: Measuring the Cost of Success," *IEEE Transactions on Systems, Man and Cybernetics A, Systems & Humans* 5, no. 5 (September 2009): 1051–1061.

Dembski, William A., and Robert J. Marks II, "The Search for a Search: Measuring the Information Cost of Higher-Level Search," *Journal of Advanced Computational Intelligence and Intelligent Informatics* 14, no. 5 (2010): 475–486.

Dembski, William A., Winston Ewert, and Robert J. Marks II, "A General Theory of Information Cost Incurred by Successful Search," in R. J. Marks II et al., eds., *Biological Information: New Perspectives* (Singapore: World Scientific, 2013), 26–63.

de Moivre, Abraham, *The Doctrine of Chances* [1718] (reprinted New York: Chelsea, 1967).

Dennett, Daniel, *Darwin's Dangerous Idea* (New York: Simon & Schuster, 1995).

Diaconis, Persi, Susan Holmes, and Richard Montgomery, "Dynamical Bias in the Coin Toss," *Society for Industrial and Applied Mathematics Review* 49, no. 2 (2007): 211–235.

D'Imperio, Mary E., *The Voynich Manuscript: An Elegant Enigma* (Fort Meade, MD: National Security Agency/Central Security Service, 1978), available online at https://www.nsa.gov/portals/75/documents/about

/cryptologic-heritage/historical-figures-publications/publications/misc
/voynich_manuscript.pdf (last accessed April 13, 2023).

Dobzhansky, Theodosius G., "Discussion of G. Schramm's paper," in Sidney
W. Fox, ed., *The Origins of Prebiological Systems and of Their
Molecular Matrices* (New York: Academic Press, 1965), 309–315.

Dornstein, Ken, *Accidentally, On Purpose: The Making of a Personal Injury
Underworld in America* (New York: St. Martin's Press, 1996).

Doudna, Jennifer A., and Samuel H. Sternberg, *A Crack in Creation: Gene
Editing and the Unthinkable Power to Control Evolution* (New York:
Mariner, 2018).

Dretske, Fred, *Knowledge and the Flow of Information* (Cambridge, MA:
MIT Press, 1981).

Dubins, Lester E., and Leonard J. Savage, *Inequalities of Stochastic
Processes: How to Gamble if You Must* (New York: Dover, 1976).

Earman, John, *Bayes or Bust? A Critical Examination of Bayesian
Confirmation Theory* (Cambridge, MA: MIT Press, 1992).

Edwards, A. W. F., "Are Mendel's Results Really Too Close?" *Biological
Review* 61 (1986): 295–312.

Eells, Ellery, *Probabilistic Causality* (Cambridge: Cambridge University
Press, 1991).

Egnor, Michael et al., "The Cerebral Windkessel and its Relevance to
Hydrocephalus: The Notch Filter Model of Cerebral Blood Flow,"
Cerebrospinal Fluid Research 3, no. 1 (2006): 1–2.

Egnor, Michael, "The Cerebral Windkessel as a Dynamic Pulsation Absorber," *Bio-Complexity* 3 (2019): 1–35.

Egnor, Michael et al., "A Quantitative Model of the Cerebral Windkessel and
Its Relevance to Disorders of Intracranial Dynamics," *Journal of
Neurosurgery: Pediatrics* (2023). In Press.

Eigen, Manfred, and Peter Schuster, *The Hypercycle* (Berlin: Springer, 1979).

ENCODE Project Consortium, "An Integrated Encyclopedia of DNA Elements in the Human Genome," *Nature* 489 (2012), 57–74.

Ethier, Stewart, and Thomas Kurtz, *Markov Processes: Characterization and
Convergence* (New York: John Wiley, 1986).

Evans, Colin, *The Casebook of Forensic Detection: How Science Solved 100
of the World's Most Baffling Crimes* (New York: Wiley, 1996).

Ewert, Winston, "The Dependency Graph of Life," *Bio-Complexity* 3 (2018):
1–27.

Ewert, Winston, William A. Dembski, and Robert J. Marks II, "Climbing the
Steiner Tree—Sources of Active Information in a Genetic Algorithm for
Solving the Euclidean Steiner Tree Problem," *Bio-Complexity* 1 (2012):
1–14.

Ewert, Winston, William A. Dembski, and Robert J. Marks II, "Algorithmic
Specified Complexity," in J. Bartlett, D. Hemser, J. Hall, eds., *Engi-*

neering and the Ultimate: An Interdisciplinary Investigation of Order and Design in Nature and Craft (Broken Arrow, OK: Blyth Institute Press, 2014), 131–149.

Ewert, Winston, Dembski, William A. Dembski, and Robert J. Marks II, "Algorithmic Specified Complexity in the Game of Life," *IEEE Transactions on Systems, Man, and Cybernetics: Systems* 45, no. 4 (2015): 584–594.

Feller, William, *An Introduction to Probability Theory and Its Applications*, vol. 1, 3rd ed. (New York: Wiley, 1968).

Feynman, Richard P., *Surely You're Joking, Mr. Feynman!* (New York: W. W. Norton, 1985).

Finkelstein, Michael O., *Quantitative Methods in Law: Studies in the Application of Mathematical Probability and Statistics to Legal Problems* (New York: Macmillan, 1978).

Finkelstein, Michael O., *Basic Concepts of Probability and Statistics in the Law* (New York: Springer-Verlag, 2009).

Finkelstein, Michael O., and Hans Levenbach, "Regression Estimates of Damages in Price-Fixing Cases," in M. H. DeGroot, S. E. Fienberg, and J. B. Kadane, eds., *Statistics and the Law* (New York: Wiley, 1986), 79–106.

Finkelstein, Michael O., and Bruce Levin, *Statistics for Lawyers*, 2nd ed. (New York: Springer-Verlag, 2001).

Fisher, Ronald A., *The Design of Experiments* (New York: Hafner, 1935).

Fisher, Ronald A., *Experiments in Plant Hybridisation* (Edinburgh: Oliver and Boyd, 1965).

Flannery, Michael A., *Intelligent Evolution: How Alfred Russel Wallace's World of Life Challenged Darwinism* (Georgetown, KY: Erasmus Press, 2020).

Franze, Kristian et al., "Müller Cells Are Living Optical Fibers in the Vertebrate Retina," *Proceedings of the National Academy of Sciences* 104, no. 20 (2007): 8287–8292.

Freedman, David, Robert Pisani, and Roger Purves, *Statistics*, 4th ed. (New York: Norton, 2007).

Gardner, Martin, "Arthur Koestler: Neoplatonism Rides Again," *World*, August 1, 1972, 87–89.

Gates, Bill, *The Road Ahead* (London: Penguin, 1996).

Gauger, Ann K., and Douglas D. Axe, "The Evolutionary Accessibility of New Enzyme Functions: A Case Study from the Biotin Pathway," *Bio-Complexity* 1 (2011): 1–17.

Gavrilets, Sergey, "Waiting Time to Parapatric Speciation," *Proceedings of the Royal Society of London B* 267 (2000): 2483–2492.

Goodman, Nelson, *Fact, Fiction, and Forecast*, 4th ed. (Cambridge, MA: Harvard University Press, 1983).

Goodwin, Brian, *How the Leopard Changed Its Spots: The Evolution of Complexity* (New York: Scribner's, 1994).

Gould, Stephen Jay, *Wonderful Life: The Burgess Shale and the Nature of History* (New York: Norton, 1989).

Graur, Dan et al., "On the Immortality of Television Sets: 'Function' in the Human Genome According to the Evolution-Free Gospel of ENCODE," *Genome Biology and Evolution* 5, no. 3 (2013): 578–590.

Greitz, Dan, "Radiological Assessment of Hydrocephalus: New Theories and Implications for Therapy," *Neurosurgical Review* 27 (2004): 145–165.

Grünwald, Peter, and Teemu Roos, "Minimum Description Length Revisited," *International Journal of Mathematics for Industry* 11, no. 1 (2020): 1930001, 1–29.

Hacking, Ian, *Logic of Statistical Inference* (Cambridge: Cambridge University Press, 1965).

Hamming, Richard W., "Error Detecting and Error Correcting Codes," *The Bell System Technical Journal* 29, no. 2 (April 1950): 147–160.

Haug, Samuel, Robert J. Marks, and William A. Dembski, "Exponential Contingency Explosion: Implications for Artificial General Intelligence," *IEEE Transactions on Systems, Man, and Cybernetics: Systems* 52, no. 5 (May 2022): 2800–2808.

Haught, John F., *God After Darwin: A Theology of Evolution* (Boulder, CO: Westview, 2000).

Hardy, G. H., and E. M. Wright, *An Introduction to the Theory of Numbers*, 5th ed. (Oxford: Clarendon Press, 1979).

Harold, Franklin, *The Way of the Cell: Molecules, Organisms and the Order of Life* (Oxford: Oxford University Press, 2001).

Harris, William H., and Judith S. Levey, eds., *The New Columbia Encyclopedia*, 4th edition (New York: Columbia University Press, 1975).

Henriques, Diana B., *The Wizard of Lies: Bernie Madoff and the Death of Trust* (New York: St. Martin's Griffin, 2011).

Hössjer, Ola, Günter Bechly, and Ann Gauger, "On the Waiting Time Until Coordinated Mutations Get Fixed in Regulatory Sequences," *Journal of Theoretical Biology* 524 (2021): 110657.

Howson, Colin, *Objecting to God* (Cambridge: Cambridge University Press, 2011).

Howson, Colin, and Peter Urbach, *Scientific Reasoning: The Bayesian Approach*, 2nd ed. (LaSalle, IL: Open Court, 1993).

Hume, David, *An Enquiry Concerning Human Understanding* [1748], ed. P. Millican (Oxford: Oxford University Press, 2008).

Hume, David, *Dialogues Concerning Natural Religion* [1779] (reprinted Buffalo, NY: Prometheus Books, 1989).

Ireland, Kenneth, and Michael Rosen, *A Classical Introduction to Modern Number Theory*, 2nd ed. (New York: Springer, 1990).

Jones, Peter W., and Peter Smith, *Stochastic Processes: An Introduction*, 3rd ed. (Boca Raton: Chapman & Hall/CRC, 2018).

Judd, John Wesley, *The Coming of Evolution: The Story of a Great Revolution in Science* [1912] (reprinted Gloucester: Echo 2010).

Jung, Carl G., *Synchronicity*, trans. R. F. C. Hull (Princeton: Princeton University Press, 1973).

Kaplan, Fred, "An Alien Concept," *Nature* 461, no. 7262 (2009): 345–346.

Kaplansky, Irving, *Set Theory and Metric Spaces*, 2nd ed. (New York: Chelsea, 1977).

Kauffman, Stuart, *The Origins of Order: Self-Organization and Selection in Evolution* (Oxford: Oxford University Press, 1993).

Kauffman, Stuart, *Investigations* (New York: Oxford University Press, 2000).

Keefe, Anthony D., and Jack W. Szostak, "Functional Proteins from a Random-Sequence Library," *Nature* 410 (2001): 715–718.

Kennedy, Gerry, and Rob Churchill, *The Voynich Manuscript: The Mysterious Code That Has Defied Interpretation for Centuries* (Rochester, Vermont: Inner Tradition, 2006).

Keynes, John Maynard, *A Tract on Monetary Reform* (London: Macmillan, 1923).

Kingsolver, Barbara, *Small Wonder* (New York: HarperCollins, 2002).

Knobloch, Eberhard, "Emile Borel as a Probabilist," in L. Krüger, L. J. Daston, and M. Heidelberger, eds., *The Probabilistic Revolution*, vol. 1 (Cambridge, MA: MIT Press, 1990), 215–233.

Kolb, Vera M., and Benton C. Clark III, *Astrobiology for a General Reader* (Newcastle: Cambridge Scholars Publishing, 2020).

Kolmogorov, Andrei, "Three Approaches to the Quantitative Definition of Information," *Problemy Peredachi Informatsii* (in translation) 1, no. 1 (1965): 3–11.

Kuhn, Thomas, *The Structure of Scientific Revolutions*, 2nd ed. (Chicago: University of Chicago Press, 1970).

Küppers, Bernd-Olaf, *Information and the Origin of Life* (Cambridge, MA: MIT Press, 1990).

Laplace, Pierre Simon de, *Essai Philosophique sur les Probabilités*, in J. R. Newman, ed., portion translated and reprinted into English in *The World of Mathematics*, 4 vols. [1814] (Redmond, WA: Tempus, 1988), 1301–1309.

Laufmann, Steve, and Howard Glicksman, *Your Designed Body* (Seattle: Discovery Institute Press, 2022).

Lehman, David, *The Perfect Murder: A Study in Detection* (New York: The Free Press, 1989).

Lehman, Joel, Jeff Clune, Dusan Misevic, et al., "The Surprising Creativity of Digital Evolution: A Collection of Anecdotes from the Evolutionary

Computation and Artificial Life Research Communities," *Artificial Life* 26, no. 2 (2020): 1–31.

Lenski, Richard E., Charles Ofria, Robert T. Pennock, and Christoph Adami, "The Evolutionary Origin of Complex Features," *Nature* 423 (May 8, 2003): 139–144.

Lenski, Richard E. et al., "Mutator Genomes Decay, Despite Sustained Fitness Gains, in a Long-Term Experiment with Bacteria," *Proceedings of the National Academy of Sciences* (published online October 10, 2017): E9026–E9035.

Lents, Nathan, *Human Errors: A Panorama of Our Glitches, from Pointless Bones to Broken Genes* (Boston: Mariner Books, 2018).

Lewis, C. S., *Miracles: A Preliminary Study*, revised edition [1960] (New York: HarperCollins eBooks, 2009).

Lewis, Michael, *The Undoing Project: A Friendship That Changed Our Minds* (New York: Norton, 2016).

Li, Ming, and Paul Vitányi, *An Introduction to Kolmogorov Complexity and Its Applications*, 4th ed. (New York: Springer, 2019).

Lloyd, Seth, "Ultimate Physical Limits to Computation," *Nature* 406 (August 2000): 1047–1054.

Lloyd, Seth, "Computational Capacity of the Universe," *Physical Review Letters* 88, no. 23 (May 2002): 237901, 1–4.

Mackie, John Leslie, "Evil and Omnipotence," *Mind* 64 (1955): 200–212.

Mackie, John Leslie, *The Cement of the Universe: A Study of Causation*, corrected paperback edition (Oxford: Clarendon, 1980).

Madani, Ali, Ben Krause, Eric R. Greene, Subu Subramanian, Benjamin P. Mohr, James M. Holton, Jose Luis Olmos Jr., Caiming Xiong, Zachary Z. Sun, Richard Socher, James S. Fraser, and Nikhil Naik, "Large Language Models Generate Functional Protein Sequences Across Diverse Families," *Nature Biotechnology* (2023): DOI:10.1038/s41587-022-01618-2.

Markopolos, Harry, *No One Would Listen: A True Financial Thriller* (New York: Wiley, 2010).

Marks, Robert J., II, Michael J. Behe, William A. Dembski, Bruce L. Gordon, and John C. Sanford, eds., *Biological Information: New Perspectives* (Singapore: World Scientific, 2013).

Marks, Robert J., II, William A. Dembski, and Winston Ewert, *Introduction of Evolutionary Informatics* (Singapore: World Scientific, 2017).

Martin, William F. et al., "The Physiology of Phagocytosis in the Context of Mitochondrial Origin," *Microbiology and Molecular Biology Reviews* 8, no. 3 (September 2017): 1–36.

Mazur, James. E., *Learning and Behavior*, 8th ed. (New York: Routledge, 2017).

McGrew, Timothy, Lydia McGrew, and Eric Vestrup, "Probabilities and the Fine-Tuning Argument: A Skeptical View," *Mind* 110, no. 440 (October 2001): 1027–1037.

McKeon, Richard, ed., *The Basic Works of Aristotle* (New York: Random House, 1941).

Medawar, Peter B., *Advice to a Young Scientist* (New York: Basic Books, 1979).

Medawar, Peter B., *The Limits of Science* (New York: Harper & Row, 1984).

Menuge, Angus, *Agents Under Fire: Materialism and the Rationality of Science* (Lanham, MD: Rowman and Littlefield, 2004).

Meyer, Stephen C., *Signature in the Cell: DNA and the Evidence for Intelligent Design* (New York: HarperOne, 2009).

Meyer, Stephen C., *Darwin's Doubt: The Explosive Origin of Animal Life and the Case for Intelligent Design* (New York: HarperOne, 2013).

Meyer, Stephen C., *Return of the God Hypothesis: Three Scientific Discoveries That Reveal the Mind Behind the Universe* (New York: HarperOne, 2021).

Milgram, Stanley, "The Small World Problem," *Psychology Today* 2 (1967): 60–67.

Mill, John Stuart, *A System of Logic: Ratiocinative and Inductive*, 8th ed. [1882] (reprinted London: Longmans, Green, and Co., 1906).

Miller, Kenneth R., *Finding Darwin's God: A Scientist's Search for Common Ground Between God and Evolution* (New York: HarperCollins, 1999).

Miller, Kenneth R., *Only a Theory: Evolution and the Battle for America's Soul* (New York: Viking, 2008).

Minnich, Scott, and Stephen C. Meyer, "Genetic Analysis of Coordinate Flagellar and Type III Regulatory Circuits in Pathogenic Bacteria," in W. Dembski, ed., *Darwin's Nemesis: Phillip Johnson and the Intelligent Design Movement* (Downers Grove, IL: InterVarsity, 2006), 214–223.

Monmonier, Mark, *How to Lie with Maps*, 3rd ed. (Chicago: University of Chicago Press, 2018).

Monod, Jacques, *Chance and Necessity* (New York: Vintage, 1972).

Moore, David S., and George P. McCabe, *Introduction to the Practice of Statistics*, 2nd ed. (New York: W. H. Freeman, 1993).

Munkres, James R., *Topology: A First Course*, 2nd ed. (Upper Saddle, N.J.: Prentice-Hall, 2000).

Nakayama, Tomohiro, Satoshi Asai, Yasuo Takahashi, Oto Maekawa, and Yasuji Kasama, "Overlapping of Genes in the Human Genome," *International Journal of Biomedical Science* 3, no. 1 (2007): 14–19.

Ocrant, Michael, "Madoff Tops Charts, Skeptics Ask How," *MAR/Hedge* 89 (May 2001): available online at https://nakedshorts.typepad.com/files/madoff.pdf (last accessed January 14, 2023).

Orgel, Leslie, *The Origins of Life* (New York: Wiley, 1973).

Pallen, Mark J., and Nicholas J. Matzke, "From *The Origin of Species* to the Origin of Bacterial Flagella," *Nature Reviews Microbiology* 4 (2006): 784–790.

Pearl, Judea, and Dana Mackenzie, *The Book of Why* (New York: Basic Books, 2018).

Pennisi, Elizabeth, "ENCODE Project Writes Eulogy for Junk DNA," *Science* 337, no. 6099 (September 7, 2012): 1159–1161.

Pennock, Robert T., "The Wizards of ID: Reply to Dembski," in R. T. Pennock, ed., *Intelligent Design Creationism and Its Critics: Philosophical, Theological, and Scientific Perspectives* (Cambridge, MA: MIT Press, 2001), 645–667.

Penrose, Roger, *The Emperor's New Mind: Concerning Computers, Minds, and the Laws of Physics* (New York: Oxford, 1989).

Peterson, Ivars, *The Jungles of Randomness: A Mathematical Safari* (New York: Wiley, 1998).

Petroski, Henry, *Invention by Design: How Engineers Get from Thought to Thing* (Cambridge, MA: Harvard University Press, 1996).

Pigliucci, Massimo, "Chance, Necessity, and the War Against Science" (review of *The Design Inference*), *BioScience* 50, no. 1 (2000): 79–81.

Planck, Max, *Scientific Autobiography and Other Papers*, trans. F. Gaynor (New York: Philosophical Library, 1949).

Plantinga, Alvin, *God, Freedom, and Evil* (Grand Rapids, MI: Eerdmans, 1977).

Plantinga, Alvin, *Warrant and Proper Function* (Oxford: Oxford University Press, 1993).

Plotkin, Joshua B., and Grzegorz Kudla, "Synonymous But Not the Same: The Causes and Consequences of Codon Bias," *Nature Reviews Genetics* 12 (2011): 32–42.

Polanyi, Michael, *Personal Knowledge: Towards a Post-Critical Philosophy* (Chicago: University of Chicago Press, 1962).

Polkinghorne, John C., *Belief in God in an Age of Science: The Terry Lectures* (New Haven: Yale University Press, 1998).

Quine, Willard V., "Naturalism; Or, Living Within One's Means," *Dialectica* 49, nos. 2–4 (1995): 251–261.

Rando, Oliver J., and Kevin J. Verstrepen, "Timescales of Genetic and Epigenetic Inheritance," *Cell* 128 (2007): 655–668.

Reeves, Mariclair A., Ann K. Gauger, and Douglas D. Axe, "Enzyme Families–Shared Evolutionary History or Shared Design? A Study of the GABA-Aminotransferase Family," *Bio-Complexity* 4 (2014): 1–16.

Reich, Eugenie Samuel, *Plastic Fantastic: How the Biggest Fraud in Physics Shook the Scientific World* (New York: St. Martin's Press, 2009).

Reid, Thomas, *Lectures on Natural Theology*, eds. J. A. Barham and J. Akins [1780] (Nashville: Influence Publishers, 2020).

Reinsel, Ruth, "Parapsychology: An Empirical Science," in P. Grim, ed., *Philosophy of Science and the Occult*, 2nd ed. (Albany, NY: State University of New York Press, 1990), 187–204.

Ricardo, Alonso, and Jack W. Szostak, "The Origin of Life on Earth," *Scientific American* (September 1, 2009), https://www.scientificamerican.com/article/origin-of-life-on-earth.

Robertson, Douglas S., "Algorithmic Information Theory, Free Will, and the Turing Test," *Complexity* 4, no. 3 (1999): 25–34.

Rosenhouse, Jason, *The Failures of Mathematical Anti-Evolutionism* (Cambridge: Cambridge University Press, 2022).

Rosenthal, Jeffrey S., "Statistics and the Ontario Lottery Retail Scandal," *Chance* 27, no. 1 (2014): 4–9.

Ross, Sheldon M., *Introduction to Probability Models*, 10th ed. (New York: Academic Press, 2010).

Royall, Richard M., *Statistical Evidence: A Likelihood Paradigm* (London: Chapman & Hall, 1997).

Rudin, Walter, *Functional Analysis*, 2nd ed. (New York: McGraw-Hill, 1991).

Rukhin, Andrew, Juan Soto, James Nechvatal, Miles Smid, Elaine Barker, Stefan Leigh, Mark Levenson, Mark Vangel, David Banks, Alan Heckert, James Dray, and San Vo, A *Statistical Test Suite for Random and Pseudorandom Number Generators for Cryptographic Applications* (Gaithersburg, MD: National Institute of Standards and Technology, 2010).

Ruse, Michael, *Monad to Man: The Concept of Progress in Evolutionary Biology* (Cambridge, MA: Harvard University Press, 1996).

Ruse, Michael, *Can a Darwinian Be a Christian? The Relationship between Science and Religion* (Cambridge: Cambridge University Press, 2001).

Russell, Bertrand, *The Problems of Philosophy* (London: Williams and Norgate, 1912).

Sagan, Carl, *Cosmos* (New York: Random House, 1980)

Sagan, Carl, *Contact: A Novel* (New York: Simon and Schuster, 1985).

Satinover, Jeffrey, *The Quantum Brain: The Search for Freedom and the Next Generation of Man* (New York: Wiley, 2001).

Savransky, Semyon, *Engineering of Creativity: Introduction to TRIZ Methodology of Inventive Problem Solving* (Boca Raton, FL: CRC Press, 2000).

Schaffer, Cullen, "A Conservation Law for Generalization Performance," in W. W. Cohen and H. Hirsh, eds., *Machine Learning: Proceedings of the Eleventh International Conference* (San Francisco: Morgan Kaufmann, 1994), 259–265.

Schneider, Thomas D, "Evolution of Biological Information," *Nucleic Acids Research* 28, no. 14 (2000): 2794–2799.

Schwartz, Barry, Edward A. Wasserman, and Steven J. Robbins, *Psychology of Learning and Behavior*, 5th ed. (New York: Norton, 2001).

Scott, Janny, "At UC San Diego: Unraveling a Research Fraud Case," *Los Angeles Times*, April 30, 1987, https://www.latimes.com/archives/la-xpm-1987-04-30-mn-2837-story.html.

Seibt, Peter, *Algorithmic Information Theory: Mathematics of Digital Information Processing* (Berlin: Springer, 2006).

Seifer, Marc J., *Wizard: The Life and Times of Nikola Tesla—Biography of a Genius* (Secaucus, NJ: Carol, 1996).

Shannon, Claude, and Warren Weaver, *The Mathematical Theory of Communication* (Urbana, IL: University of Illinois Press, 1949).

Shapiro, James A., *Evolution: A View from the 21st Century* (Upper Saddle River, NJ: FT Press Science, 2011).

Shapiro, Robert, *Origins: A Skeptic's Guide to the Creation of Life on Earth* (New York: Summit Books, 1986).

Shermer, Michael, *Why Darwin Matters: The Case Against Intelligent Design* (New York: Times Books, 2006).

Shor, Peter W., "Polynomial-Time Algorithms for Prime Factorization and Discrete Logarithms on a Quantum Computer," *SIAM Journal of Computation* 26, no. 5 (1997): 1484–1509.

Singh, Simon, *The Code Book: The Science of Secrecy from Ancient Egypt to Quantum Cryptography* (New York: Doubleday, 1999).

Smart, Nigel P., *Cryptography Made Simple* (New York: Springer, 2016).

Smith, John Maynard, *The Theory of Evolution* (London: Penguin, 1958).

Smith, Wolfgang, "Irreducible Wholeness and Dembski's Theorem," *Philos-Sophia* (March 2021), https://philos-sophia.org/irreducible-wholeness-dembski-theorem (last accessed March 14, 2023).

Snoke, David, "Systems Biology as a Research Program for Intelligent Design," *Bio-Complexity* 3 (2014): 1–7.

Sober, Elliott, "Testability," *Proceedings and Addresses of the American Philosophical Association* 73, no. 2 (1999): 47–76.

Sober, Elliott, "Intelligent Design and Probability Reasoning," *International Journal for Philosophy of Religion* 52, no. 2 (2002): 65–80.

Solomonoff, Ray J., "A Formal Theory of Inductive Inference, Part I," *Information and Control* 7 (1964a): 1–22.

Solomonoff, Ray J., "A Formal Theory of Inductive Inference, Part II," *Information and Control* 7 (1964b): 224–254.

Stalnaker, Robert C., *Inquiry* (Cambridge, MA: MIT Press, 1984).

Sternberg, Richard, "On the Roles of Repetitive DNA Elements in the Context of a Unified Genomic-Epigenetic System," *Annals of the New York Academy of Sciences* 981 (2002): 154–88.

Stoppard, Tom, *Rosencrantz and Guildenstern Are Dead* [1967] (New York: Grove Press, 2017).

Suppes, Patrick, *Probabilistic Metaphysics* (London: Blackwell, 1984).

Swinburne, Richard, *The Existence of God*, 2nd ed. (Oxford: Oxford University Press, 2004).

Takagi, Yuta A., Diep H. Nguyen, Tom B. Wexler, and Aaron D. Goldman, "The Coevolution of Cellularity and Metabolism Following the Origin of Life," *Journal of Molecular Evolution* 88 (2020): 598–617.

Thaxton, Charles B., Walter L. Bradley, and Roger L. Olsen, *The Mystery of Life's Origin: Reassessing Current Theories* (New York: Philosophical Library, 1984).

Tiessen, Axel, Paulino Pérez-Rodríguez, and Luis José Delaye-Arredondo, "Mathematical Modeling and Comparison of Protein Size Distribution in Different Plant, Animal, Fungal and Microbial Species Reveals a Negative Correlation between Protein Size and Protein Number, thus Providing Insight into the Evolution of Proteomes," *BioMed Central Research Notes* 5, no. 85 (2012): 1–22.

Timm, Torsten, and Andreas Schinner, "A Possible Generating Algorithm of the Voynich Manuscript," *Cryptologia* (published online May 25, 2019): 1–19, doi:10.1080/01611194.2019.1596999.

Townsley, Thomas D., James T. Wilson, Harrison Akers, Timothy Bryant, Salvador Cordova, T. L. Wallace, Kirk K. Durston, and Joseph E. Deweese, "PSICalc: A Novel Approach to Identifying and Ranking Critical Non-Proximal Interdependencies Within the Overall Protein Structure," *Bioinformatics Advances* (2022): 1–7.

Tribe, Larry B., "Trial by Mathematics: Precision and Ritual in the Legal Process," *Harvard Law Review* 84 (1971): 1329.

Tversky, Amos, and Daniel Kahneman, "Loss Aversion in Riskless Choice: A Reference-Dependent Model," *The Quarterly Journal of Economics* 106, no. 4 (1991): 1039–1061.

Ulam, Stanislaw M., "How to Formulate Mathematically Problems of Rate of Evolution?" in P. Moorhead and M. Kaplan, eds., *Mathematical Challenges to the Neo-Darwinian Interpretation of Evolution*, a symposium held at the Wistar Institute of Anatomy and Biology, April 25 and 26, 1966 (New York: Alan R. Liss, 1967), 21–33.

Van Hofwegen, Dustin J., Carolyn J. Hovde, and Scott A. Minnich, "Rapid Evolution of Citrate Utilization by *Escherichia coli* by Direct Selection Requires *citT* and *dctA*," *Journal of Bacteriology* 198, no. 7 (2016): 1022–1034.

Venter, J. Craig, *Life at the Speed of Light: From the Double Helix to the Dawn of Digital Life* (New York: Penguin, 2013).

Wegener, Ingo, *Complexity Theory: Exploring the Limits of Efficient Algorithms* (Berlin: Springer, 2005).

Wells, Jonathan, *Icons of Evolution: Science or Myth? Why Much of What We Teach About Evolution Is Wrong* (Washington, DC: Regnery, 2000).

Wells, Jonathan, "Using Intelligent Design Theory to Guide Scientific Research," *Progress in Complexity, Information, and Design* 3.1.2 (November 2004), available online at https://web.archive.org/web/2006 0513182316/http://www.iscid.org/papers/Wells_TOPS_051304.pdf.

Wells, Jonathan, *Zombie Science: More Icons of Evolution* (Seattle: Discovery Institute Press, 2017).

Wickramasinghe, Chandra, Kamala Wickramasinghe, and Gensuke Tokoro, *Our Cosmic Ancestry in the Stars: The Panspermia Revolution and the Origins of Humanity* (Rochester, VT: Bear & Company, 2019).

Wilkins, Adam S., "A Special Issue on Molecular Machines," *BioEssays* 25, no. 12 (December 2003): 1146.

Wilkins, John S., and Wesley R. Elsberry, "The Advantages of Theft over Toil: The Design Inference and Arguing from Ignorance," *Biology & Philosophy* 16 (2001): 709–722.

Wittgenstein, Ludwig, *Culture and Value*, ed. G. H. von Wright, trans. P. Winch (Chicago: University of Chicago Press, 1980).

Wolfram, Stephen, *What Is ChatGPT Doing ... and Why Does It Work?* (Champaign, IL: Wolfram Media, 2023), Kindle edition.

Wolpert, David H., and William G. Macready, "No Free Lunch Theorems for Optimization," *IEEE Transactions on Evolutionary Computation* 1, no. 1 (1997): 67–82.

Wright, Bradley W., Mark P. Molloy, and Paul R. Jaschke, "Overlapping Genes in Natural and Engineered Genomes," *Nature Reviews Genetics* 23 (2022): 154–168.

Wright, Ernest Vincent, *Gadsby* (Los Angeles: Wetzel, 1939).

Xue, Ruidong, Qi Ma, Matthew A. B. Baker, and Fan Bai, "A Delicate Nanoscale Motor Made by Nature—The Bacterial Flagellar Motor," *Advanced Science* 2, no. 1500129 (2015): 1–7.

Yarus, Michael, *Life from an RNA World: The Ancestor Within* (Cambridge, MA: Harvard University Press, 2011).

Yockey, Hubert P., "On the Information Content of Cytochrome *c*," *Journal of Theoretical Biology* 67, no. 3 (1977): 345–376.

Yockey, Hubert P., *Information Theory and Molecular Biology* (Cambridge: Cambridge University Press, 1992).

Yockey, Hubert P., *Information Theory, Evolution, and the Origin of Life* (Cambridge: Cambridge University Press, 2011).

Zuckoff, Mitchell, *Ponzi's Scheme: The True Story of a Financial Legend* (New York: Random House, 2005).

Zurek, Wojciech H., ed., *Complexity, Entropy and the Physics of Information* (Reading, MA: Addison-Wesley, 1990).

INDEX

Milton Keynes UK
Ingram Content Group UK Ltd.
UKHW021945101123
432363UK00005B/152